T0324519

SETS AND COMPUTATIONS

LECTURE NOTES SERIES
Institute for Mathematical Sciences, National University of Singapore

Series Editors: Chitat Chong and Kwok Pui Choi
Institute for Mathematical Sciences
National University of Singapore

ISSN: 1793-0758

Published

Vol. 23 Geometry, Topology and Dynamics of Character Varieties
edited by W Goldman, C Series & S P Tan

Vol. 24 Complex Quantum Systems: Analysis of Large Coulomb Systems
edited by Heinz Siedentop

Vol. 25 Infinity and Truth
edited by Chitat Chong, Qi Feng, Theodore A Slaman & W Hugh Woodin

Vol. 26 Notes on Forcing Axioms
by Stevo Todorcevic, edited by Chitat Chong, Qi Feng, Theodore A Slaman, W Hugh Woodin & Yue Yang

Vol. 27 *E*-Recursion, Forcing and C*-Algebras
edited by Chitat Chong, Qi Feng, Theodore A Slaman, W Hugh Woodin & Yue Yang

Vol. 28 Slicing the Turth: On the Computable and Reverse Mathematics of Combinatorial Principles
by Denis R. Hirschfeldt, edited by Chitat Chong, Qi Feng, Theodore A Slaman, W Hugh Woodin & Yue Yang

Vol. 29 Forcing, Iterated Ultrapowers, and Turing Degrees
edited by Chitat Chong, Qi Feng, Theodore A Slaman, W Hugh Woodin & Yue Yang

Vol. 30 Modular Representation Theory of Finite and *p*-Adic Groups
edited by Wee Teck Gan & Kai Meng Tan

Vol. 31 Geometric Analysis Around Scalar Curvatures
edited by Fei Han, Xingwang Xu & Weiping Zhang

Vol. 32 Mathemusical Conversations: Mathematics and Computation in Music Performance and Composition
edited by Jordan B L Smith, Elaine Chew & Gérard Assayag

Vol. 33 Sets and Computations
edited by Sy-David Friedman, Dilip Raghavan & Yue Yang

*For the complete list of titles in this series, please go to
http://www.worldscientific.com/series/LNIMSNUS

Lecture Notes Series, Institute for Mathematical Sciences,
National University of Singapore

Vol. 33

SETS AND COMPUTATIONS

Editors

Sy-David Friedman
Universität Wien, Austria

Dilip Raghavan
National University of Singapore, Singapore

Yue Yang
National University of Singapore, Singapore

World Scientific

NEW JERSEY • LONDON • SINGAPORE • BEIJING • SHANGHAI • HONG KONG • TAIPEI • CHENNAI • TOKYO

Published by

World Scientific Publishing Co. Pte. Ltd.

5 Toh Tuck Link, Singapore 596224

USA office: 27 Warren Street, Suite 401-402, Hackensack, NJ 07601

UK office: 57 Shelton Street, Covent Garden, London WC2H 9HE

Library of Congress Cataloging-in-Publication Data
Names: Friedman, Sy D., 1953– editor. | Raghavan, Dilip, editor. | Yang, Yue, 1964– editor.
Title: Sets and computations / edited by Sy-David Friedman (Universität Wien, Austria),
Dilip Raghavan (NUS, Singapore), Yue Yang (NUS, Singapore).
Description: New Jersey : World Scientific, 2017. | Series: Lecture notes series,
Institute for Mathematical Sciences, National University of Singapore ; volume 33 |
Includes bibliographical references.
Identifiers: LCCN 2017012609 | ISBN 9789813223516 (hardcover : alk. paper)
Subjects: LCSH: Set theory. | Computational complexity.
Classification: LCC QA248 .S41525 2017 | DDC 511.3/22--dc23
LC record available at https://lccn.loc.gov/2017012609

British Library Cataloguing-in-Publication Data
A catalogue record for this book is available from the British Library.

Printed in Singapore

CONTENTS

FOREWORD

The Institute for Mathematical Sciences (IMS) at the National University of Singapore was established on 1 July 2000. Its mission is to foster mathematical research, both fundamental and multidisciplinary, particularly research that links mathematics to other efforts of human endeavor, and to nurture the growth of mathematical talent and expertise in research scientists, as well as to serve as a platform for research interaction between scientists in Singapore and the international scientific community.

The Institute organizes thematic programs of longer duration and mathematical activities including workshops and public lectures. The program or workshop themes are selected from among areas at the forefront of current research in the mathematical sciences and their applications.

Each volume of the *IMS Lecture Notes Series* is a compendium of papers based on lectures or tutorials delivered at a program/workshop. It brings to the international research community original results or expository articles on a subject of current interest. These volumes also serve as a record of activities that took place at the IMS.

We hope that through the regular publication of these *Lecture Notes* the Institute will achieve, in part, its objective of reaching out to the community of scholars in the promotion of research in the mathematical sciences.

March 2017

Chitat Chong
Kwok Pui Choi
Series Editors

PREFACE

A research program on Sets and Computations was held at the IMS (Institute of Mathematical Sciences) of the NUS (National University of Singapore) from March 30 until April 30, 2015. This program not only attracted leading researchers both in Set Theory and in Computation Theory, but also placed an emphasis on recent important interactions between these two fields: computable descriptive set theory, computation theory for uncountable structures, forcing in computational complexity theory, complexity theory for computations on sets, infinite-time Turing machines, randomness beyond the hyperarithmetic and much more.

Our program was well-timed, as these connections have only become apparent in the last few years. And of course, we look forward to seeing them develop further in the near and long-term future.

The overall program structure reflected the above through a series of lectures by the participants:

Weeks 1–2 (30 March–10 April 2015): Set Theory (ideals, cardinal characteristics, set-theoretic topology, forcing axioms, Fraïssé limits, very large cardinals, singular cardinal combinatorics, combinatorial set theory, reflection principles and proper forcing). Lectures by Chodounsky, Dimonte, Farkas, A. Fischer, V. Fischer, Fuchino, Goldstern, Medini, Peng, Sakai, Shi, Sinapova, Solecki, Steprans, Todorcevic, Törnquist, Usuba, Verner, Wu, Yorioka and S. Zhang.

Week 3 (13–17 April 2015): Interactions between Set Theory and Computation Theory (effective descriptive set theory, computable cardinal characteristics, computational complexity in set theory, effective theory of Polish group actions, infinite-time Turing machines and computation on uncountable structures). Lectures by Arai, Beckmann, Buss, Kihara, Montalban, Nies, Selivanov, Schweber, Weiermann, Welch, Wilken and Wong.

Weeks 4–5 (20–30 April 2015): Computation Theory (the metamathematics of Ramsey's theorem, Turing degrees, variants of Hilbert's tenth problem, computation theory on the reals, c.e. equivalence relations, reverse recursion theory on admissible ordinals, computable categoricity, normal numbers and random graphons). Lectures by Harizanvo, Harrison-Trainor, W. Li, Miller, Morozov, Reimann, Shafer, Shore, Slaman, Sorbi, Yokoyama and Yu.

The current volume collects 10 papers resulting from the program, including 4 on Set Theory, 4 on Interactions and 2 on Computation Theory. We are very happy with the success of the program and with the excellent papers that have resulted from it.

We also would like to thank Professor Chitat Chong, IMS Director, for enabling us to carry out this program and to the friendly and efficient IMS staff for all of their help, both with the program and with this volume. And of course, the financial support of the IMS was essential for attracting so many international researchers to the program.

March 2017

Sy-David Friedman
Universität Wien

Dilip Raghavan
National University of Singapore

Yue Yang
National University of Singapore
Volume Editors

Q

Jörg Brendle[a]
Graduate School of System Informatics
Kobe University
Rokko-dai 1-1, Nada-ku
Kobe 657-8501, Japan
brendle@kurt.scitec.kobe-u.ac.jp

A Q-set is an uncountable set of reals all of whose subsets are relative G_δ sets. We prove that, for an arbitrary uncountable cardinal κ, there is consistently a Q-set of size κ whose square is not Q. This answers a question of A. Miller.

1. Introduction

A *Q-set* is an uncountable set of reals in which every subset is a relative G_δ set. If X is a Q-set, then clearly $2^{|X|} = \mathfrak{c}$. In particular, the existence of Q-sets implies $2^{\omega_1} = \mathfrak{c}$. On the other hand, under Martin's Axiom MA every uncountable sets of reals of cardinality $< \mathfrak{c}$ is Q [3]. Przymusiński [7] proved that if there is a Q-set, then there is a Q-set of size $\aleph_1$ all of whose finite powers are Q as well. Furthermore, by the above, it is clear that under MA all finite powers of Q-sets are again Q. This left open the question of whether there can consistently exist a Q-set whose square is not Q. In [1], Fleissner claimed the consistency of a stronger statement, namely, that there is a Q-set of size $\aleph_2$ while no square of a set of reals of size $\aleph_2$ is Q. However, Miller (Theorem 8 in [5]) observed that Fleissner's argument was flawed. More explicitly, he showed that if there is a Q-set of size $\aleph_2$, then there is a set of reals $X = \{x_\alpha : \alpha < \omega_2\}$ such that the set of points

[a]Partially supported by Grant-in-Aid for Scientific Research (C) 15K04977, Japan Society for the Promotion of Science. I also acknowledge partial support from Michael Hrušák's grants, CONACyT grant no. 177758 and PAPIIT grant IN-108014, during my stay at UNAM in spring 2015, when this research was started.

"above the diagonal", $\{(x_\alpha, x_\beta) : \alpha < \beta < \omega_2\}$, is a relative G_δ set in the square X^2. While this does not contradict the statement of Fleissner's result (whose correctness remains open as far as we know), it does contradict his method of proof, for he claimed that in his model, for every set of reals $X = \{x_\alpha : \alpha < \omega_2\}$, the set $\{(x_\alpha, x_\beta) : \alpha < \beta < \omega_2\}$ was not a relative G_δ in X^2.

Here we show:

Theorem: *Let κ be an arbitrary uncountable cardinal. It is consistent that there is a Q-set of size κ whose square is not Q.*

This answers a question of A. Miller (Problem 7.16 in [6]; see also [5], after the proof of Theorem 8 on p. 32).

Our model is, in a sense, very natural: we first add κ Cohen reals c_α, $\alpha < \kappa$, and then turn the set $C = \{c_\alpha : \alpha < \kappa\}$ into a Q-set in a finite support iteration of length κ^+, going through all subsets of C by a book-keeping argument, and turning them into relative G_δ's by ccc forcing (see, e.g., [4], Section 5, for this forcing). To have control over the names for conditions arising in the iteration, we present a recursive definition of this forcing with finitary conditions. Incidentally, this model is the same as the one used by Fleissner in [1], for the case $\kappa = \omega_2$, though our description is somewhat different. Furthermore, it is similar to the one used by Fleissner and Miller in [2], for the case $\kappa = \omega_1$, so that our technique also shows that the square of their Q-set, which is concentrated on the rationals, is not a Q-set. The main point is, of course, to prove that C^2 is not a Q-set. To this end, we show that the set of points "above the diagonal", $\{(c_\alpha, c_\beta) : \alpha < \beta < \kappa\}$, is not a relative G_δ set in C^2. Unlike for Fleissner's work [1], there is no contradiction with Miller's result (Theorem 8 in [5]), because we prove a much weaker statement.

Acknowledgment. I thank Michael Hrušák for bringing this problem to my attention and for suggesting the model. I am also grateful to the referee for pointing out the reference [2].

2. Proof of the Theorem

Assume the ground model V satisfies GCH. We perform a finite support iteration $(\mathbb{P}_\gamma : \gamma \leq \kappa^+)$ of ccc forcing. As usual, elements of $\mathbb{P}_\gamma$ are functions with domain γ. Also, a book-keeping argument will hand us down a sequence $(\dot{A}_\gamma : \gamma < \kappa^+)$ of $\mathbb{P}_\gamma$-names of subsets of κ such that for every $\mathbb{P}_{\kappa^+}$-name $\dot{A}$ for a subset of κ there are $\gamma < \kappa^+$ such that $\mathbb{P}_\gamma$ forces that $\dot{A} = \dot{A}_\gamma$. Since this is a standard argument we will omit details. The recursive definition of the $\mathbb{P}_\gamma$ is as follows:

- $\mathbb{P}_0$ is the trivial forcing (as usual).
- $\mathbb{P}_1$ is $\mathbb{C}_\kappa$, the forcing for adding κ Cohen reals: if $p \in \mathbb{P}_1$, then $p(0) = (\sigma^p_\alpha : \alpha \in F^p)$ where $F^p \subseteq \kappa$ is finite and $\sigma^p_\alpha \in 2^{<\omega}$ for $\alpha \in F^p$; the ordering is given by $q \leq p$ if $F^q \supseteq F^p$ and $\sigma^q_\alpha \supseteq \sigma^p_\alpha$ for all $\alpha \in F^p$.
- If γ is a limit ordinal, $\mathbb{P}_\gamma$ consists of all functions p with $\mathrm{dom}(p) = \gamma$ such that there is $0 < \delta < \gamma$ with $p \restriction \delta \in \mathbb{P}_\delta$ and $p(\epsilon) = \emptyset$ for all ϵ with $\delta \leq \epsilon < \gamma$; the ordering is given by $q \leq p$ if $q \restriction \delta \leq p \restriction \delta$ for all $\delta < \gamma$.
- If $\gamma \geq 1$, $\mathbb{P}_{\gamma+1}$ consists of all functions p with $\mathrm{dom}(p) = \gamma + 1$, $p \restriction \gamma \in \mathbb{P}_\gamma$ and $p(\gamma)$ is a finite (possibly empty) subset of $(2^{<\omega} \cup \kappa) \times \omega$ such that
  - if $(\alpha, n) \in p(\gamma)$ for some $\alpha \in \kappa$ and $n \in \omega$, then $\alpha \in F^p$ and $p \restriction \gamma \Vdash \alpha \notin \dot{A}_\gamma$,
  - if $(\alpha, n), (\sigma, n) \in p(\gamma)$ for some $\alpha \in \kappa$, $\sigma \in 2^{<\omega}$ and $n \in \omega$, then σ^p_α and σ are incompatible in $2^{<\omega}$;

  the ordering is given by $q \leq p$ if $q \restriction \gamma \leq p \restriction \gamma$ and $q(\gamma) \supseteq p(\gamma)$.

For $\gamma \leq \kappa^+$ and $p \in \mathbb{P}_\gamma$, let $\mathrm{supp}(p) = \{0\} \cup \{\gamma > 0 : p(\gamma) \neq \emptyset\}$ denote the *support* of p. First of all, let us note that this is indeed an iteration, that is, $\mathbb{P}_\gamma \lessdot \mathbb{P}_\delta$ for $\gamma < \delta$. This is straightforward, for if $p \in \mathbb{P}_\delta$ and $q' \in \mathbb{P}_\gamma$ with $q' \leq p \restriction \gamma$, then $q \in \mathbb{P}_\delta$ defined by $q \restriction \gamma = q'$ and $q(\epsilon) = p(\epsilon)$ for $\gamma \leq \epsilon < \delta$ is a common extension of q' and p.

Here is a sufficient criterion for compatibility of conditions.

Lemma 2.1: *Assume $p, q \in \mathbb{P}_\gamma$ are such that*

- $\sigma^p_\alpha = \sigma^q_\alpha$ *for all* $\alpha \in F^p \cap F^q$,
- *for all* $\delta \in \mathrm{supp}(p) \cap \mathrm{supp}(q)$ *with* $\delta > 0$ *and all* $(\sigma, n) \in 2^{<\omega} \times \omega$,

$$(\sigma, n) \in p(\delta) \Longleftrightarrow (\sigma, n) \in q(\delta).$$

Then p and q are compatible with a canonical common extension r given by

- $\mathrm{supp}(r) = \mathrm{supp}(p) \cup \mathrm{supp}(q)$,
- $F^r = F^p \cup F^q$,
- $\sigma^r_\alpha = \begin{cases} \sigma^p_\alpha & \text{if } \alpha \in F^p \\ \sigma^q_\alpha & \text{if } \alpha \in F^q, \end{cases}$

- $r(\delta) = p(\delta) \cup q(\delta)$ *for all* $\delta \in \mathrm{supp}(r)$ *with* $\delta > 0$.

Proof: By induction on $\delta \leq \gamma$ we simultaneously prove that $r{\restriction}\delta$ as defined in the lemma is indeed a condition and that $r{\restriction}\delta \leq p{\restriction}\delta, q{\restriction}\delta$ holds. For $\delta = 1$, this is straightforward by assumption. The limit step is also clear.

So suppose this has been proved for $\delta \geq 1$, and we show it for $\delta + 1$. If $(\alpha, n) \in r(\delta)$ for some α and n, then, without loss of generality, we may assume $(\alpha, n) \in p(\delta)$. Thus $p{\restriction}\delta \Vdash \alpha \notin \dot{A}_\delta$ and, since $r{\restriction}\delta \leq p{\restriction}\delta$ by induction hypothesis, we also have $r{\restriction}\delta \Vdash \alpha \notin \dot{A}_\delta$, as required.

Next assume (α, n) and (σ, n) belong to $r(\delta)$ for some α, σ and n. If both belong to either $p(\delta)$ or $q(\delta)$, there is nothing to show. So we may assume without loss of generality $(\alpha, n) \in p(\delta)$ and $(\sigma, n) \in q(\delta)$. In particular, $\delta \in \mathrm{supp}(p) \cap \mathrm{supp}(q)$ and, by assumption, $(\sigma, n) \in p(\delta)$ follows. Since p is a condition, $\sigma^r_\alpha = \sigma^p_\alpha$ and σ are incompatible, as required.

Thus we have proved that $r{\restriction}\delta + 1 \in \mathbb{P}_{\delta+1}$, and $r{\restriction}\delta + 1 \leq p{\restriction}\delta + 1, q{\restriction}\delta + 1$ now follows easily. This completes the induction and the proof of the lemma.

□

This lemma presents a basic pattern of how to define a common extension of two conditions and then show by induction that the object defined really is a condition. We shall use this pattern several times, see Lemma 2.3 and Claim 2.10 below. An immediate consequence is:

Corollary 2.2: $\mathbb{P}_\gamma$ *is ccc (and even satisfies Knaster's condition) for every* $\gamma \leq \kappa^+$.

Proof: Let $\{p_\zeta : \zeta < \omega_1\} \subseteq \mathbb{P}_\gamma$. By a straightforward Δ-system argument we see that we may assume that for any $\zeta < \xi < \omega_1$, $p = p_\zeta$ and $q = p_\xi$ satisfy the assumptions of the previous lemma. Hence the p_ζ are pairwise compatible. □

Furthermore, compatible conditions sort of have a "minimal" extension.

Lemma 2.3: *Assume* $p, q \in \mathbb{P}_\gamma$ *are compatible with common extension* r. *Then there is a condition* $s = s^{p,q,r} \in \mathbb{P}_\gamma$ *with* $r \leq s \leq p, q$ *such that*

- $\mathrm{supp}(s) = \mathrm{supp}(p) \cup \mathrm{supp}(q)$,
- $F^s = F^p \cup F^q$,
- $\sigma^s_\alpha = \sigma^r_\alpha$ *for all* $\alpha \in F^s$,
- $s(\delta) = p(\delta) \cup q(\delta)$ *for all* $\delta \in \mathrm{supp}(s)$ *with* $\delta > 0$.

Proof: By induction on $\delta \leq \gamma$ we simultaneously prove that $s\restriction\delta$ as defined in the lemma is indeed a condition and that $r\restriction\delta \leq s\restriction\delta \leq p\restriction\delta, q\restriction\delta$ holds. For $\delta = 1$ this is obvious. The limit step is also straightforward.

So assume this has been proved for $\delta \geq 1$. First suppose that $(\alpha, n) \in s(\delta)$ for some α and n. Without loss of generality we may assume that $(\alpha, n) \in p(\delta)$. Thus $p\restriction\delta \Vdash \alpha \notin \dot{A}_\delta$. Since $s\restriction\delta \leq p\restriction\delta$ by induction hypothesis, also $s\restriction\delta \Vdash \alpha \notin \dot{A}_\delta$, as required.

Next suppose $(\alpha, n), (\sigma, n)$ are in $s(\delta)$ for some α, σ and n. Since $r \leq p, q$, $r(\delta) \supseteq s(\delta)$, and $(\alpha, n), (\sigma, n) \in r(\delta)$ follows. In particular, $\sigma^s_\alpha = \sigma^r_\alpha$ and σ are incompatible in $2^{<\omega}$, as required.

Thus $s\restriction\delta + 1 \in \mathbb{P}_{\delta+1}$ and $r\restriction\delta + 1 \leq s\restriction\delta + 1 \leq p\restriction\delta + 1, q\restriction\delta + 1$ is now obvious. This completes the induction and the proof of the lemma. □

In particular, two conditions p, q are compatible iff they have a common extension s with $\mathrm{supp}(s) = \mathrm{supp}(p) \cup \mathrm{supp}(q)$, $F^s = F^p \cup F^q$, $\sigma^s_\alpha \supseteq \sigma^p_\alpha, \sigma^q_\alpha$ for all $\alpha \in F^s$, and $s(\delta) = p(\delta) \cup q(\delta)$ for all $\delta \in \mathrm{supp}(s)$. (Lemma 2.3 will not be needed but gives some motivation for Claim 2.10 below.)

Let $\dot{G}$ be the name for the $\mathbb{P}_{\kappa^+}$-generic filter. For $0 < \gamma < \kappa^+$ and $n \in \omega$ define $\mathbb{P}_{\kappa^+}$-names (more explicitly, $\mathbb{P}_{\gamma+1}$-names)

$$\dot{U}_{\gamma,n} = \bigcup\{[\sigma] : (\sigma, n) \in p(\gamma) \text{ for some } p \in \dot{G}\} \text{ and } \dot{H}_\gamma = \bigcap_{n\in\omega} \dot{U}_{\gamma,n}.$$

Clearly, the $\dot{U}_{\gamma,n}$ are names for open sets, and $\dot{H}_\gamma$ is the name for a G_δ set. Also let

$$\dot{c}_\alpha = \bigcup\{\sigma : \sigma = \sigma^p_\alpha \text{ for some } p \in \dot{G}\}$$

be the canonical name for the Cohen real added in coordinate α of stage 1 of the forcing, for each $\alpha < \kappa$.

Lemma 2.4: *(i) Assume $n \in \omega, \alpha < \kappa, \gamma < \kappa^+$ and $p \in \mathbb{P}_{\kappa^+}$ are given such that $p\restriction\gamma \Vdash_\gamma \alpha \in \dot{A}_\gamma$. Then there are $q \leq p$ and $\sigma \subsetneq \sigma^q_\alpha$ such that $(\sigma, n) \in q(\gamma)$.*

(ii) Assume $\alpha < \kappa, \gamma < \kappa^+$ and $p \in \mathbb{P}_{\kappa^+}$ are given such that $p\restriction\gamma \Vdash_\gamma \alpha \notin \dot{A}_\gamma$. Then there are $q \leq p$ and $n \in \omega$ such that $(\alpha, n) \in q(\gamma)$.

Proof: (i) Assume n, α, γ, p are given as required. Then clearly $(\alpha, n) \notin p(\gamma)$. For $\beta \in F^p$ extend σ^p_β to σ^q_β such that they are pairwise incompatible. This defines $q\restriction 1$ (in particular, $F^q = F^p$). For $\delta > 0$ let

$$q(\delta) = \begin{cases} p(\delta) & \text{if } \delta \neq \gamma \\ p(\delta) \cup \{(\sigma^q_\alpha, n)\} & \text{if } \delta = \gamma \end{cases}$$

(in particular, $\text{supp}(q) = \text{supp}(p) \cup \{\gamma\}$). It is easy to see that q is a condition and that it strengthens p.

(ii) Assume α, γ, p are given as required. Let $F^q = F^p \cup \{\alpha\}$ and let σ^q_α be either σ^p_α (if $\alpha \in F^p$) or arbitrary (otherwise). Let n be large enough so that no (σ, n) belongs to $p(\gamma)$. For $\delta > 0$ let

$$q(\delta) = \begin{cases} p(\delta) & \text{if } \delta \neq \gamma \\ p(\delta) \cup \{(\alpha, n)\} & \text{if } \delta = \gamma \end{cases}$$

(in particular, $\text{supp}(q) = \text{supp}(p) \cup \{\gamma\}$). Again, it is easy to see that q is a condition and that it strengthens p. □

Corollary 2.5: *(i) Assume $\alpha < \kappa, \gamma < \kappa^+$ and $p \in \mathbb{P}_{\kappa^+}$ are such that $p{\restriction}\gamma \Vdash_\gamma \alpha \in \dot{A}_\gamma$. Then $p{\restriction}\gamma + 1 \Vdash_{\gamma+1} \dot{c}_\alpha \in \dot{H}_\gamma$.*

(ii) Assume $\alpha < \kappa, \gamma < \kappa^+$ and $p \in \mathbb{P}_{\kappa^+}$ are such that $p{\restriction}\gamma \Vdash_\gamma \alpha \notin \dot{A}_\gamma$. Then $p{\restriction}\gamma + 1 \Vdash_{\gamma+1} \dot{c}_\alpha \notin \dot{H}_\gamma$.

Proof: (i) This is immediate from Lemma 2.4 (i).

(ii) Let $q \leq p$ be arbitrary. By Lemma 2.4 (ii), there are $r \leq q$ and $n \in \omega$ with $(\alpha, n) \in r(\gamma)$. It suffices to show that $r \Vdash \dot{c}_\alpha \notin \dot{U}_{\gamma,n}$.

Assume that G is $\mathbb{P}_{\kappa^+}$-generic with $r \in G$ and $c_\alpha \in U_{\gamma,n}$. By definition of $U_{\gamma,n}$, there are $r' \leq r$ and $\sigma \in 2^{<\omega}$ such that $r' \in G$, $(\sigma, n) \in r'(\gamma)$ and $\sigma \subseteq c_\alpha$. Hence there is $r'' \leq r'$ such that $r'' \in G$ and $\sigma \subseteq \sigma^{r''}_\alpha$. Since $(\alpha, n), (\sigma, n) \in r''(\gamma)$, this contradicts the definition of a condition. Hence we must have $c_\alpha \notin U_{\gamma,n}$, as required. □

As a consequence, we see that

$$\Vdash_{\gamma+1} \dot{H}_\gamma \cap \{\dot{c}_\alpha : \alpha < \kappa\} = \{\dot{c}_\alpha : \alpha \in \dot{A}_\gamma\}.$$

Thus we obtain:

Corollary 2.6: $\Vdash_{\kappa^+}$ "$\{\dot{c}_\alpha : \alpha < \kappa\}$ *is a Q-set*".

To see that the square of the set of Cohen reals is not a Q-set, it clearly suffices to establish the following:

Main Lemma 2.7: $\Vdash_{\kappa^+}$ "$\{(\dot{c}_\alpha, \dot{c}_\beta) : \alpha < \beta < \kappa\}$ is not a relative G_δ-set".

Proof: Assume that $\dot{V}_n$, $n \in \omega$, are $\mathbb{P}_{\kappa^+}$-names of open sets in the plane $(2^\omega)^2$ such that $\{(\dot{c}_\alpha, \dot{c}_\beta) : \alpha < \beta < \kappa\} \subseteq \bigcap_{n\in\omega} \dot{V}_n$ is forced by the trivial condition. We shall find $\beta < \alpha < \kappa$ such that the trivial condition forces $(\dot{c}_\alpha, \dot{c}_\beta) \in \bigcap_{n\in\omega} \dot{V}_n$. This is clearly sufficient.

There are $\dot{S}_n \subseteq (2^{<\omega})^2$ such that for every $n \in \omega$, $\dot{V}_n = \bigcup\{[\sigma] \times [\tau] : (\sigma, \tau) \in \dot{S}_n\}$ is forced. For each $n \in \omega$ and each $(\sigma, \tau) \in (2^{<\omega})^2$, let $\{p^j_{n,\sigma,\tau} : j \in \omega\}$ be a maximal antichain of conditions in $\mathbb{P}_{\kappa^+}$ deciding the statement $(\sigma, \tau) \in \dot{S}_n$.

Let $Z = \{z_i : i \in \omega\}$ be a countable set disjoint from κ. We say that $\mathsf{t} = (\Gamma^{\mathsf{t}}, \Delta^{\mathsf{t}}, \bar{D}^{\mathsf{t}} = (D^{\mathsf{t}}_\gamma : \gamma \in \Gamma^{\mathsf{t}} \setminus \{0\}), \bar{E}^{\mathsf{t}} = (E^{\mathsf{t}}_\gamma; \gamma \in \Gamma^{\mathsf{t}} \setminus \{0\}), \bar{\tau}^{\mathsf{t}} = (\tau^{\mathsf{t}}_\zeta : \zeta \in \Delta^{\mathsf{t}}))$ is a *pattern* if

- $\Gamma^{\mathsf{t}} \subseteq \kappa^+$ is finite, $0 \in \Gamma^{\mathsf{t}}$,
- $\Delta^{\mathsf{t}} \subseteq \kappa \cup Z$ is finite and $\Delta^{\mathsf{t}} \cap Z$ is an initial segment of Z, i.e., $\Delta^{\mathsf{t}} \cap Z = \{z_i : i < k\}$ for some k,
- $D^{\mathsf{t}}_\gamma \subseteq 2^{<\omega} \times \omega$ is finite for $\gamma \in \Gamma^{\mathsf{t}}$,
- $E^{\mathsf{t}}_\gamma \subseteq \Delta^{\mathsf{t}} \times \omega$ is finite for $\gamma \in \Gamma^{\mathsf{t}}$,
- $\tau^{\mathsf{t}}_\zeta \in 2^{<\omega}$ for $\zeta \in \Delta^{\mathsf{t}}$.

Usually we will omit the superscript t. Let PAT denote the collection of all patterns.

Let $X \subseteq \kappa^+$ and $Y \subseteq \kappa$. Assume $0 \in X$. Also let $p \in \mathbb{P}_{\kappa^+}$. We say that $(\Gamma, \Delta, \bar{D}, \bar{E}, \bar{\tau}) \in \mathsf{PAT}$ is the *(X, Y)-pattern of p* if $\Gamma \subseteq X$, $\Delta \subseteq Y \cup Z$ and for some (necessarily unique) one-to-one function $\varphi = \varphi^p : \Delta \to \kappa$ with $\varphi{\restriction}(\Delta \cap Y) = \mathrm{id}$ and $\varphi{\restriction}(\Delta \cap Z) : Z \to \kappa \setminus Y$ order-preserving (i.e., if $i < j$ and $z_i, z_j \in \Delta \cap Z$, then $\varphi(z_i) < \varphi(z_j)$), the following hold:

- $\mathrm{supp}(p) \cap X = \Gamma$,
- $F^p = \{\varphi(\zeta) : \zeta \in \Delta\}$ (in particular, $F^p \cap Y = \Delta \cap Y$ and $F^p \setminus Y = \{\varphi(\zeta) : \zeta \in \Delta \cap Z\}$),
- $\sigma^p_{\varphi(\zeta)} = \tau_\zeta$ for $\zeta \in \Delta$,
- for $\gamma \in \Gamma$ and $(\sigma, n) \in 2^{<\omega} \times \omega$:

$$(\sigma, n) \in p(\gamma) \Longleftrightarrow (\sigma, n) \in D_\gamma,$$

- for $\gamma \in \Gamma$ and $(\zeta, n) \in \Delta \times \omega$:

$$(\varphi(\zeta), n) \in p(\gamma) \Longleftrightarrow (\zeta, n) \in E_\gamma.$$

Clearly each condition has a unique (X, Y)-pattern.

Let $\chi \geq (2^{\kappa^+})^+$, and let $M \prec H(\chi)$ be a countable elementary submodel containing κ, $\mathbb{P}_{\kappa^+}$, Z, PAT, all $\dot{V}_n$, $\dot{S}_n$, and $\{p^j_{n,\sigma,\tau} : j \in \omega, (\sigma, \tau) \in (2^{<\omega})^2\}$. Choose any ordinals β, α with $M \cap \omega_1 \leq \beta < \alpha < \omega_1$. We shall see that β and α are as required, that is, that $\Vdash (\dot{c}_\alpha, \dot{c}_\beta) \in \bigcap_{n \in \omega} \dot{V}_n$. By a standard density argument, it suffices to prove the following:

Main Claim 2.8: For all $n \in \omega$ and all $p \in \mathbb{P}_{\kappa^+}$ there are $q \leq p$ and $\sigma_0, \tau_0 \in 2^{<\omega}$ such that $\sigma_0 \subseteq \sigma^q_\alpha$, $\tau_0 \subseteq \sigma^q_\beta$ and $q \Vdash (\sigma_0, \tau_0) \in \dot{S}_n$.

Proof: Fix $n \in \omega_2$. Let $p \in \mathbb{P}_{\kappa^+}$ be given. By going over to a stronger condition, if necessary, we may assume $\alpha, \beta \in F^p$. Let $X = M \cap \kappa^+$ and $Y = M \cap \kappa$. Let $\mathrm{t} = (\Gamma, \Delta, \bar{D}, \bar{E}, \bar{\tau})$ be the (X, Y)-pattern of p. Put $X_0 := \Gamma$ and $Y_0 := \Delta \cap \kappa$. Then we see that $X_0 = \Gamma = \mathrm{supp}(p) \cap X = \mathrm{supp}(p) \cap M$ and $Y_0 = \Delta \cap Y = F^p \cap Y = F^p \cap M$. In particular, t is also the (X_0, Y_0)-pattern of p. Furthermore, unlike X and Y, X_0 and Y_0 belong to M, and so does t. Since $\alpha \in F^p \setminus Y$, there is $i \in \omega$ such that $\varphi^p(z_i) = \alpha$. So

$$H_\chi \models \exists p' \in \mathbb{P}_{\kappa^+}\ \exists \alpha' < \omega_1\ (\mathrm{t} \text{ is the } (X_0, Y_0)\text{-pattern of } p' \text{ and } \varphi^{p'}(z_i) = \alpha')$$

because this statement is true for $p' = p$ and $\alpha' = \alpha$. By elementarity we obtain

$$M \models \exists p' \in \mathbb{P}_{\kappa^+}\ \exists \alpha' < \omega_1\ (\mathrm{t} \text{ is the } (X_0, Y_0)\text{-pattern of } p' \text{ and } \varphi^{p'}(z_i) = \alpha').$$

Let $p' \in M \cap \mathbb{P}_{\kappa^+}$ and $\alpha' < M \cap \omega_1 \leq \beta$ be the witnesses.

Claim 2.9: *p and p' are compatible, with common extension p'' given canonically according to Lemma 2.1.*

Proof: We need to verify the two conditions in Lemma 2.1. Since $F^{p'} \subseteq M \cap \kappa = Y$ and $F^p \cap Y = F^p \cap Y_0 = \Delta \cap Y_0 = F^{p'} \cap Y_0$, we see that $F^p \cap F^{p'} = \Delta \cap Y_0$ and $\sigma^p_\zeta = \tau_\zeta = \sigma^{p'}_\zeta$ for all $\zeta \in \Delta \cap Y_0$. Similarly, since $\mathrm{supp}(p') \subseteq M \cap \kappa^+ = X$ and $\mathrm{supp}(p) \cap X = \mathrm{supp}(p) \cap X_0 = \Gamma = \mathrm{supp}(p') \cap X_0$, we have that $\mathrm{supp}(p) \cap \mathrm{supp}(p') = \Gamma$ and, for $\gamma \in \Gamma$,

$$(\sigma, n) \in p(\gamma) \Longleftrightarrow (\sigma, n) \in D_\gamma \Longleftrightarrow (\sigma, n) \in p'(\gamma).$$

Hence the conditions are indeed satisfied. □

Let $\psi = \varphi^{p'} \circ (\varphi^p)^{-1}$. Clearly ψ maps F^p one-to-one and onto $F^{p'}$, and we have $\alpha' = \psi(\alpha)$. We also note for later use that for all $\zeta \in F^p$, $\sigma^{p''}_\zeta = \sigma^p_\zeta = \tau_{(\varphi^p)^{-1}(\zeta)} = \tau_{(\varphi^{p'})^{-1}(\psi(\zeta))} = \sigma^{p'}_{\psi(\zeta)} = \sigma^{p''}_{\psi(\zeta)}$, and for all $\gamma \in \Gamma$ and $\zeta \in F^p$,

$$(\zeta, n) \in p(\gamma) \Longleftrightarrow ((\varphi^p)^{-1}(\zeta), n) \in E_\gamma \Longleftrightarrow (\psi(\zeta), n) \in p'(\gamma).$$

Since $\Vdash (\dot{c}_{\alpha'}, \dot{c}_\beta) \in \dot{V}_n$, there are $\tilde{p} \leq p''$ and $\sigma_0, \tau_0 \in 2^{<\omega}$ such that

$$\tilde{p} \Vdash (\dot{c}_{\alpha'}, \dot{c}_\beta) \in [\sigma_0] \times [\tau_0] \subseteq \dot{V}_n,$$

that is, $\sigma_0 \subseteq \sigma^{\tilde{p}}_{\alpha'}$, $\tau_0 \subseteq \sigma^{\tilde{p}}_\beta$ and $\tilde{p} \Vdash (\sigma_0, \tau_0) \in \dot{S}_n$.

By construction, for some $j \in \omega$, the condition $r = p^j_{n,\sigma_0,\tau_0} \in M$ is compatible with $\tilde{p}$. In particular, r must also force $(\sigma_0, \tau_0) \in \dot{S}_n$. Furthermore we know that $\mathrm{supp}(r) \subseteq X$ and $F^r \subseteq Y$. By strengthening $\tilde{p}$, if necessary, we may assume that $\tilde{p} \leq r$.

Claim 2.10: *p and r have a common extension q such that $\sigma^q_\alpha = \sigma^{\tilde{p}}_{\alpha'}$ and $\sigma^q_\beta = \sigma^{\tilde{p}}_\beta$.*

Note that we know already that p and r are compatible because they have common extension $\tilde{p}$; however, here we construct a different common extension q (in general $\sigma^{\tilde{p}}_\alpha$ and σ^q_α are distinct).

Proof: As in the proofs of Lemmata 2.1 and 2.3, we define q and then show by induction on γ that $q \restriction \gamma \in \mathbb{P}_\gamma$ and that $q \restriction \gamma$ extends both $p \restriction \gamma$ and $r \restriction \gamma$.

- $\mathrm{supp}(q) = \mathrm{supp}(p) \cup \mathrm{supp}(r)$,
- $F^q = F^p \cup F^r$,
- for $\zeta \in F^q$, $\sigma^q_\zeta = \begin{cases} \sigma^{\tilde{p}}_\zeta & \text{if } \zeta \in F^r \text{ or } \zeta = \beta \\ \sigma^{\tilde{p}}_{\psi(\zeta)} & \text{if } \zeta \in F^p \setminus F^r \text{ and } \zeta \neq \beta, \end{cases}$
- for $\gamma \in \mathrm{supp}(q)$ with $\gamma > 0$, $q(\gamma) = p(\gamma) \cup r(\gamma)$.

For the induction, first let $\gamma = 1$. If $\zeta \in F^r$, then by $\tilde{p} \leq r$, $\sigma^r_\zeta \subseteq \sigma^{\tilde{p}}_\zeta = \sigma^q_\zeta$. If, additionally, $\zeta \in F^p$, then by $\tilde{p} \leq p''$, $\sigma^p_\zeta = \sigma^{p''}_\zeta \subseteq \sigma^{\tilde{p}}_\zeta = \sigma^q_\zeta$. If $\zeta = \beta$, we similarly have $\sigma^p_\beta = \sigma^{p''}_\beta \subseteq \sigma^{\tilde{p}}_\beta = \sigma^q_\beta$. Finally, if $\zeta \in F^p \setminus F^r$ with $\zeta \neq \beta$, then, using again $\tilde{p} \leq p''$, we see that $\sigma^p_\zeta = \sigma^{p'}_{\psi(\zeta)} = \sigma^{p''}_{\psi(\zeta)} \subseteq \sigma^{\tilde{p}}_{\psi(\zeta)} = \sigma^q_\zeta$. Thus $q \restriction 1 \leq p \restriction 1, r \restriction 1$, as required.

Note, in particular, that $\sigma^q_\alpha = \sigma^{\tilde{p}}_{\psi(\alpha)} = \sigma^{\tilde{p}}_{\alpha'}$.

As usual, the limit step of the induction is trivial. Let us assume we have proved the statement for some $\gamma \geq 1$, and let us prove it for $\gamma + 1$. First assume that $(\zeta, n) \in q(\gamma)$. Then we see that $q \restriction \gamma \Vdash \zeta \notin \dot{A}_\gamma$ as in the proofs of Lemmata 2.1 and 2.3.

Hence assume $(\zeta, n), (\sigma, n) \in q(\gamma)$. As in the proof of Lemma 2.1, we may assume that $\gamma \in \mathrm{supp}(p) \cap \mathrm{supp}(r)$. In particular $\gamma \in X_0 = \Gamma$. First suppose $(\zeta, n) \in r(\gamma)$ and $(\sigma, n) \in p(\gamma)$. Then $\zeta \in F^r \subseteq F^{\tilde{p}}$. Since $\tilde{p} \leq p, r$, we see that $(\zeta, n), (\sigma, n) \in \tilde{p}(\gamma)$, and since $\tilde{p}$ is a condition, σ and $\sigma^{\tilde{p}}_\zeta = \sigma^q_\zeta$ are incompatible, as required.

Next suppose $(\zeta, n) \in p(\gamma)$ and $(\sigma, n) \in r(\gamma)$. So $\zeta \in F^p$. We split into cases according to whether $\zeta = \beta$ or not. First assume $\zeta \neq \beta$. Then, using $\tilde{p} \leq p'' \leq p'$, we see $(\psi(\zeta), n) \in p'(\gamma) \subseteq p''(\gamma) \subseteq \tilde{p}(\gamma)$ and $(\sigma, n) \in \tilde{p}(\gamma)$.

Since $\tilde{p}$ is a condition, $\sigma^{\tilde{p}}_{\psi(\zeta)}$ and σ are incompatible. If $\zeta \in F^r$, then $\zeta \in Y_0$, and $\psi(\zeta) = \zeta$ follows. Hence $\sigma^q_\zeta = \sigma^{\tilde{p}}_\zeta = \sigma^{\tilde{p}}_{\psi(\zeta)}$. If $\zeta \notin F^r$, then $\sigma^q_\zeta = \sigma^{\tilde{p}}_{\psi(\zeta)}$ by definition. In either case, we see that σ^q_ζ and σ are incompatible.

Now assume $\zeta = \beta$. Since $\tilde{p} \leq p, r$, we have $(\beta, n), (\sigma, n) \in \tilde{p}(\gamma)$, and $\sigma^q_\beta = \sigma^{\tilde{p}}_\beta$ and σ are incompatible because $\tilde{p}$ is a condition.

This shows that $q{\restriction}\gamma+1$ is a condition. Clearly $q{\restriction}\gamma+1 \leq p{\restriction}\gamma+1, r{\restriction}\gamma+1$. This completes the induction. □

Thus $q \leq p$, $\sigma_0 \subseteq \sigma^q_\alpha$, $\tau_0 \subseteq \sigma^q_\beta$ and q forces that (σ_0, τ_0) belongs to $\dot{S}_n$, and the proof of the main claim is complete. □

This completes the proof of the main lemma and of the theorem. □

References

1. W.G. Fleissner, *Squares of Q sets*, Fund. Math. **118** (1983), 223-231.
2. W.G. Fleissner and A.W. Miller, *On Q sets*, Proc. Amer. Math. Soc. **78** (1980), 280-284.
3. D.A. Martin and R.M. Solovay, *Internal Cohen extensions*, Ann. Math. Logic **2** (1970), 143-178.
4. A.W. Miller, *Descriptive Set Theory and Forcing*, Lecture Notes in Logic **4**, Springer 1995.
5. A.W. Miller, *A hodgepodge of sets of reals*, Note di Matematica **27** (2007), 25-39.
6. A.W. Miller, *Some interesting problems*, version of April 2015.
7. T. Przymusiński, *The existence of Q-sets is equivalent to the existence of strong Q-sets*, Proc. Amer. Math. Soc. **79** (1980), 626-628.

A FRAÏSSÉ APPROACH TO THE POULSEN SIMPLEX

Clinton Conley[a]
Department of Mathematical Sciences, Wean Hall 6113
Carnegie Mellon University, Pittsburgh, PA 15213, USA
clintonc@andrew.cmu.edu

Asger Törnquist[b]
Department of Mathematical Sciences, University of Copenhagen
Universitetsparken 5, 2100 Copenhagen, Denmark
asgert@math.ku.dk

We consider the Fraïssé class of finite dimensional simplices, and show that the Fraïssé limit is the so-called Poulsen simplex, a separable simplex with dense extremal boundary which was constructed by E. T. Poulsen in 1961.

1. Introduction

This paper presents an attempt at connecting the *Poulsen simplex*, a well-known object in convexity theory, with *Fraïssé theory*, a well-known limit construction in model theory.

Convexity theory is a part of analysis concerned with compact convex sets in topological vector spaces. Choquet simplices (or just simplices) are compact convex sets of particular interest because each point in a Choquet simplex is the centre of mass of a unique probability measure supported on the *extremal boundary* (see §2) of the simplex. In finite dimension, examples include triangles and tetrahedrons, as every point inside a triangle or

[a]Clinton Conley was supported by the NSF grant DMS-1500906.
[b]Asger Törnquist was supported by a Sapere Aude fellowship (level 2) from Denmark's Natural Sciences Research Council, and by the DNRF Niels Bohr Professorship of Lars Hesselholt.

tetrahedron is the unique convex combination of the vertices. Circles and spheres are, on the other hand, not simplices.

In infinite dimension, the geometrical and topological properties of simplices can be far more intricate. The Poulsen simplex is a particular example of a simplex whose behaviour is impossible in finite dimension: It is a separable simplex, yet the extremal boundary is a dense G_δ set in the Poulsen simplex. The question if a simplex with these properties could exist had been raised by Choquet, and Poulsen answered Choquet's question by giving a construction in [9].

It turns out that up to affine homeomorphism there is only one separable simplex in which the extremal boundary is dense, a fact that was proved in the paper [6]. Along the way, the authors of [6] prove that the Poulsen simplex has a number of homogeneity properties, which eventually facilitates a proof of uniqueness that a logician would recognize as a back-and-forth argument.

This raises the question if the Poulsen simplex can be constructed as some sort of Fraïssé limit, so that existence, uniqueness and homogeneity properties of the Poulsen simplex are linked more naturally.

This note presents such a construction and ensuing proofs of uniqueness. We take the route, á la Kadison, of going to the natural dual category of *complete order unit spaces*. This has several advantages, the most immediate of which is working with *direct* limits as opposed to inverse limits. However, in principle everything we do could be rephrased in terms of inverse limits, and dual ("projective") Fräisse limits in the spirit of [2].

The paper is organized as follows: §2 presents the required background on simplices and Kadison's dual theory of complete order unit spaces; In §3 we show that the class of finite-dimensional complete order unit spaces form a Fraïssé class, and that (the completion of) the resulting limit Fraïssé limit is the order unit space associated to the Poulsen simplex; and uniqueness of the Poulsen simplex in terms of its defining topological properties is established. §4 contains a few open problems.

2. Background on Convexity Theory

This section recalls the elementary notions of convexity theory that we need; for further background, [8] provides an accessible reference; a more systematic development can be found in [1] or the encyclopedic [7].

Let K be a compact convex subset of some locally convex (real) vector space. We let $A(K)$ denote the set of all continuous affine real-valued $f : K \to \mathbb{R}$, which we equip with the sup-norm, denoted $\|\cdot\|_\infty$. One easily verifies that $A(K)$ is then a closed subspace of $(C_{\mathbb{R}}(K), \|\cdot\|_\infty)$, the Banach space of real-valued continuous functions on K, and so $A(K)$ is itself a Banach space. We let $A^*(K)$ denote the dual space of $A(K)$, i.e. the space of real-valued bounded functionals on $A(K)$.

The *extremal boundary* of K, denoted $\partial_e K$, consists of the points $x \in K$ that cannot be written as a non-trivial convex combination of points in K. The following result is as old as water, but we include a pleasantly short proof.

Proposition 2.1: *If K is metrizable, then $\partial_e K$ is a G_δ set.*

Proof: Fix a compatible metric d on K. For each $\varepsilon > 0$ and $\frac{1}{2} > \delta > 0$, let

$$A_{\varepsilon,\delta} = \{(x, y, z, t) \in K \times K \times K \times [0,1] : d(y, z) \geq \varepsilon \wedge t \in [\delta, 1-\delta] \wedge x = ty + (1-t)z\}.$$

Then $A_{\varepsilon,\delta}$ is closed, and so compact, whence

$$\mathrm{proj}_X A_{\varepsilon,\delta} = \{x : (\exists y, z, t)(x, y, z, t) \in A_{\varepsilon,\delta}\}$$

is closed. Since

$$x \notin \partial_e K \iff (\exists \varepsilon, \delta \in \mathbb{Q}_+) x \in \mathrm{proj}_X(A_{\varepsilon,\delta})$$

it follows that $K \setminus \partial_e K$ is F_σ, as required. □

If μ is a Borel probability measure on K, we say that μ *represents* $x_0 \in K$ if for all $f \in A(K)$ we have

$$f(x_0) = \int f(x) d\mu(x).$$

A compact convex set K is a *simplex* if every $x \in K$ is represented by a unique Borel probability measure supported on $\partial_e K$.

2.1. *Order unit spaces*

An *order unit space* is a normed, ordered vector space over A with a distinguished vector $e \in A$ such that (see pp. 68–69 in [1]):

(1) $a \leq 0$ iff $(\forall n \in \mathbb{N})\ na \leq e$.

(2) $\|a\| = \inf\{\lambda > 0 : -\lambda e \leq a \leq \lambda e\}$.

We note that (2) implies that order preserving maps between order unit spaces are contractions.

A *state* on (A, e) is a positive continuous functional $\phi : A \to \mathbb{R}$ such that $\|\phi\| = \phi(e) = 1$. The *state space* $S(A, e)$ (written $S(A)$ if e is clear from the context) is the set of all states. Note that $S(A, e)$ is a compact convex subset of A^* (the space of all continuous functionals) when this space is given the weak-$*$ topology (in the usual sense of real Banach spaces).

A typical example of a (complete) order unit space is $C_{\mathbb{R}}(X)$, where X is some compact set (the unit is the constant 1 function, the norm is the sup norm, $\|\cdot\|_\infty$). If K is a compact convex set then $A(K)$ is an order unit space. In the case that K is finite dimensional with n vertices then $A(K) \simeq (\mathbb{R}^n, \|\cdot\|_\infty)$. Given K a compact convex set, we define a state $\phi_x \in S(A(K))$ by $\phi_x(f) = f(x)$. The next result is due to Kadison, but see [1].

Theorem 2.2: *The category of compact convex sets and the category of complete order unit spaces are dual, and that map $K \mapsto A(K)$ is a contravariant functor, with inverse $A \mapsto S(A)$. Moreover, the map $x \mapsto \phi_x$ is an affine homeomorphism between K and $S(A(K))$, and the map $\rho_f : A \to A(S(A)) : \phi \mapsto \phi(f)$ is an isomorphism of complete order unit spaces.*

Above, we have defined a Choquet simplex to be a convex compact set where every point is represented uniquely by a normalized boundary measure. Usually, one defines a Choquet simplex to be a compact convex set where $A^*(K)$ is a vector lattice (i.e., each pair of vectors in $A^*(K)$ have a least upper bound and a greatest lower bound). The equivalence of our definition with this one is then a theorem, due to Choquet, see [1]. In terms of $A(K)$, we also have the following characterization: K is a Choquet simplex iff $A(K)$ has the Riesz interpolation property (see Corollary II.3.11 in [1]).

2.2. *Direct, inductive, and inverse limits*

Let A_i, $i \in \mathbb{N}$ be a sequence of order unit spaces, and let $\phi_{j,i} : A_i \to A_j$, $j > i$, be a family of order unit space monomorphisms such that for all

$k > j > i$ we have $\phi_{k,j} \circ \phi_{j,i} = \phi_{k,i}$ (this identity also explains why the order of the indices on $\phi_{j,i}$ was chosen as it was). Then there is an order unit space A_∞ and o.u.s. monomorphisms $\phi_{\infty,j} : A_j \to A_\infty$ such that for all $j > i$ we have $\phi_{\infty,j} \circ \phi_{j,i} = \phi_{\infty,i}$. We call A_∞ the *direct* limit of the system $(A_i, \phi_{i+1,i})$ (the existence of A_∞ can be seen by a standard argument). The space A_∞ is essentially uniquely determined by the maps $\phi_{\infty,i}$. However, A_∞ need not be a complete order unit space even if the A_i are. By the *inductive* limit of the system $(A_i, \phi_{j,i})$ we mean the completion of the order unit space A_∞ (the operations and ordering extend naturally).

The natural dual construction of the above for compact convex sets correspond to forming the *inverse limit* of a family $(K_i, f_{j,i})$ of compact convex sets K_i with respect to the affine, continuous surjections $f_{j,i} : K_i \to K_j$, $j > i$.

A remarkable fact is the following theorem of Lazar and Lindenstrauss:

Theorem 2.3: *Every metrizable Choquet simplex can be represented as an inverse limit of finite dimensional simplices. Equivalently, every separable complete order unit space A with the Riesz interpolation property can be represented as an inductive limit $(\mathbb{R}^{n_i}, \phi_{j,i})$.*

This is in fact just a special case of a more general theorem about the representation of Banach spaces whose dual are L^1 spaces. However, it serves as motivation for the following discussion.

The standard basis in $\mathbb{R}^n$ will be denoted by $(e_j^n)_{1\leq j\leq n}$. An order unit map $\phi : \mathbb{R}^n \to A$, where A is an order unit space, is determined by the partition of unity $(f_j)_{1\leq j\leq n} = (\phi(e_j^n))_{1\leq j\leq n}$ in A. If $A = A(K)$ for some compact convex space K, then we will call such a partition of unity *peaked* if $\max_K f_j = 1$ for all $j \leq n$. It is not hard to see that ϕ is isometric precisely when the partion of unity $(\phi(e_j^n))$ is peaked.

In particular, any isometric order unit embedding $\phi : \mathbb{R}^n \to \mathbb{R}^{n+1}$ must, after rearranging the basis, have the form

$$\phi(e_j^n) = e_j^{n+1} + a_j e_{n+1}^{n+1}$$

where $0 \leq a_j \leq 1$ for $1 \leq j \leq n$, and $\sum_{j=1}^n a_j = 1$. We will call this the *standard form* (with coefficients $(a_j)_{1\leq j\leq n}$) of an order unit map $\mathbb{R}^n \to \mathbb{R}^{n+1}$. It is easy to see that if $\psi : \mathbb{R}^n \to \mathbb{R}^{n+k}$ is an order unit map $n, k \in$

$\mathbb{N}$, then there are intermediate order unit maps $\psi_0, \ldots, \psi_{k-1}$, where $\psi_i : \mathbb{R}^{n+i} \to \mathbb{R}^{n+i+1}$, $0 \leq i \leq k-1$, such that

$$\psi = \psi_{k-1} \circ \cdots \circ \psi_0.$$

After possibly rearranging the bases, each map ψ_i has the above form

$$\phi_i(e_j^{n+i}) = e_j^{n+i+1} + a_{j,i} e_{n+i+1}^{n+i+1}.$$

It follows by the Lazar-Lindenstrauss theorem that every separable, complete order unit space A can be represented as an inductive limit of the form $(\mathbb{R}^i, \psi_i)$, with ψ_i on the standard form with coefficients $(a_{j,i})_{1 \leq j \leq i}$. The matrix $(a_{j,i})_{1 \leq j \leq i, i \in \mathbb{N}}$ is called a *representing matrix* for A (and $S(A)$.) If, conversely, we are given a matrix of elements $(a_{j,i})_{1 \leq j \leq i, \iota \in \mathbb{N}}$ satisfying $0 \leq a_{j,i} \leq 1$ and $\sum_{j=1}^{i} a_{j,i} = 1$ for all $i \in \mathbb{N}$, then it is clear that this matrix determines a corresponding system $(\mathbb{R}^i, \psi_i)$.

3. Existence and Uniqueness of the Poulsen Simplex

3.1. *Amalgamation*

Let $\mathcal{C}$ be a category of objects, and $\Phi_{\mathcal{C}}$ be a class of morphisms among the objects in $\mathcal{C}$. We will say that $\Phi_{\mathcal{C}}$ has the *amalgamation property* if whenever $A, B, C \in \mathcal{C}$ and $\phi, \psi \in \Phi_{\mathcal{C}}$ are morphisms $\phi : A \to B$, $\psi : A \to C$, then there is $D \in \mathcal{C}$ and $\phi' : B \to D$ and $\psi' : C \to D$ such that $\phi' \circ \phi = \psi' \circ \psi$.

Lemma 3.1: *Let $\mathcal{C} = \{\mathbb{R}^n : n \in \mathbb{N}\}$, and let Φ be the class of all order unit maps between elements of $\mathcal{C}$. Then Φ has the amalgamation property.*

Proof: Let $\phi : \mathbb{R}^n \to \mathbb{R}^m$ and $\psi : \mathbb{R}^n \to \mathbb{R}^k$ be order unit morphisms. After re-arranging the basis, we may assume they have the standard form. Define $\phi' : \mathbb{R}^m \to \mathbb{R}^m \oplus \mathbb{R}^k$ by

$$\phi'(e_j^m) = \begin{cases} e_j^m + \psi(e_j^n) & \text{if } j \leq n \\ e_j^m & \text{otherwise} \end{cases}$$

and define $\psi' : \mathbb{R}^k \to \mathbb{R}^m \oplus \mathbb{R}^k$ by

$$\psi'(e_j^m) = \begin{cases} e_j^k + \phi(e_j^n) & \text{if } j \leq n \\ e_j^k & \text{otherwise.} \end{cases}$$

Then $\phi' \circ \phi(e_j^n) = \phi(e_j^n) \oplus \psi(e_j^n) = \psi' \circ \psi(e_j^n)$, as required. □

The proof clearly shows that if $S \subseteq \mathbb{R}$ is an additive subgroup containing 1 then the set $\Phi_S \subseteq \Phi$, consisting of those maps $\phi : \mathbb{R}^n \to \mathbb{R}^m$ (some $n, m \in \mathbb{N}$) with $\phi(e_i^n) \in S^m$, also satisfies amalgamation. It follows that if $\Phi_0 \subseteq \Phi$ is a countable set of order unit maps, then there is a countable $\tilde{\Phi}_0 \supseteq \Phi_0$ which is closed under amalgamation.

Definition 3.2: For $\varepsilon > 0$, and order unit maps $\phi, \psi : \mathbb{R}^n \to \mathbb{R}^m$, we will write $\phi \sim_\varepsilon \psi$ ("ϕ ε-approximates ψ") if $\|\phi(e_i^n) - \psi(e_i^n)\| < \varepsilon$ for all $1 \leq i \leq n$.

A set $\Phi_0 \subseteq \Phi$ is said to be *dense* if for all $\phi \in \Phi$ and $\varepsilon > 0$ there is $\psi \in \Phi_0$ such that $\psi \sim_\varepsilon \phi$.

It is clear that $\Phi_\mathbb{Q}$ and Φ_D, where D denotes the dyadic rationals, are countable dense subsets of Φ.

3.2. *Generic inductive systems*

Definition 3.3: Let $(\mathbb{R}^{n_i}, \phi_i)_{i \in \mathbb{N}}$ be a directed system of order unit spaces and maps.

1. We will say that this system has the *extension property* (EP) if whenever $\varepsilon > 0$ and $\psi : \mathbb{R}^n \to \mathbb{R}^m$ and $\gamma : \mathbb{R}^n \to \mathbb{R}^{n_i}$ are given there is $j \geq i$ and $\gamma' : \mathbb{R}^m \to \mathbb{R}^{n_j}$ such that $\|\phi_{j,i} \circ \gamma(e_k^n) - \gamma' \circ \psi(e_k^n)\| < \varepsilon$ for all $k \leq n$. I.e., the diagram

$$\begin{array}{ccc} \mathbb{R}^n & \xrightarrow{\psi} & \mathbb{R}^m \\ \gamma \downarrow & \circlearrowleft^{\varepsilon} & \downarrow \gamma' \\ \mathbb{R}^{n_i} & \xrightarrow[\phi_{j,i}]{} & \mathbb{R}^{n_j} \end{array}$$

ε-commutes (which is indicated by the symbol $\circlearrowleft^{\varepsilon}$).

2. The system $(\mathbb{R}^{n_i}, \phi_i)$ has the *genericity property* (GP) if for all $\psi : \mathbb{R}^{n_i} \to \mathbb{R}^m$ and $\varepsilon > 0$ there is $j \geq i$ and $\gamma' : \mathbb{R}^m \to \mathbb{R}^{n_j}$ such that $\gamma' \circ \psi \sim_\varepsilon \phi_{j,i}$. I.e., the diagram

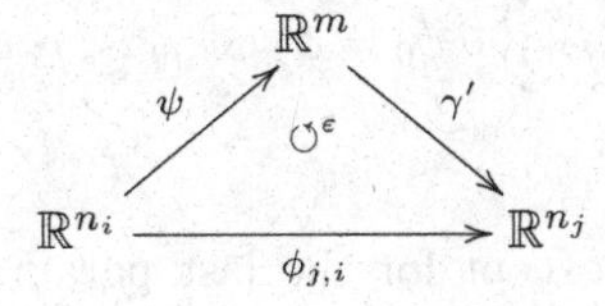

ε-commutes.

Lemma 3.4: *A system $(\mathbb{R}^{n_i}, \phi_i)$ has the EP iff it has the GP.*

Proof: If $(\mathbb{R}^{n_i}, \phi_i)$ has the EP and $\psi : \mathbb{R}^{n_i} \to \mathbb{R}^m$ and $\varepsilon > 0$ are given, simply let $\gamma : \mathbb{R}^{n_i} \to \mathbb{R}^{n_i}$ and applying the EP to (ψ, γ).

Conversely, if $(\mathbb{R}^{n_i}, \phi_i)$ has the GP and $\psi : \mathbb{R}^n \to \mathbb{R}^m$, $\gamma : \mathbb{R}^n \to \mathbb{R}^{n_i}$, and $\varepsilon > 0$ are given, then amalgamate ψ and γ to obtain $\bar{\psi} : \mathbb{R}^{n_i} \to \mathbb{R}^{m+n_i}$ and $\bar{\gamma} : \mathbb{R}^m \to \mathbb{R}^{m+n_i}$ such that $\bar{\psi} \circ \psi = \bar{\gamma} \circ \gamma$. By the GP there is $j \geq i$ and $\gamma' : \mathbb{R}^{m+n_i} \to \mathbb{R}^{n_j}$ such that $\gamma' \circ \bar{\psi} \sim_\varepsilon \phi_{j,i}$. But clearly then we also have that $\phi_{j,i} \circ \gamma \sim_\varepsilon \gamma' \circ \bar{\psi} \circ \psi$.

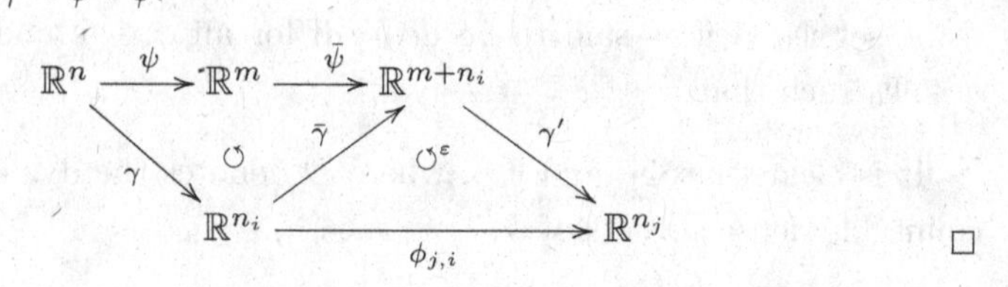

□

Theorem 3.5: *There is a directed system with the EP (and the GP).*

Proof: Let $\Phi_0 \subseteq \Phi$ be a countable dense subclass of Φ which is closed under the amalgamation property. Let $(\psi_i, \gamma_i)_{i \in \mathbb{N}}$ enumerate (with infinite repetition) all pairs $\psi_i : \mathbb{R}^n \to \mathbb{R}^m$, $\gamma_i : \mathbb{R}^n \to \mathbb{R}^k$, where $n < k, m$ range over $\mathbb{N}$. We will define by recursion a directed system $(\mathbb{R}^{n_i}, \phi_i)$ with the property that

(∗) If $\gamma_{i_0} : \mathbb{R}^n \to \mathbb{R}^{n_{i_1}}$ for some $i_0, i_1 < j$ then there is $i_1 < j_1 \leq j$ and $\gamma' : \mathbb{R}^m \to \mathbb{R}^{j_1}$ such that $\gamma' \circ \psi_{i_0} = \phi_{j_1, i_1} \circ \gamma_{i_0}$.

Suppose $(\mathbb{R}^{n_i}, \psi_i)_{i \leq j}$ has been constructed satisfying (∗) and that $\gamma_j : \mathbb{R}^n \to \mathbb{R}^{n_{i_1}}$ for some $i_1 \leq j$, and $\psi_j : \mathbb{R}^n \to \mathbb{R}^m$ for some $m \geq n$. By amalgamating the pair $(\phi_{j,i_1} \circ \gamma_j, \psi_j)$ we then obtain $n_{j+1} > n_j$ and $\phi_j : \mathbb{R}^{n_j} \to \mathbb{R}^{n_{j+1}}$ and $\gamma' : \mathbb{R}^m \to \mathbb{R}^{n_{j+1}}$ such that $\phi_j \circ \gamma_j = \gamma' \circ \psi_j$, as required.

Now let $\varepsilon > 0$ and $\psi : \mathbb{R}^n \to \mathbb{R}^m$ and $\gamma : \mathbb{R}^n \to \mathbb{R}^{n_i}$ for some $i \in \mathbb{N}$. Let $\psi_0, \gamma_0 \in \Phi_0$ be such that $\psi_0 \sim_{\varepsilon/2} \psi$ and $\gamma_0 \sim_{\varepsilon/2} \gamma$. It follows from (∗) there is $n_j > n_i$ and $\gamma' : \mathbb{R}^m \to \mathbb{R}^{n_j}$ such that $\gamma' \circ \psi_0 = \phi_{j,i} \circ \gamma_0$. Then

$$\gamma' \circ \psi \sim_{\varepsilon/2} \gamma' \circ \psi_0 = \phi_{j,i} \circ \gamma_0 \sim_{\varepsilon/2} \phi_{j,i} \circ \gamma,$$

whence $\gamma' \circ \psi \sim_\varepsilon \phi_{j,i} \circ \gamma$. □

The previous proof (except for the last paragraph) looks exactly like the proof of the existence of the Fraïssé limit in classical Fraïssé theory, and this is of course no coincidence. Next we show that the (completed) limit object is unique, using an argument that a model theorist would call a *back-and-forth* argument, and a functional analysts or operator algebraists would call an (approximate) *intertwining argument*, see e.g. [10].

Theorem 3.6: *If* $(\mathbb{R}^{n_i}, \phi_i)$ *and* $(\mathbb{R}^{m_j}, \psi_j)$ *are directed systems with the GP then* $A = \overline{\lim}(\mathbb{R}^{n_i}, \phi_i)$ *and* $B = \overline{\lim}(\mathbb{R}^{m_j}, \psi_j)$ *are isomorphic.*

Proof: Fix a summable sequence $(\varepsilon_i)_{i\in\mathbb{N}}$ of positive tolerances. Using the GP, we can define strictly increasing sequences $(i_k)_{k\in\mathbb{N}}$ and $(j_k)_{k\in\mathbb{N}}$, and order unit maps $\gamma_k : \mathbb{R}^{n_{i_k}} \to \mathbb{R}^{m_{j_k}}$ and $\delta_k : \mathbb{R}^{m_{j_k}} \to \mathbb{R}^{n_{i_{k+1}}}$ such that $\delta_k \circ \gamma_k \sim_{\varepsilon_{2k-1}} \phi_{n_{i_k}, n_{i_{k+1}}}$ and $\gamma_{k+1} \circ \delta_k \sim_{\varepsilon_{2k}} \psi_{m_{j_k}, m_{j_{k+1}}}$. In other words, we will have the following approximate intertwining:

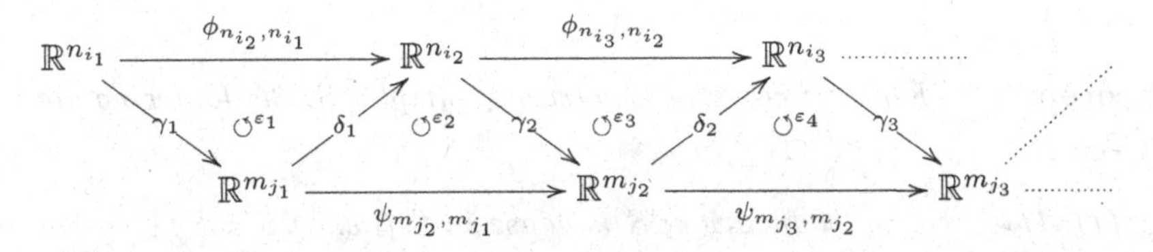

Then for each n_{i_l} and $k \leq n_{i_l}$, the sequence $a_p^{n_{i_l},k} = \psi_{\infty,m_{j_p}} \circ \gamma_p \circ \phi_{n_{i_p},n_{i_l}}$, $(p \geq l)$ is Cauchy and the elements $a_\infty^{n_{i_l},k} = \lim_{p\to\infty} a_p^{n_{i_l},k}$, $1 \leq k \leq n_{i_l}$ forms a partition of unity in B. Define $\gamma_{\infty,n_{i_l}} : \mathbb{R}^{n_{i_l}} \to B$ by $\gamma_{\infty,n_{i_l}}(e_k^{n_{i_l}}) = a_\infty^{n_i,l,k}$. Then we have that $\gamma_{\infty,n_{i_p}} \circ \phi_{n_{i_p},n_{i_l}} = \gamma_{\infty,n_{i_l}}$, and so we can define an order unit monomorphism $\gamma : \lim(\mathbb{R}^{n_i}, \phi_i) \to B$ by $\gamma(\phi_{\infty,n_{i_l}}(x)) = \gamma_{\infty,n_{i_l}}(x)$. Since γ is isometric we can extend it to an order unit morphism $\bar{\gamma} : A \to B$.

It only remains to show that $\mathrm{ran}(\bar{\gamma}) = B$. For this it is enough to show that $\mathrm{ran}(\gamma)$ is dense in B. For this, note that $\psi_{\infty,m_{j_l}} \sim_{\varepsilon_p} \psi_{\infty,m_{j_{p+1}}} \circ \gamma_{p+1} \circ \delta_p \circ \psi_{m_{j_p},m_{j_l}}$, and so

$$\psi_{\infty,m_{j_l}} \sim_{\sum_{i=p}^{\infty} \varepsilon_i} \gamma_{\infty,n_{i_{p+1}}} \circ \delta_p \circ \psi_{m_{j_p},m_{j_l}},$$

which implies that $\mathrm{ran}(\gamma)$ is dense in B. □

Before stating our main theorem (Theorem 3.9), let us introduce the following property which is free of reference to inductive limit representations, and at the same time evokes Gurarij's well-known Banach space.

Definition 3.7: An order unit space A is said to have the *Gurarij property* if for all $\varepsilon > 0$, $\psi : \mathbb{R}^n \to \mathbb{R}^m$ and $\gamma : \mathbb{R}^n \to A$ there is $\gamma' : \mathbb{R}^m \to A$ such that

$$\|\gamma' \circ \psi(e_j^n) - \gamma(e_j^n)\| < \varepsilon$$

for all $1 \leq j \leq n$.

Lemma 3.8: *Let* $(\mathbb{R}^{n_i}, \phi_i)$ *be an inductive system. Then* $\overline{\lim}(\mathbb{R}^{n_i}, \phi_i)$ *has the Gurarij property iff* $(\mathbb{R}^{n_i}, \phi_i)$ *has the GP.*

Proof: This is a straightforward exercise in using the definitions. □

The next theorem finally connects all the dots: The Poulsen simplex is uniquely characterized by being a Fraïssé limit, as well as by its topological properties.

Theorem 3.9: *For a (non-trivial) metrizable simplex* S*, the following are equivalent:*

(1) The extremal boundary of S *is dense in* S *(i.e.,* $\overline{\partial_e S} = S$*).*
(2) Any (equivalently, some) inductive limit representation $(\mathbb{R}^{n_i}, \phi_i)$ *of* $A(S)$ *has the GP.*
(3) $A(S)$ *has the Gurarij property.*

We will need the following Lemma:

Lemma 3.10: *Let* S *be a simplex, and let* $(\mathbb{R}^{n_i}, \phi_i)$ *be an inductive limit representation* $A(S) = \overline{\lim}(\mathbb{R}^{n_i}, \phi_i)$*, with limit maps* $\phi_{\infty,i} : \mathbb{R}^{n_i} \to A(S)$*. For each* i *and* $1 \leq j \leq n_i$*, suppose* $x_j^{n_i}$ *is a peak point of* $\phi_{\infty,i}(e_j^{n_i})$ *in* $\partial_e S$*. Then the set* $P = \{x_j^{n_i} : i \in \mathbb{N} \wedge 1 \leq j \leq n_i\}$ *is dense in* $\partial_e S$*.*

Proof: Let us first point out that if $f \in A(S)$ and $M_f = \{x \in S : f(x) = \max_S f\}$, then $M_f \cap \partial_e S \neq \emptyset$. Indeed, if $x \in M_f$ and μ is a measure on $\partial_e S$ representing x, then the fact that $f(x) = \int f(y) d\mu(y)$ implies that $f(y) = \max_S f$ for μ-a.a. y. (By the same token, the set M_f is easily seen to be a face of S.)

Now fix $\tilde{x} \in \partial_e S$, and assume for a contradiction that $\tilde{x} \notin \overline{P}$. Let U be an open neighborhood of $\tilde{x}$ such $U \cap P = \emptyset$, and pick for each $i \in \mathbb{N}$ a point $x_i \in \langle x_1^{n_i}, \ldots, x_{n_i}^{n_i} \rangle$ such that $x_i \to \tilde{x}$. Write $x_i = \sum_{j=1}^{n_i} a_j^{n_i} x_j^{n_i}$ as a convex combination, so that the measure $\mu_i = \sum_{j=1}^{n_i} a_j^{n_i} \delta_{x_j^{n_i}}$ has $x_j^{n_i}$ as a barycenter. After possibly going to a subsequence, we may assume that μ_i converges to some measure μ. The barycenter of μ must be $\tilde{x}$, and since $\tilde{x}$ is a boundary point this means that $\mu = \delta_{\tilde{x}}$. Thus $\mu(U) = 1$, while $\mu_i(U) = 0$ for all i, contradicting that $\mu_i \to \mu$. □

Proof: Let $(\mathbb{R}^{n_i}, \phi_i)$ be an inductive limit representation of $A(S)$, and let $\phi_{\infty,i} : \mathbb{R}^{n_i} \to A(S)$ be the associated maps into the limit. For simplicity, we identify $\mathbb{R}^{n_i}$ with its image in $A(S)$ under $\phi_{\infty,i} : \mathbb{R}^{n_i} \to A(S)$.

(1) $\Longrightarrow$ (2). Let $\gamma : \mathbb{R}^{n_i} \to \mathbb{R}^{n_i+1}$ be of the standard form, $\gamma(e_j^{n_i}) = e_j^{n_i+1} + a_j e_{n_i+1}^{n_i+1}$, and let $\varepsilon > 0$. Fix for each $1 \leq j \leq n_i$ a peak point $x_j \in S$ of $e_j^{n_i}$, and let $\tilde{x} = \sum_{j=1}^{n_i} a_j x_j$. Consider then the open neighborhood of $\tilde{x}$ defined by $U = \{x \in S : |(\forall j \leq n_i) e_j^{n_i}(x) - e_j^{n_i}(\tilde{x})| < \varepsilon\}$. Since $\overline{\partial_e S} = S$, it follows from Lemma 3.10 that there is some $k > i$, a $1 \leq l \leq n_k$, and $x_0 \in \partial_e S$ such that $e_l^{n_k}(x_0) = 1$ and $x_0 \in U$. Since we may write

$$e_j^{n_i} = \sum_{p=1}^{n_k} \langle \phi_{k,i}(e_j^{n_i}), e_p^{n_k} \rangle e_p^{n_k},$$

which is valid in $A(S)$, it follows by evaluating at x_0 that $|\langle \phi_{k,i}(e_j^{n_i}), e_l^{n_k} \rangle - a_j| < \varepsilon$ (note that $e_j^{n_i}(\tilde{x}) = a_j$.) Define $\psi : \mathbb{R}^{n_i+1} \to \mathbb{R}^{n_k}$ by letting

$$\psi(e_j^{n_i+1}) = \phi_{k,i}(e_j^{n_i}) - \langle \phi_{k,i}(e_j^{n_i}), e_l^{n_k} \rangle e_l^{n_k}$$

for $j \leq n_i$ and $\psi(e_{n_i+1}^{n_i+1}) = e_l^{n_k}$. Then

$$\psi \circ \gamma(e_j^{n_i}) = \phi_{k,i}(e_j^{n_i}) + (a_j - \langle \phi_{k,i}(e_j^{n_i}), e_l^{n_k} \rangle) e_l^{n_k},$$

whence $\|\phi_{k,i}(e_j^{n_i}) - \psi \circ \gamma(e_j^{n_i})\|_\infty < \varepsilon$, as required.

(2) $\Longrightarrow$ (1). Suppose now that $(\mathbb{R}^{n_i}, \phi_i)$ has the GP, and let $x_0 \in S$. Fix $\varepsilon > 0$ and $i \in \mathbb{N}$, and consider the neighborhood $U = \{x \in S : (\forall j \leq n_i)|e_j^{n_i}(x) - e_j^{n_i}(x_0)| < \varepsilon\}$. Let $a_j = e_j^{n_i}(x_0)$, let $\gamma : \mathbb{R}^{n_i} \to \mathbb{R}^{n_i+1}$ be the order unit map $\gamma(e_j^{n_i}) = e_j^{n_i+1} + a_j e_{n_i+1}^{n_i+1}$, and let $\psi : \mathbb{R}^{n_i+1} \to \mathbb{R}^{n_k}$ be such that $\psi \circ \gamma \sim_\varepsilon \phi_{k,i}$. Let $x \in \partial_e S$ be a peak point of $\phi_{\infty,k} \circ \psi(e_{n_i+1}^{n_i+1})$. Then $\phi_{\infty,k} \circ \psi \circ \gamma(e_j^{n_i})(x) = a_j$ and so since $\psi \circ \gamma \sim_\varepsilon \phi_{k,i}$ we have

$$|e_j^{n_i}(x_0) - e_j^{n_i}(x)| = |\phi_{\infty,k} \circ \psi \circ \gamma(e_j^{n_i})(x) - e_j^{n_i}(x)| \leq \|\psi \circ \gamma(e_j^{n_i}) - \phi_{k,i}(e_j^{n_i})\|_\infty < \varepsilon$$

for all $j \leq n_i$, as required. □

From now on we denote the Poulsen simplex by $\mathbb{P}$.

Proposition 3.11: *The Poulsen simplex is surjectively universal: For every metrizable simplex S there is a affine continuous surjection $\sigma : \mathbb{P} \to S$.*

Proof: We just need to show that there is an order unit space embedding of $A(S)$ into $A(\mathbb{P})$. For this, fix a direct limit system $(\mathbb{R}^{n_i}, \psi_i)$ representing $A(S)$ and $(\mathbb{R}^{m_i}, \phi_i)$ representing $A(\mathbb{P})$, and use the EP repeatedly to build the desired embedding. □

A slightly more elaborate argument along the same lines shows:

Proposition 3.12: *$\mathbb{P}$ has the following universality property: If S_1 and S_2 are metrizable simplices and $\rho : S_2 \twoheadrightarrow S_1$, $\sigma : \mathbb{P} \twoheadrightarrow S_1$ are continuous affine surjections, then there is $\theta : \mathbb{P} \to S_2$ such that $\sigma = \rho \circ \theta$. This property characterizes $\mathbb{P}$ up to isomorphism.*

Remark 3.13: The approach to the Poulsen simplex presented here is closely related to the approach to the uniqueness of the Gurarij space presented in [2], and the category theoretic/axiomatic approach to this type of problem developed in [5].

4. Questions

This paper is at best one scratch on the surface of the Poulsen simplex as seen from the point of view of Fraïssé theory. There are many questions that can now be asked, but answering these may not be so easy.

Lindenstrauss, Olsen and Sternfeld proved in [6] that the Poulsen simplex has the following universality property: Every metrizable simplex is isomorphic (i.e., affinely homeomorphic) to a face of the Poulsen simplex.

Question 4.1: Can this "face universality" property of the Poulsen simplex be derived from the Fraïssé theoretic approach?

Lindenstrauss, Olsen and Sternfeld go on to prove that any isomorphism between *proper* faces of the Poulsen simplex can be extended to an automorphism of $\mathbb{P}$.

Question 4.2: Can this "facial homogeneity" be proved from the Fraïssé theoretic approach to $\mathbb{P}$?

Another line of questions that are natural to consider are questions about the automorphism group of $\mathbb{P}$, i.e., the group $\mathrm{Aut}(\mathbb{P})$ of all affine homeomorphisms of $\mathbb{P}$ onto itself. This is a Polish group that sits inside the group of all autohomeomorphisms of $\mathbb{P}$ onto itself.

The group $\mathrm{Aut}(\mathbb{P})$ acts continuously on $\mathbb{P}$ itself, of course, and does so by affine transformations. This action has no fixed point, whence:

Proposition 4.3: $\mathrm{Aut}(\mathbb{P})$ *is not amenable.*

Sketch of Proof: If $x \in \mathbb{P}$ were a fixed point, then take μ to be a boundary (probability) measure representing x. Find a compact subset $K \subseteq \partial_e \mathbb{P}$ such that $\mu(K) > \frac{1}{2}$. The affine complement (see [6]) of the face generated by K is isomorphic to $\mathbb{P}$ itself, whence contains of a set K' homeomorphic to K. Now find an automorphism α of $\mathbb{P}$ which carries K to K' (by homogeneity of proper faces such exists). Since x is fixed, α must also leave μ invariant (as it is the unique measure representing x), but that is impossible since $\mu(K) > \frac{1}{2}$, $\alpha(K) = K'$, and $K \cap K' = \emptyset$. □

Of course, this also rules out that $\mathrm{Aut}(\mathbb{P})$ is extremely amenable, in the sense discussed at length in [3]. However, it may still be possible to calculate the universal minimal flow, and so we ask:

Question 4.4: What is the universal minimal flow of $\mathrm{Aut}(\mathbb{P})$? Is it separable? Can it be explicitly calculated?

Another property that is often of interest for large Polish groups is the automatic continuity property. For a number of large Polish groups, it can be shown that *any* algebraic homomorphism into a separable group must be continuous (see e.g. [4] or [11] and references therein), and we say such a group has the "automatic continuity property". In [4], the properties of Fraïssé theoretic representations are linked via the notion of "ample generics" to automatic continuity, and so we ask:

Question 4.5: Does $\mathrm{Aut}(\mathbb{P})$ have the automatic continuity property?

References

1. Erik M. Alfsen, *Compact convex sets and boundary integrals*, Springer-Verlag, New York, 1971, Ergebnisse der Mathematik und ihrer Grenzgebiete, Band 57. MR 0445271 (56 #3615)

2. Trevor Irwin and Sławomir Solecki, *Projective Fraïssé limits and the pseudo-arc*, Transactions of the American Mathematical Society **358** (2006), no. 7, 3077–3096.
3. Alexander S. Kechris, Vladimir G. Pestov, and Stevo Todorcevic, *Fraïssé limits, ramsey theory, and topological dynamics of automorphism groups*, Geometric and Functional Analysis **15** (2005), no. 1, 106–189.
4. Alexander S. Kechris and Christian Rosendal, *Turbulence, amalgamation, and generic automorphisms of homogeneous structures*, Proceedings of the London Mathematical Society **94** (2007), no. 2, 302–350.
5. Wiesław Kubiś, *Fraïssé sequences: category-theoretic approach to universal homogeneous structures*, Annals of Pure and Applied Logic **165** (2014), no. 11, 1755–1811.
6. Joram Lindenstrauss, Gunnar Olsen, and Y. Sternfeld, *The poulsen simplex*, Annales de l'institut Fourier **28** (1978), no. 1, 91–114.
7. Jaroslav Lukeš, Jan Malý, Ivan Netuka, and Jiří Spurný, *Integral representation theory*, de Gruyter Studies in Mathematics, vol. 35, Walter de Gruyter & Co., Berlin, 2010, Applications to convexity, Banach spaces and potential theory. MR 2589994 (2011e:46002)
8. Robert R. Phelps, *Lectures on Choquet's theorem*, second ed., Lecture Notes in Mathematics, vol. 1757, Springer-Verlag, Berlin, 2001. MR 1835574 (2002k:46001)
9. Ebbe T. Poulsen, *A simplex with dense extreme points*, Annales de l'institut Fourier **11** (1961), 83–87.
10. M. Rørdam, *Classification of nuclear C^*-algebras*, Encyclopaedia of Math. Sciences, vol. 126, Springer-Verlag, Berlin, 2002.
11. Marcin Sabok, *Automatic continuity for isometry groups*, arXiv preprint arXiv:1312.5141 (2013).

ON THE SET-GENERIC MULTIVERSE

Sy-David Friedman*, Sakaé Fuchino† and Hiroshi Sakai‡

*Kurt Gödel Research Center for Mathematical Logic
Universität Wien
Vienna, Austria
sdf@logic.univie.ac.at

†*Graduate School of System Informatics*
Kobe University
Kobe, Japan
fuchino@diamond.kobe-u.ac.jp

‡*Graduate School of System Informatics*
Kobe University
Kobe, Japan
hsakai@people.kobe-u.ac.jp

The forcing method is a powerful tool to prove the consistency of set-theoretic assertions relative to the consistency of the axioms of set theory. Laver's theorem and Bukovský's theorem assert that set-generic extensions of a given ground model constitute a quite reasonable and sufficiently general class of standard models of set-theory.

In Sections 2 and 3 of this note, we give a proof of Bukovsky's theorem in a modern setting (for another proof of this theorem see [4]). In Section 4, we check that the multiverse of set-generic extensions can be treated as a collection of countable transitive models in a conservative extension of ZFC. The last section then deals with the problem of the existence of infinitely-many independent buttons, which arose in the modal-theoretic approach to the set-generic multiverse by J. Hamkins and B. Loewe [12].

2010 Mathematical Subject Classification: 03E40 03E70
Keywords: Philosophy of set theory, forcing, multiverse, Laver's theorem, Bukovský's theorem.

Contents

1. The Category of Forcing Extensions as the Set-Theoretic Multiverse

The forcing method is a powerful tool to prove the consistency of set-theoretic (i.e., mathematical) assertions relative to (the consistency of) the axioms of set theory. If a sentence σ in the language $\mathcal{L}_{\mathrm{ZF}}$ of set theory is proved to be relatively consistent with the axioms of set theory (ZFC) by some forcing argument then it is so in the sense of the strictly finitist standpoint of Hilbert: the forcing proof can be recast into an algorithm $\mathcal{A}$ such that, if a formal proof $\mathcal{P}$ of a contradiction from ZFC + σ is ever given, then we can transform $\mathcal{P}$ with the help of $\mathcal{A}$ to another proof of a contradiction from ZFC or even ZF alone.

The "working set-theorists" however prefer to see their forcing arguments not as mere discussions concerning manipulations of formulas in a formal system but rather concerning the "real" mathematical universe in which they "live". Forcing for them is thus a method of extending the universe of set theory where they originally "live" (the ground model, usually denoted as "V") to many (actually more than class many in the sense of V) different models of set theory called generic extensions of V. Actually, a family of generic extensions is constructed for certain V-definable partial orderings $\mathbb{P}$. Each such generic extension is obtained first by fixing a so-called generic filter G which is a filter over $\mathbb{P}$, sitting outside V with a "generic" sort of transcendence over V, and then by adding G to V to generate a new structure — the generic extension $V[G]$ of V — which is also a model of ZFC. Often this process of taking generic extensions over some model of set theory is even repeated transfinitely-many times. As a result, a set-theorist performing forcing constructions is seen to live in many different models of set theory simultaneously. This is manifested in many technical expositions of forcing where the reader very often finds narratives beginning with phrases like: "Working in $V[G]$, ...", "Let $\alpha < \kappa$ be such that x is in the αth intermediate model $V[G_\alpha]$ and ...", "Now returning to V, ...", etc., etc.

Although this "multiverse" view of forcing is in a sense merely a modus loquendi, it is worthwhile to study the possible pictures of this multiverse per se. Some initial moves in this direction have been taken e.g. in [1], [2], [5], [6], [7], [8], [11], [12], [21], [24] etc. The term "multiverse" probably originated in work of Woodin in which he considered the "set-generic multiverse", the "class" of set-theoretic universes which forms the closure of the given initial universe V under set-generic extension and set-generic ground models. Sometimes we also have to consider the constellations of the set-generic multiverse where V cannot be reconstructed as a set-generic extension of some of or even any of the proper inner models of V. To deal with such cases it is more convenient to consider the expanded generic multiverse where we also assume that the multiverse is also closed under the construction of definable inner models.

The set-generic universe should be distinguished from the "class-generic multiverse", defined in the same way but with respect to class-forcing extensions and ground models, as well as inner models of class-generic extensions that are not themselves class-generic (see [5]). It is even possible to go beyond class-forcing by considering forcings whose conditions are classes, so-called hyperclass forcings (see [6]). The broadest point of view with regard to the multiverse is expressed in [7], where the "hyperuniverse" is taken to consist of *all* universes which share the same ordinals as the initial universe (which is taken to be countable to facilitate the construction of new universes). The hyperuniverse is closed under all notions of forcing.

In this article, we restrict our attention to the set-generic multiverse. The well-posedness of questions regarding the set-generic multiverse is established by the theorems of Laver and Bukovský which we discuss in Section 2. These theorems show that the set-generic extensions and set-generic ground models of a given universe represent a "class" of models with a natural characterization.

The straightforward formulation of the set-generic multiverse requires the notion of "class" of classes which cannot be treated in the usual framework of ZF set theory, but, as emphasized at the beginning, theorems about the set-generic multiverse are actually meta-theorems about ZFC. However we can also consider a theory which is a conservative extension of ZFC in which set-generic extensions and set-generic ground models are real objects in the theory and the set-generic multiverse a definable class. In Section 4, we consider such a system and show that it is a conservative extension of ZFC.

The multiverse view sometimes highlights problems which would never have been asked in the conventional context of forcing constructions (see [11]). As one such example we consider in Section 5 the problem of the existence of infinitely many independent buttons (in the sense of [12]).

2. Laver's Theorem and Bukovský's Theorem

In the forcing language, we often have to express that a certain set is already in the ground model, e.g. in a statement like: $p \Vdash_{\mathbb{P}}$ " ... $\dot{x}$ is in V and ... ". In such situations we can always find a large enough ordinal ξ such that the set in question should be found in that level of the cumulative hierarchy in the ground model. So we can reformulate a statement like the one above into something like $p \Vdash_{\mathbb{P}}$ " ... $\dot{x} \in \check{V}_\xi$ and ... " which is a legitimate expression in the forcing language.

This might be one of the reasons why it is proved only quite recently that the ground model is always definable in an arbitrary set-generic extension:

Theorem 2.1: (R. Laver [17], **H. Woodin** [23]) *There is a formula $\varphi^*(x, y)$ in $\mathcal{L}_{\mathrm{ZF}}$ such that, for any transitive model V of ZFC and set-generic extension $V[G]$ of V there is $a \in V$ such that, for any $b \in V[G]$*

$$b \in V \quad \Leftrightarrow \quad V[G] \models \varphi^*(a, b). \qquad \square$$

An important corollary of Laver's theorem is that a countable transitive model of ZFC can have at most countably many ground models for set forcing.

Bukovský's theorem gives a natural characterization of inner models M of V such that V is a set-generic extension of M.[a] Note that, by Laver's theorem Theorem 2.1, such an M is then definable in V. However the inner model M of V may be introduced as a class in the sense of von Neumann-Bernays-Gödel class theory (NBG) and in such a situation the definability of M in V may not be immediately clear.

Let us begin with the following observation concerning κ-c.c. generic extensions. We shall call a partial ordering *atomless* if each element of it has at least two extensions which are incompatible with each other.

[a]In the terminology of [8], M is a ground of V.

Lemma 2.2: *Let κ be a regular uncountable cardinal. If $\mathbb{P}$ is a κ-c.c. atomless partial ordering, then $\mathbb{P}$ adds a new subset of $2^{<\kappa}$.*

Proof. Without loss of generality, we may assume that $\mathbb{P}$ consists of the positive elements of a κ-c.c. atomless complete Boolean algebra. Note that $\mathbb{P}$ adds new subsets of On since $\mathbb{P}$ adds a new set (e.g. the $(V, \mathbb{P})$-generic set). Suppose that $\dot{S}$ is a $\mathbb{P}$-name of a new subsfet of On. Let θ be a sufficiently large regular cardinal and let $M \prec \mathcal{H}(\theta)$ be such that

(2.1) $|M| \leq 2^{<\kappa}$;

(2.2) ${}^{<\kappa}M \subseteq M$ and

(2.3) $\mathbb{P}, \dot{S}, \kappa \in M$.

Let $\dot{T}$ be a $\mathbb{P}$-name such that $\Vdash_{\mathbb{P}}$ "$\dot{T} = \dot{S} \cap M$". By (2.1), it is enough to show the following, where V denotes the ground model:

Claim 2.2.1: $\Vdash_{\mathbb{P}}$ "$\dot{T} \notin V$".

$\vdash$ Otherwise there would be $p \in \mathbb{P}$ and $T \in V$, $T \subseteq$ On such that

(2.4) $p \Vdash_{\mathbb{P}}$ "$\dot{T} = \check{T}$".

We show in the following that then we can construct a strictly decreasing sequence $\langle q_\alpha : \alpha < \kappa \rangle$ in $\mathbb{P} \cap M$ such that

(2.5) $p \leq_{\mathbb{P}} q_\alpha$ for all $\alpha < \kappa$.

But since $\{q_\alpha \cdot -q_{\alpha+1} : \alpha < \kappa\}$ is then a pairwise disjoint subset of $\mathbb{P}$, this contradicts the κ-c.c. of $\mathbb{P}$.

Suppose that $\langle q_\alpha : \alpha < \delta \rangle$ for some $\delta < \kappa$ has been constructed. If δ is a limit, let $q_\delta = \prod_{\alpha<\delta} q_\alpha$. Then we have $p \leq_{\mathbb{P}} q_\delta$ and $q_\delta \leq_{\mathbb{P}} q_\alpha$ for all $\alpha < \delta$. Since $\langle q_\alpha : \alpha < \delta \rangle \in M$ by (2.2), we also have $q_\delta \in M$.

If $\delta = \beta+1$, then, since $M \models$"q_β does not decide $\dot{S}$" by the elementarity of M, there are $\xi \in \mathrm{On} \cap M$ and $q, q' \in \mathbb{P} \cap M$ with $q, q' \leq_{\mathbb{P}} q_\beta$ such that $q \Vdash_{\mathbb{P}}$ "$\xi \in \dot{S}$" and $q' \Vdash_{\mathbb{P}}$ "$\xi \notin \dot{S}$". At least one of them, say q, must be incompatible with p. Then $q_\delta = q_\beta \cdot -q$ is as desired. $\dashv$ (Claim 2.2.1)

□ (Lemma 2.2)

Note that, translated into the language of complete Boolean algebras, the lemma above just asserts that no κ-c.c. atomless Boolean algebra $\mathbb{B}$ is $(2^{<\kappa}, 2)$-distributive.

Suppose now that we work in NBG, V is a transitive model of ZF and M an inner model of ZF in V (that is M is a transitive class $\subseteq V$ with

$(M, \in) \models$ ZF). For a regular uncountable cardinal κ in M, we say that M *κ-globally covers V* if for every function f (in V) with $\mathrm{dom}(f) \in M$ and $\mathrm{rng}(f) \subseteq M$, there is a function $g \in M$ with $\mathrm{dom}(g) = \mathrm{dom}(f)$ such that $f(i) \in g(i)$ and $M \models |\, g(i) \,| < \kappa$ for all $i \in \mathrm{dom}(f)$.

Theorem 2.3: (L. Bukovský, [3])[b] *Suppose that V is a transitive model of ZFC, $M \subseteq V$ an inner model of ZFC and κ is a regular uncountable cardinal in M. Then M κ-globally covers V if and only if V is a κ-c.c. set-generic extension of M.*

As the referee of the paper points out, this theorem can be formulated more naturally in the von Neumann-Bernays-Gödel class theory (NBG) since in the framework of ZFC this theorem can only be formulated as a meta-theorem, that is, as a collection of theorems consisting corresponding statements for each formula which might define an inner model M.

Proof of Theorem 2.3: If V is a κ-c.c. set-generic extension of M, say by a partial ordering $\mathbb{P} \in M$ with $M \models$"$\mathbb{P}$ has the κ-c.c.", then it is clear that M κ-globally covers V (for f as above, let $\dot{f} \in M$ be a $\mathbb{P}$-name of f and g be defined by letting $g(\alpha)$ to be the set of all possible values $\dot{f}(\alpha)$ may take).

The proof of the converse is done via the following Lemma 2.4. Note that, by Grigorieff's theorem (see Corollary 2.6 below), the statement of this Lemma is a consequence of Bukovský's theorem:

Lemma 2.4: *Suppose that M is an inner model of a transitive model V of ZFC such that M κ-globally covers V for some κ regular uncountable in M. Then for any $A \in V$, $A \subseteq \mathrm{On}$, $M[A]$ is*[c] *a κ-c.c. set-generic extension of M.*

Note that it can happen easily that $M[A]$ is not a set generic extrension of M. For example, $0^{\#}$ exists and $M = L$, then $M[0^{\#}]$ is not a set-generic extension of M.

---

[b]Tadatoshi Miyamoto told us that James Baumgartner independently proved this theorem in an unpublished note using infinitary logic.

[c]$M[A]$ may be defined by $M[A] = \bigcup_{\alpha \in \mathrm{On}} L(V_\alpha^M \cup \{A\})$. $M[A]$ is a model of ZF: this can be seen easily e.g. by applying Theorem 13.9 in [13]. If M also satisfies AC then $M[A]$ satisfies AC as well since, in this case, it is easy to see that a well-ordering of $(V_\alpha)^M \cup \{A\}$ belongs to $M[A]$ for all $\alpha \in \mathrm{On}$.

We first show that Theorem 2.3 follows from Lemma 2.4. Assume that M κ-globally covers V. We have to show that V is a κ-c.c. set-generic extension of M. In V, let λ be a regular cardinal such that $\lambda^{<\kappa} = \lambda$ and $A \subseteq \mathrm{On}$ be a set such that

$$(2.6) \quad (\mathcal{P}(\lambda))^{M[A]} = (\mathcal{P}(\lambda))^V.$$

Then, by Lemma 2.4, $M[A]$ is a κ-c.c. generic extension of M and hence we have $M[A] \models$ "κ is a regular cardinal". Actually we have $M[A] = V$. Otherwise there would be a $B \in V \setminus M[A]$ with $B \subseteq \mathrm{On}$. Since $M[A]$ κ-globally covers $M[A][B]$, we may apply Lemma 2.4 on this pair and conclude that $M[A][B]$ is a (non trivial) κ-c.c. generic extension of $M[A]$. By Lemma 2.2, there is a new element of $\mathcal{P}((2^{<\kappa})^{M[A]}) \subseteq \mathcal{P}(\lambda)$ in $M[A][B]$. But this is a contradiction to (2.6). □ (Theorem 2.3)

Proof of Lemma 2.4: We work in M and construct a κ-c.c. partial ordering $\mathbb{P}$ such that $M[A]$ is a $\mathbb{P}$-generic extension over M.

Let $\mu \in \mathrm{On}$ be such that $A \subseteq \mu$ and let $\mathcal{L}_\infty(\mu)$ be the infinitary sentential logic with atomic sentences

$$(2.7) \quad \text{“}\alpha \in \dot{A}\text{” for } \alpha \in \mu$$

and the class of sentences closed under $\neg$ and $\bigvee$ where $\neg$ is to be applied to a formula and $\bigvee$ to an arbitrary set of formulas. To be specific let us assume that the atomic sentences "$\alpha \in \dot{A}$" for $\alpha \in \mu$ are coded by the sets $\langle \alpha, 0\rangle$ for $\alpha \in \mu$, the negation $\neg\varphi$ by $\langle \varphi, 1\rangle$ and the infinitary disjunction $\bigvee\Phi$ by $\langle \Phi, 2\rangle$. We regard the usual disjunction $\vee$ of two formulas as a special case of $\bigvee$ and other logical connectives like "$\bigwedge$", "$\wedge$", "$\rightarrow$" as being introduced as abbreviations of usual combinations of $\neg$ and $\bigvee$. For a sentence $\varphi \in \mathcal{L}_\infty(\mu)$ and $B \subseteq \mu$, we write $B \models \varphi$ when φ holds if each atomic sentence of the form "$\alpha \in \dot{A}$" in φ is interpreted by "$\alpha \in B$" and logical connectives in φ are interpreted in canonical way. For a set Γ of sentences, we write $B \models \Gamma$ if $B \models \psi$ for all $\psi \in \Gamma$. For $\Gamma \subseteq \mathcal{L}_\infty(\mu)$ and φ, we write $\Gamma \models \varphi$ if $B \models \Gamma$ implies $B \models \varphi$ for all $B \subseteq \mu$ (in V).

Let $\vdash$ be a notion of provability for $\mathcal{L}_\infty(\mu)$ in some logical system which is correct (i.e. $\Gamma \vdash \varphi$ always implies $\Gamma \models \varphi$),[d] upward absolute (i.e. $M \subseteq N$ and $M \models$ "$\Gamma \vdash \varphi$" always imply $N \models$ "$\Gamma \vdash \varphi$" for any transitive models M, N of ZF) and sufficiently strong (so that all the arguments used below work for this $\vdash$). In Section 3 we introduce one such deductive system (as well as

[d]More precisely, we assume that ZF proves the correctness of $\vdash$.

an alternative approach without using such a deduction system, based on Lévy Absoluteness).

Let $\lambda = \max\{\kappa, \mu^+\}$ and $\mathcal{L}_\lambda(\mu) = \mathcal{L}_\infty(\mu) \cap (V_\lambda)^M$. Let $f \in V$ be a mapping $f : \left(\mathcal{P}(\mathcal{L}_\lambda(\mu))\right)^M \setminus \{\emptyset\} \to \left(\mathcal{L}_\lambda(\mu)\right)^M$ such that, for any $\Gamma \in \left(\mathcal{P}(\mathcal{L}_\lambda(\mu))\right)^M \setminus \{\emptyset\}$, we have $f(\Gamma) \in \Gamma$ and $A \models f(\Gamma)$ if $A \models \bigvee\!\!\bigvee \Gamma$. Since M κ-globally covers V, there is a $g \in M$ with $g : \left(\mathcal{P}(\mathcal{L}_\lambda(\mu))\right)^M \setminus \{\emptyset\} \to \mathcal{P}_{<\kappa}\left(\mathcal{L}_\lambda(\mu)\right)^M$ such that $f(\Gamma) \in g(\Gamma) \subseteq \Gamma$ for all $\Gamma \in (\mathcal{P}(\mathcal{L}_\lambda(\mu)))^M \setminus \{\emptyset\}$.

In M, let

(2.8) $T = \{\bigvee\!\!\bigvee \Gamma \to \bigvee\!\!\bigvee g(\Gamma) \,:\, \Gamma \in \mathcal{P}(\mathcal{L}_\lambda(\mu)) \setminus \{\emptyset\}\}$.

Note that $M[A] \models$ "$A \models T$". It follows that T is consistent with respect to our deduction system (in V). In M, let

(2.9) $\mathbb{P} = \{\varphi \in \mathcal{L}_\lambda(\mu) \,:\, T \nvdash \neg\varphi\}$

and for φ, $\psi \in \mathbb{P}$, let

(2.10) $\varphi \leq_{\mathbb{P}} \psi \;\Leftrightarrow\; T \vdash \varphi \to \psi$.

Claim 2.4.1: For $\varphi \in \mathcal{L}_\lambda(\mu)$, if $A \models \varphi$ then we have $\varphi \in \mathbb{P}$. In particular, "$\alpha \in \dot{A}$" $\in \mathbb{P}$ for all $\alpha \in A$ and "$\neg(\alpha \in \dot{A})$" $\in \mathbb{P}$ for all $\alpha \in \mu \setminus A$.

$\vdash$ Suppose $A \models \varphi$. We have to show $T \nvdash \neg\varphi$: If $T \vdash \neg\varphi$ in M, then we would have $V \models$ "$T \vdash \neg\varphi$". Since $A \models T$ in V, it follows that $A \models \neg\varphi$. This is a contradiction. $\dashv$ (Claim 2.4.1)

Claim 2.4.2: For φ, $\psi \in \mathbb{P}$, φ and ψ are compatible if and only if

(2.11) $T \nvdash \neg(\varphi \wedge \psi)$.

Note that (2.11) is equivalent to

(2.12) $T \nvdash \neg\varphi \vee \neg\psi \quad (\Leftrightarrow\; T \nvdash \varphi \to \neg\psi)$.

$\vdash$ Suppose that φ, $\psi \in \mathbb{P}$ are compatible. By the definition of $\leq_{\mathbb{P}}$ this means that there is $\eta \in \mathbb{P}$ such that $T \vdash \eta \to \varphi$ and $T \vdash \eta \to \psi$. For this η we have $T \vdash \eta \to (\varphi \wedge \psi)$. Since $T \nvdash \neg\eta$ by the consistency of T, it follows that $T \nvdash \neg(\varphi \wedge \psi)$.

Conversely if $T \nvdash \neg(\varphi \wedge \psi)$. Then $(\varphi \wedge \psi) \in \mathbb{P}$. Since $T \vdash (\varphi \wedge \psi) \to \varphi$ and $T \vdash (\varphi \wedge \psi) \to \psi$, we have $(\varphi \wedge \psi) \leq_{\mathbb{P}} \varphi$ and $(\varphi \wedge \psi) \leq_{\mathbb{P}} \psi$. Thus φ and ψ are compatible with respect to $\leq_{\mathbb{P}}$. $\dashv$ (Claim 2.4.2)

Claim 2.4.3: $\mathbb{P}$ has the κ-c.c.

$\vdash$ Suppose that $\Gamma \subseteq \mathbb{P}$ is an antichain. Since $|\,g(\Gamma)\,| < \kappa$, it is enough to show that $g(\Gamma) = \Gamma$. Suppose otherwise and let $\varphi_0 \in \Gamma \setminus g(\Gamma)$. Since "$\bigvee\!\!\!\bigvee\Gamma \rightarrow \bigvee\!\!\!\bigvee g(\Gamma)$" $\in T$ and $\vdash \varphi_0 \rightarrow \bigvee\!\!\!\bigvee\Gamma$, we have

(2.13) $T \vdash \varphi_0 \rightarrow \bigvee\!\!\!\bigvee g(\Gamma)$.

It follows that there is $\varphi \in g(\Gamma)$ such that φ_0 and φ are compatible. This is because otherwise we would have $T \vdash \varphi_0 \rightarrow \neg\varphi$ for all $\varphi \in g(\Gamma)$ by Claim 2.4.2. Hence $T \vdash \varphi_0 \rightarrow \bigwedge\!\!\!\bigwedge\{\neg\varphi : \varphi \in g(\Gamma)\}$ which is equivalent to $T \vdash \varphi_0 \rightarrow \neg\bigvee\!\!\!\bigvee g(\Gamma)$. From this and (2.13), it follows that $T \vdash \neg\varphi_0$. But this is a contradiction to the assumption that $\varphi_0 \in \mathbb{P}$.

Now, since Γ is pairwise incompatible, it follows that $\varphi_0 = \varphi \in g(\Gamma)$. This is a contradiction to the choice of φ_0. $\dashv$ (Claim 2.4.3)

In V, let $G(A) = \{\varphi \in \mathbb{P} : A \models \varphi\}$. By Claim 2.4.1, we have $G(A) = \{\varphi \in \mathcal{L}_\lambda(\mu) : A \models \varphi\}$ and A is definable from $G(A)$ over M as $\{\alpha \in \mu :$ "$\alpha \in \dot{A}$" $\in G(A)\}$. Thus we have $M[G(A)] = M[A]$.

Hence the following two Claims prove our Lemma:

Claim 2.4.4: $G(A)$ is a filter in $\mathbb{P}$.

$\vdash$ Suppose that $\varphi \in G(A)$ and $\varphi \leq_{\mathbb{P}} \psi$. Since this means that $A \models \varphi$ and $T \vdash \varphi \rightarrow \psi$, it follows that $A \models \psi$. That is, $\psi \in G(A)$.

Suppose now that φ, $\psi \in G(A)$. This means that

(2.14) $A \models \varphi$ and $A \models \psi$.

Hence we have $A \models \varphi \wedge \psi$. By Claim 2.4.1, it follows that $(\varphi \wedge \psi) \in \mathbb{P}$, that is, $T \nvdash \neg(\varphi \wedge \psi)$. Thus φ and ψ are compatible by Claim 2.4.2. $\dashv$ (Claim 2.4.4)

Claim 2.4.5: $G(A)$ is $\mathbb{P}$-generic.

$\vdash$ Working in M, suppose that Γ is a maximal antichain in $\mathbb{P}$. By Claim 2.4.3, we have $|\,\Gamma\,| < \kappa$ and hence we have $\bigvee\!\!\!\bigvee\Gamma \in \mathcal{L}_\lambda(\mu)$ and hence $\bigvee\!\!\!\bigvee\Gamma \in \mathbb{P}$: For $\varphi \in \Gamma$, since $\varphi \in \mathbb{P}$ we have $T \nvdash \neg\varphi$ and $\vdash \varphi \rightarrow \bigvee\!\!\!\bigvee\Gamma$. It follows $T \nvdash \bigvee\!\!\!\bigvee\Gamma$.

Moreover we have $T \vdash \bigvee\!\!\!\bigvee\Gamma$: Otherwise $\neg\bigvee\!\!\!\bigvee\Gamma$ would be an element of $\mathbb{P}$ incompatible with every $\varphi \in \Gamma$. A contradiction to the maximality of Γ.

Hence $A \models \bigvee\!\!\!\bigvee\Gamma$ and thus there is $\varphi \in \Gamma$ such that $A \models \varphi$. That is, $\varphi \in G(A)$. $\dashv$ (Claim 2.4.5)

$\square$ (Lemma 2.4)

The proof of Theorem 2.3 from Lemma 2.4 relies on Lemma 2.2 and the Axiom of Choice is involved both in the statement and the proof of Lemma 2.2.

On the other hand, Lemma 2.4 can be proved without assuming the Axiom of Choice in M: It suffices to eliminate choice from the proof of Claim 2.4.5.

Proof of Claim 2.4.5 without the Axiom of Choice in M: Working in M, suppose that D is a dense subset of $\mathbb{P}$. Then $A \models \bigvee\!\!\!\bigvee D$: Otherwise we would have $T \nvdash \bigvee\!\!\!\bigvee D$. Since

(2.15) $T \vdash \bigvee\!\!\!\bigvee D \leftrightarrow \bigvee\!\!\!\bigvee g(D)$,

it follows that $T \nvdash \bigvee\!\!\!\bigvee g(D)$. Since $\bigvee\!\!\!\bigvee g(D) \in \mathcal{L}_\lambda(\mu)$, this implies $\neg\bigvee\!\!\!\bigvee g(D) \in \mathbb{P}$. Since D is dense in $\mathbb{P}$ there is $\varphi_0 \in D$ such that $T \vdash \varphi_0 \rightarrow \neg\bigvee\!\!\!\bigvee g(D)$. By (2.15), it follows that $T \vdash \varphi_0 \rightarrow \neg\bigvee\!\!\!\bigvee D$. On the other hand, since $\varphi_0 \in D$ we have $T \vdash \varphi_0 \rightarrow \bigvee\!\!\!\bigvee D$. Hence we have $T \vdash \neg\varphi_0$ which is a contradiction to $\varphi_0 \in \mathbb{P}$.

Thus there is $\varphi_1 \in D$ such that $A \models \varphi_1$, that is, $\varphi_1 \in G(A)$.

□ (Claim 2.4.5 without AC in M)

The next corollary follows immediately from this remark:

Corollary 2.5: *Work in NBG. Suppose that V is a model of ZFC and M is an inner model of V (of ZF) such that M κ-globally covers V. If $V = M[A]$ for some set $A \subseteq \mathrm{On}$ then V is a κ-c.c. set-generic extension of M.* □

We do not know if Corollary 2.5 is false without the added assumption that V is $M[A]$ for a set of ordinals A.

More generally, it seems to be open if there is a characterisation of the set-generic extensions of an arbitrary model of ZF; or at least of such extensions given by partial orders which are well-ordered in the ground model.

Grigorieff's theorem can be also obtained by a modification of the proof of Theorem 2.3.

Corollary 2.6: (S. Grigorieff [10]) *Suppose that M is an inner model of a model V of ZFC and V is a set-generic extension of M. Then any inner model N of V (of ZFC) with $M \subseteq N$ is a set-generic extension of M and hence definable in V. Also, for such N, V is a set-generic extension of N.*

If V is κ-c.c. set-generic extension of M in addition, then N is a κ-c.c. set-generic extension of M and V is a κ-c.c. set-generic extension of N. ❑

Similarly to Theorem 2.3, we can also characterize generic extensions obtained via a partial ordering of cardinality $\leq \kappa$.

For M and V as above, we say that V is *κ-decomposable* into M if for any $a \in V$ with $a \subseteq M$, there are $a_i \in M$, $i \in \kappa$ such that $a = \bigcup_{i<\kappa} a_i$.

Theorem 2.7: *Suppose that V is a transitive model of ZFC and M an inner model of ZFC definable in V and κ is a cardinal in M. Then V is a generic extension of M by a partial ordering in M of size $\leq \kappa$ (in M) if and only if M κ^+-globally covers V and V is κ-decomposable into M.*

Proof. If V is a generic extension of M by a generic filter G over a partial ordering $\mathbb{P} \in M$ of size $\leq \kappa$ (in M) then $\mathbb{P}$ has the κ^+-c.c. and hence M κ^+-globally covers V by Theorem 2.1. V is κ-decomposable into M since, for any $a \in V$ with $a = \dot{a}^G$, we have $a = \bigcup\{\{m \in M : p \Vdash_{\mathbb{P}} \text{“}m \in \dot{a}\text{”}\} : p \in G\}$.

Suppose now that M κ^+-globally covers V and V is κ-decomposable into M. By Theorem 2.3, there is a κ^+-c.c. partial ordering $\mathbb{P}$ in M and a $\mathbb{P}$-generic filter G over M such that $V = M[G]$. Without loss of generality, we may assume that $\mathbb{P}$ consists of the positive elements of a complete Boolean algebra $\mathbb{B}$ (in M).

By κ-decomposability, G can be decomposed into κ sets $G_i \in M$, $i < \kappa$. Without loss of generality, we may assume that $\mathbb{1}_{\mathbb{P}}$ forces this fact. So letting $\dot{G}$ be the standard name of G and $\dot{G}_i$, $i < \kappa$ be names of G_i, $i < \kappa$ respectively, we may assume

(2.16) $\Vdash_{\mathbb{P}} \text{“}\dot{G} = \bigcup_{i<\kappa} \dot{G}_i\text{”}.$

Working in M, let $X_i \subseteq \mathbb{P}$ be a maximal pairwise incompatible set of conditions p which decide $\dot{G}_i$ to be $G_{i,p} \in M$ for each $i < \kappa$. By the κ^+-c.c. of $\mathbb{P}$, we have $|\,X_i\,| \leq \kappa$. Clearly, we have $p \leq_{\mathbb{P}} \prod^{\mathbb{B}} G_{i,p}$ for all $i < \kappa$ and $p \in X_i$. Let $\mathbb{P}' = \bigcup\{X_i : i < \kappa\}$. Then $|\,\mathbb{P}'\,| \leq \kappa$.

Claim 2.7.1: $\mathbb{P}'$ is dense in $\mathbb{P}$.

$\vdash$ Suppose $p \in \mathbb{P}$. Then there is $q \leq p$ such that q decides some $\dot{G}_i$ to be $G_{i,q}$ and $p \in G_{i,q}$. Let $r \in X_i$ be compatible with q. Then we have $r \leq_{\mathbb{P}} \prod^{\mathbb{B}} G_{i,r} = \prod^{\mathbb{B}} G_{i,q} \leq p$. $\dashv$ (Claim 2.7.1)

Thus V is a $\mathbb{P}'$-generic extension over M. ❑ (Theorem 2.7)

3. A Formal Deductive System for $\mathcal{L}_\infty(\mu)$

In the proof of Lemma 2.4, we used a formal deductive system of $\mathcal{L}_\infty(\mu)$ without specifying exactly which system we are using. It is enough to consider a system of deduction which contains all logical axioms we used in the course of the proof together with modus ponens and some infinitary deduction rules like:

$$\frac{\varphi_i \to \psi, \quad i \in I}{\bigvee\!\!\!\bigvee\{\varphi_i : i \in I\} \to \psi}$$

What we need for such a system is that its correctness and upward absoluteness hold while we do not make use of any version of completeness of the system.

Formal deduction systems for infinitary logics have been studied extensively in 1960s and 1970s, see e.g. [14], [15], [20]. Nevertheless, to be concrete, we shall introduce below such a deductive system S for $\mathcal{L}_\infty(\mu)$.

One peculiar task for us here is that we have to make our deduction system S such that S does not rely on AC so that we can apply it in an inner model M which does not necessarily satisfy AC to obtain Corollary 2.5.

Recall that we have introduced $\mathcal{L}_\infty(\mu)$ as the smallest class containing the sets $\langle\alpha, 0\rangle$, $\alpha \in \mu$ as the codes of the prediactes "$\alpha \in \dot{A}$" for $\alpha \in \mu$ and closed with respect to $\langle\varphi, 1\rangle$ for $\varphi \in \mathcal{L}_\infty(\mu)$ and $\langle\Phi, 2\rangle$ for all sets $\Phi \subseteq \mathcal{L}_\infty(\mu)$ where $\langle\varphi, 1\rangle$ and $\langle\Phi, 2\rangle$ represent $\neg\varphi$ and $\bigvee\!\!\!\bigvee\Phi$ respectively. Here, to be more precise about the role of the infinite conjunction we add the infinitary logical connective $\bigwedge\!\!\!\bigwedge$, and assume that $\bigwedge\!\!\!\bigwedge\Phi$ is coded by $\langle\Phi, 3\rangle$ and thus $\mathcal{L}_\infty(\mu)$ is also closed with respect to $\langle\Phi, 3\rangle$ for all sets $\Phi \subseteq \mathcal{L}_\infty(\mu)$.

The axioms of S consist of the following formulas:

(A1) $\varphi(\varphi_0, \varphi_1, ..., \varphi_{n-1})$

for each tautology $\varphi(A_0, A_1, ..., A_{n-1})$ of (finitary) propositional logic and $\varphi_0, \varphi_1, ..., \varphi_{n-1} \in \mathcal{L}_\infty(\mu)$;

(A2) $\varphi \to \bigvee\!\!\!\bigvee\Phi$ and $\bigwedge\!\!\!\bigwedge\Phi \to \varphi$

for any set $\Phi \subset \mathcal{L}_\infty(\mu)$ and $\varphi \in \Phi$;

(A3) $\neg(\bigwedge\!\!\!\bigwedge\Phi) \leftrightarrow \bigvee\!\!\!\bigvee\{\neg\varphi : \varphi \in \Phi\}$ and
$\neg(\bigvee\!\!\!\bigvee\Phi) \leftrightarrow \bigwedge\!\!\!\bigwedge\{\neg\varphi : \varphi \in \Phi\}$

for any set $\Phi \subseteq \mathcal{L}_\infty(\mu)$; and

(A4) $\varphi \wedge (\bigvee\!\!\bigvee \Psi) \leftrightarrow \bigvee\!\!\bigvee \{\varphi \wedge \psi : \psi \in \Psi\}$ and
$\varphi \vee (\bigwedge\!\!\bigwedge \Psi) \leftrightarrow \bigwedge\!\!\bigwedge \{\varphi \vee \psi : \psi \in \Psi\}$
for any $\varphi \in \mathcal{L}_\infty(\mu)$ and any set $\Psi \subseteq \mathcal{L}_\infty(\mu)$.

Deduction Rules:

(Modus Ponens) $$\frac{\{\varphi, \varphi \to \psi\}}{\psi}$$

(R1) $$\frac{\{\varphi \to \psi : \varphi \in \Phi\}}{\bigvee\!\!\bigvee \Phi \to \psi}$$ **(R2)** $$\frac{\{\varphi \to \psi : \psi \in \Psi\}}{\varphi \to \bigwedge\!\!\bigwedge \Psi}$$

A proof of $\varphi \in \mathcal{L}_\infty(\mu)$ from $\Gamma \subseteq \mathcal{L}_\infty(\mu)$ is a labeled tree $\langle \mathsf{T}, f \rangle$ such that

(3.1) $\mathsf{T} = \langle \mathsf{T}, \leq \rangle$ is a tree growing upwards with its root r_0 and T with $(\leq)^{-1}$ is well-founded;

(3.2) $f : \mathsf{T} \to \mathcal{L}_\infty(\mu)$;

(3.3) $f(r_0) = \varphi$;

(3.4) if $t \in \mathsf{T}$ is a maximal element then either $f(t) \in \Gamma$ or t is one of the axioms of S;

(3.5) if $t \in \mathsf{T}$ and $P \subseteq \mathsf{T}$ is the set of all immediate successors of t, then
$$\frac{\{f(p) : p \in P\}}{f(t)}$$
is one of the deduction rules.

We have to stress here that, in (3.5), we do not assume that the function f is one-to-one since otherwise we have to choose a proof for each formula in the set in the premises of **(R1)** and **(R2)**. Thus, for example, we can deduce $T \vdash \bigwedge\!\!\bigwedge \Phi$ in S from $T \vdash \varphi$ for all $\varphi \in \Phi$ without appealing to AC.

Now the proof of the following is an easy exercise:

Proposition 3.1: *(1) For any $B \subseteq \mu$, $T \subseteq \mathcal{L}_\infty(\mu)$ and $\varphi \in \mathcal{L}_\infty(\mu)$, if $T \vdash \varphi$ and $B \models T$, then we have $B \models \varphi$.*

(2) For transitive models M, N of ZF such that M is an inner model of N, if $M \models$ "$\langle \mathsf{T}, f \rangle$ is a proof of φ in $\mathcal{L}_\infty(\mu)$", then

$N \models$ "$\langle \mathsf{T}, f \rangle$ is a proof of φ in $\mathcal{L}_\infty(\mu)$".

Proof. (1): By induction on cofinal subtrees of a fixed proof $\langle \mathsf{T}, f \rangle$ of φ.
(2): Clear by definition. ❑ (Proposition 3.1)

An alternative setting to the argument by means of a deductive system is to make use of the following definition of $M \models$ "$\Gamma \vdash \varphi$" in the proof of Lemma 2.4:

> $M \models$ "$\Gamma \vdash \varphi$" iff for any $B \subseteq \mu$ in some set-forcing extension $M[G]$ of M, $M[G] \models B \models \psi$ for all $\psi \in \Gamma$ always implies $M[G] \models B \models \varphi$.

Note that this is definable in M using the forcing relation definable on M. It remains to verify that this notion has the desired degree of absoluteness. Actually we can easily prove the full absoluteness, that is, if N is a transitive model containing M with the same ordinals as those of M then, for Γ, $\varphi \in M$ with $M \models \Gamma \subseteq \mathcal{L}_\infty(\mu)$ and $M \models \varphi \in \mathcal{L}_\infty(\mu)$, $\Gamma \vdash \varphi$ holds in M iff $\Gamma \vdash \varphi$ holds in N.

First suppose that $B \subseteq \mu$ is a set of ordinals in a set-generic extension $N[G]$ of N such that B witnesses the failure of $\Gamma \vdash \varphi$ in N. Let x be a real which is generic over N for the Lévy collapse of a sufficiently large ν to ω such that Γ and μ become countable in the generic extension $N[x]$. Then x is also Lévy generic over M and $M[x]$ is a submodel of $N[x]$. By Lévy Absoluteness, it follows that that there exists $B' \subseteq \mu$ in $M[x]$ which also witnesses the failure of $\Gamma \vdash \varphi$ in M.

Conversely, suppose that $\Gamma \vdash \varphi$ holds in N and let $B \subseteq \mu$ be a set of ordinals in a set-generic extension $M[G]$ of M such that B witnesses the failure of $\Gamma \vdash \varphi$ in M. Then B also belongs to an extension of M which is generic for the Lévy collapse of sufficently large ν to ω; choose a condition p in this forcing which forces the existence of such a B. Now if x is Lévy-generic over N and contains the condition p, we see that there is a counterexample to $\Gamma \vdash \varphi$ in N witnessed in $N[x]$, contrary to our assumption.

With both of the interpretations of $\vdash$ we can check that the arguments in Section 2 go through.

4. An Axiomatic Framework for the Set-Generic Multiverse

In this section, we consider some possible axiomatic treatments of the set-generic multiverse. Such axiomatic treatments are also discussed e.g. in [9], [19], [22]. We introduce a conservative extension MZFC of ZFC in which we can treat the multiverse of set-generic extensions of models of ZFC as a collection of countable transitive models. This system or some further extension of it (which can possibly also treat tame class forcings) may be used as a basis for direct formulation of statements concerning the multiverse.

The language $\mathcal{L}_{\mathrm{MZF}}$ of the axiom system MZFC consists of the ϵ-relation symbol '$\in$', and a constant symbol 'v' which should represent the countable transitive "ground model".

The axiom system MZFC consists of

(4.1) all axioms of ZFC;

(4.2) "v is a countable transitive set";

(4.3) "$\mathrm{v} \models \varphi$" for all axioms φ of ZFC.

By (4.1), MZFC proves the (unique) existence of the closure $\mathcal{M}$ of "$\{\mathrm{v}\}$" under forcing extension and definable "inner model" of "ZF" (here 'ZF' is set in quotation marks since we can only argue in metamathematics that such "inner model" satisfies each instance of replacement). Note that $\mathcal{M} \subseteq \mathcal{H}_{\aleph_1}$. Here "inner model" is actually phrased in $\mathcal{L}_{\mathrm{ZF}}$ as "transitive almost universal subset closed under Gödel operations". If we had $\mathrm{v} \models \mathrm{ZFC}$, we would have $w \models \mathrm{ZF}$ for any inner model w of v in this sense by Theorem 13.9 in [13]. In MZFC, however, we have only $\mathrm{v} \models \varphi$ for each axiom φ of ZFC (in the meta-mathematics). Nevertheless, for all such "inner model" w and hence for all $w \in \mathcal{M}$, we have $w \models \varphi$ for all axiom φ of ZF by the proof of Theorem 13.9 in [13] and the Forcing Theorem. Apparently, this is enough to consider $\mathcal{M}$ in this framework as the set-generic multiverse.

Similarly, we can also start from any extension of ZFC (e.g. with some additional large cardinal axiom) and make $\mathcal{M}$ closed under some more operations such as some well distinguished class of class forcing extensions.

The following theorem shows that we do not increase the consistency strength by moving from ZFC to MZFC.

Theorem 4.1: *MZFC is a conservative extension of ZFC: for any sentence ψ in $\mathcal{L}_{\mathrm{ZF}}$, we have* $\mathrm{ZFC} \vdash \psi \iff \mathrm{MZFC} \vdash \psi$. *In particular, MZFC is equiconsistent with ZFC.*

Proof. "$\Rightarrow$" is trivial.

For "$\Leftarrow$", suppose that $\mathrm{MZFC} \vdash \psi$ for a formula ψ in $\mathcal{L}_{\mathrm{ZF}}$. Let $\mathcal{P}$ be a proof of ψ from MZFC and let T be the finite fragment of ZFC consisting of all axioms φ of ZFC such that $\mathrm{v} \models \varphi$ appears in $\mathcal{P}$. Let $\Phi(x)$ be the formula in $\mathcal{L}_{\mathrm{ZF}}$ saying

$$\text{"}x \text{ is a countable transitive set and } x \models \bigwedge\!\!\!\bigwedge T\text{"}.$$

By the Deduction Theorem, we can recast $\mathcal{P}$ to a proof of $\mathrm{ZFC} \vdash \forall x(\Phi(x) \rightarrow$

ψ). On the other hand we have ZFC $\vdash \exists x \Phi(x)$ (by the Reflection Principle, Downward Löwenheim-Skolem Theorem and Mostowski's Collapsing Theorem). Hence we obtain a proof of ψ from ZFC alone. □ (Theorem 4.1)

It may be a little bit disappointing if each set-theoretic universe in the multiverse seen from the "meta-universe" is merely a countable set. Of course if M is an inner model of a model W of ZFC (i.e. M is a model which is a transitive class $\subseteq W$ and $M, W \models$ ZFC) there are always partial ordering $\mathbb{P}$ in M for which there is no $(M, \mathbb{P})$-generic set in W (e.g. any partial ordering collapsing a cardinal of W cannot have its generic set in W).

However, if we are content with a meta-universe which is not a model of full ZFC, we can work with the following setting where each of the "elements" of the set-generic multiverse is an inner model of a meta-universe: starting from a model V of ZFC with an inaccessible cardinal κ, we generically extend it to $W = V[G]$ by Lévy collapsing κ to ω_1. Letting $M = \mathcal{H}(\kappa)^V$, we have $M \models$ ZFC and M is an inner model of $W = \mathcal{H}(\kappa)^{V[G]} = \mathcal{H}(\omega_1)^{V[G]}$. $W \models$ ZFC $-$ the Power Set Axiom and for any partial ordering $\mathbb{P}$ in M there is a $(M, \mathbb{P})$-generic set in W. Thus an NBG-type theory of W with a new unary predicate corresponding to M can be used as a framework of the theory for the set-generic multiverse (which is obtained by considering all the set-generic grounds of M, and then all the set generic extensions of them, etc.) as a "class" of classes in W. A setting similar to this idea was also discussed in [19].

5. Independent Buttons

The multiverse view sometimes highlights problems which would be never asked in the conventional context of forcing constructions. The existence of infinitely many independent buttons which arose in connection with the characterization of the modal logic of the set-generic multiverse (see [12]) is one such question.

A sentence φ in $\mathcal{L}_{\rm ZF}$ is said to be a *button* (for set-genericity) if any set-generic extension $V[G]$ of the ground model V has a further set-generic extension $V[G][H]$ such that φ holds in all set-generic extensions of $V[G][H]$. Let us say that a button φ is pushed in a set-generic extension $V[G]$ if φ holds in all further set-generic extensions $V[G][H]$ of $V[G]$ (including $V[G]$ itself).

Formulas φ_n, $n \in \omega$ are *independent buttons*, if φ_n, $n \in \omega$ are unpushed buttons and for any set-generic extension $V[G]$ of the ground model V and any $X \subseteq \omega$ in $V[G]$,

(5.1) if $\{n \in \omega : V[G] \models \varphi_n \text{ is pushed}\} \subseteq X$ then there is a set-generic extension $V[G][H]$ such that $\{n \in \omega : V[G][H] \models \varphi_n \text{ is pushed}\} = X$.

In [12], it is claimed that formulas b_n, $n \in \omega$ form an infinite set of independent buttons over $V = L$ where b_n is a formula asserting: "${\omega_n}^L$ is not a cardinal". This is used to prove that the principles of forcing expressible in the modal logic of the set-theoretic multiverse as a Kripke frame where modal operator $\Box$ is interpreted as:

(5.2) $M \models \Box\varphi \Leftrightarrow$ in all set-generic extensions $M[G]$ of M we have $M[G] \models \varphi$

coincides with the modal theory S4.2 (Main Theorem 6 in [12]).

Unfortunately, it seems that there is no guarantee that (5.1) holds in an arbitrary set-generic extension $V[G]$ for these b_n, $n \in \omega$.

In the following, we introduce an alternative set of infinitely many formulas which are actually independent buttons for any ground model of ZFC+ "GCH below $\aleph_\omega$" + "$\aleph_n = \aleph_n^L$ for all $n \in \omega$" which can be used as b_n, $n \in \omega$ in [12].

We first note that, for Main Theorem 6 in [12] we actually need only the existence of an arbitrary finite number of independent buttons. In the case of $V = L$ the following formulas can be used for this: Let ψ_n be the statement that $\aleph_n^L$ is a cardinal and the L-least $\aleph_n^L$-Suslin tree T_n^L in L (i.e., the L-least normal tree of height $\aleph_n^L$ with no antichain of size $\aleph_n^L$ in L) is still $\aleph_n^L$-Suslin. If M is a set-generic (or arbitrary) extension of L in which the button $\neg\psi_n$ has not been pushed, then by forcing with T_n^L over M we push this button and do not affect any of the other unpushed buttons $\neg\psi_m$, $m \neq n$, as this forcing is $\aleph_n$-distributive and has size $\aleph_n$. Rittberg [18] also found independent buttons under $V = L$.

Now we turn to a construction of infinitely many independent buttons for which we even do not need the existence of Suslin trees. For $n \in \omega$, let φ_n be the statement:

(5.3) there is an injection from ${\aleph_{n+2}}^L$ to $\mathcal{P}({\aleph_n}^L)$.

Note that φ_n is pushed in a set-generic extension $V[G]$ if and only if it holds in $V[G]$. Thus φ_n for each $n \in \omega$ is a button provided that φ_n does not hold in the ground model. We show that these φ_n, $n \in \omega$ are independent buttons (over any ground model where they are unpushed — e.g., when $V = L$).

Suppose that we are working in some model W of ZFC. In W, let $A = \{n \in \omega : \Box\varphi_n \text{ holds}\}$ and $B \subseteq \omega$ be arbitrary with $A \subseteq B$. It is enough to prove the following

Proposition 5.1: *We can force (over W) that φ_n holds for all $n \in B$ and $\neg\varphi_n$ for all $n \in \omega \setminus B$.*

Proof. In W, let $\kappa_n = |\aleph_n{}^L|$ for $n \in \omega$. We use the notation of [16] on the partial orderings with partial functions and denote with $\mathrm{Fn}(\kappa, \lambda, \mu)$ the set of all partial functions from κ to λ with cardinality $< \mu$ ordered by reverse inclusion. By Δ-System Lemma, it is easy to see that $\mathrm{Fn}(\kappa, \lambda, \mu)$ has the $(\lambda^{<\mu})^+$-c.c. Let

$$(5.4) \quad \mathbb{P}_n = \begin{cases} \mathrm{Fn}(\kappa_{n+2}, 2, \kappa_n) & \text{if } n \in B \setminus A \\ \mathbb{1} & \text{otherwise.} \end{cases}$$

Let $\mathbb{P} = \prod_{n \in \omega} \mathbb{P}_n$ be the full support product of $\mathbb{P}_n$, $n \in \omega$. Then we clearly have $\Vdash_{\mathbb{P}}$ "φ_n" for all $n \in B$. Thus to show that $\mathbb{P}$ creates a generic extension as desired, it is enough to show that $\Vdash_{\mathbb{P}}$ "$\neg\varphi_n$" for all $n \in \omega \setminus B$.

Suppose that

$$(5.5) \quad n \in \omega \setminus B.$$

Then we have

$$(5.6) \quad \mathbb{P}_n = \mathbb{1}.$$

Since φ_n does not hold in W, we have $\kappa_n < \kappa_{n+1} < \kappa_{n+2}$ and $2^{\kappa_n} = \kappa_{n+1}$ in W. By (5.6), $\mathbb{P}$ factors as $\mathbb{P} \sim \mathbb{P}(< n) \times \mathbb{P}(> n)$ where $\mathbb{P}(< n) = \prod_{k<n} \mathbb{P}_k$ and $\mathbb{P}(> n) = \prod_{k>n} \mathbb{P}_k$.

We show that both $\mathbb{P}(> n)$ and $\mathbb{P}(< n)$ over $\mathbb{P}(> n)$ do not add any injection from κ_{n+2} into $\mathcal{P}(\kappa_n)$.

$\mathbb{P}(> n)$ is κ_{n+1}-closed. Thus it does not add any new subsets of κ_n. So if it added an injection from κ_{n+2} into $\mathcal{P}(\kappa_n)$ then it would collapse the cardinal κ_{n+2}. Since $\mathbb{P}(> n)$ further factors as $\mathbb{P}(> n) \sim \mathbb{P}_{n+1} \times \prod_{k>n+1} \mathbb{P}_k$ and $\prod_{k>n+1} \mathbb{P}_k$ is κ_{n+2}-closed the only way $\mathbb{P}(> n)$ could collapse κ_{n+2} would be if $\mathbb{P}_{n+1}$ did so. But then, since $\mathbb{P}_{n+1}$ has the $(2^{<\kappa_{n+1}})^+$-c.c. with $(2^{<\kappa_{n+1}})^+ = (2^{\kappa_n})^+$, we would have $2^{\kappa_n} \geq \kappa_{n+2}$. This is a contradiction to the choice (5.5) of n. So $\mathbb{P}(> n)$ forces φ_n to fail.

In the rest of the proof, we work in $W^{\mathbb{P}(>n)}$ and show that $\mathbb{P}(<n)$ does not add any injection from κ_{n+2} into $\mathcal{P}(\kappa_n)$. Note that, by κ_{n+1}-closedness of $\mathbb{P}(>n)$, we have $\mathrm{Fn}(\kappa_{m+2}, 2, \kappa_m)^W = \mathrm{Fn}(\kappa_{m+2}, 2, \kappa_m)^{W^{\mathbb{P}(>n)}}$ for $m < n$.

We have the following two cases:

Case I. $n - 1 \in A \cup (\omega \setminus B)$. Then $\mathbb{P}(<n) \sim \mathbb{P}(<m)$ for some $m < n$ and $\mathbb{P}(<m)$ has the $(2^{\kappa_{m-1}})^+$-c.c. with $(2^{\kappa_{m-1}})^+ \leq \kappa_n$.

Case II. $n - 1 \in B \setminus A$. Then $2^{<\kappa_{n-1}} = \kappa_n$ and $\mathbb{P}(<n)$ has the κ_{n+1}-c.c.

In both cases the partial ordering $\mathbb{P}(<n)$ has κ_{n+1}-c.c. and hence the cardinals κ_{n+1} and κ_{n+2} are preserved. Since $\mathbb{P}(<n)$ has at most cardinality $2^{\kappa_{n-1}} \cdot \kappa_{n+1} = \kappa_{n+1}$, it adds at most ${\kappa_{n+1}}^{\kappa_n} = \kappa_{n+1}$ new subsets of κ_n and thus the size of $\mathcal{P}(\kappa_n)$ remains unchanged. This shows that $\Vdash_{\mathbb{P}}$ "$\neg\varphi_n$".

□ (Proposition 5.1)

Acknowledgments

The authors wish to thank Dr. Toshimichi Usuba and the anonymous referee for valuable comments on this article.

The first author would like to thank the FWF (Austrian Science Fund) for its support through project number P 28420.

The second author is supported by Grant-in-Aid for Scientific Research (C) No. 21540150 and Grant-in-Aid for Exploratory Research No. 26610040 of the Ministry of Education, Culture, Sports, Science and Technology Japan (MEXT).

The third author is supported by Grant-in-Aid for Young Scientists (B) No. 23740076 of the Ministry of Education, Culture, Sports, Science and Technology Japan (MEXT).

References

1. Tatiana Arrigoni and Sy-David Friedman, Foundational implications of the inner model hypothesis, Annals of Pure and Applied Logic, Vol. 163, (2012), 1360–66.
2. Tatiana Arrigoni and Sy-David Friedman, The hyperuniverse program, Bulletin of Symbolic Logic 19, No. 1, (2013), 77–96.
3. Lev Bukovský, Characterization of generic extensions of models of set theory, Fundamenta Mathematica 83 (1973), 35–46.
4. Lev Bukovský, Generic Extensions of Models of ZFC, a lecture note of a talk at the Novi Sad Conference in Set Theory and General Topology, Novi Sad, August 18–21, (2014).

5. Sy-David Friedman, Strict genericity, in *Models, Algebras and Proofs*, proceedings of the 1995 Latin American Logic Symposium, (1999), 129–139.
6. Sy-David Friedman, *Fine Structure and Class Forcing*, de Gruyter series in logic and its applications, volume 3, (2000).
7. Sy-David Friedman, Internal consistency and the inner model hypothesis, Bulletin of Symbolic Logic, Vol. 12, No. 4, December (2006), 591–600.
8. Gunter Fuchs, Joel David Hamkins and Jonas Reitz, Set Theoretic Geology, Annals of Pure and Applied Logic, Vol. 166, Iss. 4 (2015), 464–501.
9. Victoria Gitman and Joel Hamkins, A natural model of the multiverse axioms, Nortre Dame Journal of Formal Logic, Vol. 51, (4), (2010), 475–484.
10. Serge Grigorieff, Intermediate Submodels and Generic Extensions in Set Theory, The Annals of Mathematics, Second Series, Vol. 101, No. 3 (1975), 447–490.
11. Joel David Hamkins, The set-theoretical multiverse, Review of Symbolic Logic, Vol. 5, (2012), 416–449.
12. Joel David Hamkins and Benedikt Löwe, The modal logic of forcing, Transactions of the American Mathematical Society Vol. 360, No. 4, (2008), 1793–1817.
13. Thomas Jech, *Set Theory*, The Third Millennium Edition, Springer (2001/2006).
14. Carol Karp, *Languages with Expressions of Infinite Length*, North-Holland, (1964).
15. H. Jerome Keisler, *Model Theory for Infinitary Logic*, North-Holland (1974).
16. Kenneth Kunen, *Set Theory, An Introduction to Independence Proofs*, North-Holland (1980).
17. Richard Laver, Certain very large cardinals are not created in small forcing extensions, Annals of Pure and Applied Logic 149 (2007) 1–6.
18. Colin Jakob Rittberg, On the modal logic of forcing, Diploma Thesis, (2010).
19. John R. Steel, Gödel's program, in: Juliette Kennedy (ed.), *Interpreting Gödel: Critical Essays*, Cambridge University Press (2014), 153–179.
20. Gaishi Takeuti, *Proof Theory*, 2nd Ed., North-Holland, (1987).
21. Toshimichi Usuba, The downward directed grounds hypothesis and large large cardinals, preprint.
22. Jouko Väänaän, Multiverse Set Theory and Absolutely Undecidable Propositions, in: Juliette Kennedy (ed.), *Interpreting Gödel: Critical Essays*, Cambridge University Press (2014), 180–208.
23. W. Hugh Woodin, Recent developments on Cantor's Continuum Hypothesis, in *Proceedings of the Continuum in Philosophy and Mathematics, 2004*, Carlsberg Academy, Copenhagen, November (2004).
24. W. Hugh Woodin, The realm of the infinite, in: Michael Heller and W. Hugh Woodin (eds.), *Infinity: New Research Frontiers*, Cambridge University Press, (2011).
25. W. Hugh Woodin, The Continuum Hypothesis, the generic multiverse of sets, and the Ω conjecture, in: J. Kennedy and R. Kossak (eds.), *Set Theory, Arithmetic, and Foundations of Mathematics*, ASL lecture notes in Logic, Cambridge Univ. Press (2011), 13-42.

REAL GAMES AND STRATEGICALLY SELECTIVE COIDEALS

Paul B. Larson
Department of Mathematics, Miami University, Oxford, Ohio 45056 USA
larsonpb@miamioh.edu

Dilip Raghavan
Department of Mathematics, National University of Singapore
Singapore 119076
dilip.raghavan@protonmail.com

We introduce a notion of strategically selective coideal, and show that tall strategically selective coideals do not exist under $\mathrm{AD}_{\mathbb{R}}$, generalizing a classical theorem of Mathias. We discuss some issues involved in generalizing this result to semiselective coideals.

1. The Main Theorem

A set $\mathcal{C} \subseteq \mathcal{P}(\omega)$ is a *coideal* if $\mathcal{P}(\omega) \setminus \mathcal{C}$ is an ideal containing Fin, the collection of finite subsets of ω. A *fast diagonalization* of a sequence $\langle A_i : i < \omega \rangle \in \mathcal{P}(\omega)^\omega$ is a set $E = \{e_i : i \in \omega\}$ (listed in increasing order) such that $e_0 \in A_0$ and $e_{i+1} \in A_{e_i}$ for all $i \in \omega$. A coideal $\mathcal{C}$ is

- *tall* if for no infinite $A \subseteq \omega$ is $\mathcal{C} \cap \mathcal{P}(A) = \mathcal{P}(A) \setminus \mathrm{Fin}$, and
- *selective* if every $\subseteq$-decreasing sequence $\langle A_i : i < \omega \rangle \in \mathcal{C}^\omega$ has a fast diagonalization in $\mathcal{C}$.

The axiom $\mathrm{AD}_{\mathbb{R}}$ asserts the determinacy of all games of length ω where the players play subsets of ω. Woodin has shown the consistency of $\mathrm{AD}_{\mathbb{R}}$ relative to large cardinals (see Theorem 9.3 of [10], for instance). The axiom DC is a weak form of the Axiom of Choice, asserting that each tree of height ω without terminal nodes has an infinite branch. The axiom $\mathrm{DC}_{\mathbb{R}}$ is the restriction of DC to trees on $\mathbb{R}$; $\mathrm{DC}_{\mathbb{R}}$ is easily seen to be a consequence of $\mathrm{AD}_{\mathbb{R}}$. Mathias [8] showed that $\mathrm{AD}_{\mathbb{R}}$ implies that there are no tall selective coideals on ω. In this note, we give a generalization of this fact. Our proof consists of combining Mathias's proof with results of Solovay and Woodin.

Given a coideal $\mathcal{C}$, we let $\mathcal{G}_\mathcal{C}$ be the game of length ω in which players I and II choose the members of a $\subseteq$-decreasing sequence of elements of $\mathcal{C}$, with II winning if the sequence constructed has a fast diagonalization in $\mathcal{C}$.

Definition 1.1: A coideal $\mathcal{C}$ is *strategically selective* if player I does not have a winning strategy in $\mathcal{G}_\mathcal{C}$.

It follows almost immediately from the definitions that a selective coideal is strategically selective, as in this case all runs of the game $\mathcal{G}_\mathcal{C}$ are won by II. As defined by Farah [3], a coideal $\mathcal{C}$ is *semiselective* if whenever $A \in \mathcal{C}$ and $\mathcal{D}_i$ $(i \in \omega)$ are dense open subsets of the partial order $(\mathcal{C}, \subseteq)$ below A, there exist sets $A_i \in \mathcal{D}_i$ $(i \in \omega)$ and a fast diagonalization of $\langle A_i : i \in \omega\rangle$ in $\mathcal{C}$. As we shall see, the Axiom of Choice implies that the notions of strategically selective and semiselective are equivalent.

Assuming $\mathrm{AD}_\mathbb{R}$, strategically selective coideals are easily seen to be semiselective; we don't know if the reverse implication holds. Uniformization is the statement that for each $U \subseteq \mathbb{R}\times\mathbb{R}$, there is a partial function $f : \mathbb{R} \to \mathbb{R}$ with the property that for each $x \in \mathbb{R}$, if there is a y such that (x, y) is in U, then x is in the domain of f and $(x, f(x))$ is in U. Uniformization is an easy consequence of $\mathrm{AD}_\mathbb{R}$, via a game in which each player plays once.

Theorem 1.2: *If Uniformization holds, then every strategically selective codideal is semiselective.*

Proof: Suppose that $\mathcal{C}$ is a coideal which is not semiselective, and let $\mathcal{D}_i$ $(i \in \omega)$ and $A \in \mathcal{C}$ witness this. Fix a strategy Σ in $\mathcal{G}_\mathcal{C}$ for player I where I plays a subset of A in $\mathcal{D}_0$ as his first move, and, for any $i > 0$, I plays in response to any sequence of length $2i$ an element of $\mathcal{D}_i$ contained in the last move made by player II. Uniformization implies that there exists such a strategy. Now suppose that $\bar{A} = \langle A_i : i < \omega\rangle$ is a run of $\mathcal{G}_\mathcal{C}$ where I has played according to Σ, and suppose toward a contradiction that $\{e_i : i \in \omega\}$ is a fast diagonalization of $\bar{A}$ in $\mathcal{C}$. Then $\{e_{2i+1} : i \in \omega\}$ and $\{e_{2i} : i \in \omega\}$ are each fast diagonalizations of $\langle A_{2i} : i \in \omega\rangle$, and at least one of these two sets is in $\mathcal{C}$, giving a contradiction. □

Given a coideal $\mathcal{C}$, we let $P_\mathcal{C}$ denote the partial order of mod-I containment on $\mathcal{C}$, where $I = \mathcal{P}(\omega) \setminus \mathcal{C}$. Farah [3] has shown that when $\mathcal{C}$ is semiselective, forcing with $P_{\mathcal{C}}$ adds a selective ultrafilter (see Theorem 2.1 below).

The following is our main theorem.

Theorem 1.3: *If* $\mathrm{AD}_{\mathbb{R}}$ *holds, then there are no tall strategically selective coideals on* ω.

Question 1.4 below is open, as far as we know. Corollary 3.4 of [2] states a result which would imply a positive answer to the question. However, there appears to be a gap in the proof, corresponding roughly to the issue of obtaining functions a and b as in the statement of Proposition 2.4 below in the context of $\mathrm{AD}_{\mathbb{R}}$ (i.e., without using the Axiom of Choice).

Question 1.4: Does $\mathrm{AD}_{\mathbb{R}}$ imply that there are no tall semiselective coideals on ω?

We begin with Solovay's result on the existence of a normal fine measure on $\mathcal{P}_{\aleph_1}(\mathbb{R})$ under $\mathrm{AD}_{\mathbb{R}}$. Here *normality* of a measure μ on $\mathcal{P}_{\aleph_1}(\mathbb{R})$ means that if $A \in \mu$ and f is a function on A such that $f(\sigma)$ is a nonempty subset of σ for each $\sigma \in A$, then there is a real x which is in $f(\sigma)$ for μ-many σ.

Lemma 1.5 (Solovay [9]). *If* $\mathrm{AD}_{\mathbb{R}}$ *holds, then there is normal fine measure on* $\mathcal{P}_{\aleph_1}(\mathbb{R})$.

Proof: Given $A \subseteq \mathcal{P}_{\aleph_1}(\mathbb{R})$, consider the game $\mathcal{G}_m(A)$ where players I and II pick alternately pick finite sets of reals s_i ($i \in \omega$) and I wins if $\bigcup\{s_i : i \in \omega\} \in A$. Let μ be the set of A for which I has a winning strategy in $\mathcal{G}_m(A)$. That μ is fine (i.e., contains the set of supersets of each countable set of reals) is immediate. That it is an ultrafilter follows from running two strategies against one another. Normality follows the fact that given a family of games $\mathcal{G}_x$ indexed by reals for which II has a winning strategy, there is a function picking such a strategy for each game, induced by the game where I first picks x and then I and II play $\mathcal{G}_x$. Fixing f as in the statement of normality and supposing that player II has a winning strategy for each payoff set of the form $F_x = \{\sigma \mid x \in f(\sigma)\}$, we can build a countable set of reals $\sigma \in \mathrm{dom}(f)$ which results from a run of $\mathcal{G}_m(F_x)$ according to a winning strategy for player II, for each $x \in \sigma$, giving a contradiction. $\square$

Given a set of ordinals S and a formula ϕ, let us write $A_{S,\phi}$ for the set $\{x \in (\omega^\omega)^{<\omega} : L[S,x] \models \phi(S,x)\}$. Formally extending a definition due to

Woodin, we say that the pair (S, ϕ) is an ∞-*Borel code* for the set $A_{S,\phi}$, and we say that a set $B \subseteq (\omega^\omega)^{<\omega}$ is ∞-*Borel* if there exists such a pair (S, ϕ) with $A_{S,\phi} = B$ (i.e., if B has an ∞-Borel code). A tree on the ordinals projecting to a subset of ω^ω is an example of an ∞-Borel code, but the assumption that every set of reals is ∞-Borel is weaker than the assumption that every set of reals is the projection of a tree on the ordinals. The statement that every subset of $(\omega^\omega)^{<\omega}$ is ∞-Borel is easily seen to be equivalent to the assertion that every subset of ω^ω is ∞-Borel, which in turn is part of Woodin's axiom AD^+ (see [4]).

The following theorem is unpublished.

Theorem 1.6 (Woodin). *If* $\mathrm{AD}_\mathbb{R}$ *holds, then every set of reals is* ∞-*Borel.*

Recall that *Mathias forcing* Q_U relative to an ultrafilter U consists of pairs (s, A), where s is finite subset of ω and $A \in U$, with the order $(s, A) \geq (t, B)$ if $s \subseteq t$, $B \subseteq A$ and $t \setminus s \subseteq A$. The following is due to Mathias ([8], Theorem 2.0).

Theorem 1.7 (Mathias). *Suppose that* M *is a model of* $ZF + \mathrm{DC}_\mathbb{R}$ *and that* U *is a selective ultrafilter in* M. *Then a set* $x \subseteq \omega$ *is* M-*generic for* Q_U *if and only if* $x \setminus y \in \mathrm{Fin}$ *for all* $y \in U$.

The following proof puts together the facts listed above. The ultraproduct construction in the proof is taken from the proof of Theorem 9.39 from [11], except that we use ∞-Borel codes instead of trees on the ordinals.

Proof: [Proof of Theorem 1.3.] Suppose that $\mathrm{AD}_\mathbb{R}$ holds, and that $\mathcal{C}$ is a tall strategically selective coideal. Fix a winning strategy Σ for player II in $\mathcal{G}_\mathcal{C}$. By Theorem 1.6, there are formulas ϕ and ψ, and sets of ordinals S and T such that

$$\mathcal{C} \times \Sigma = A_{S,\phi}$$

and

$$(\omega^\omega)^{<\omega} \setminus (\mathcal{C} \times \Sigma) = A_{T,\psi}.$$

Let μ be a normal fine measure on $\mathcal{P}_{\aleph_1}(\mathbb{R})$. Let (M, E) be the ultraproduct $\prod_{\sigma \in \mathcal{P}_{\aleph_1}(\mathbb{R})} L(\sigma, S, T)/\mu$ constructed inside $L(\mathbb{R})[S, T, \mu]$. Then

- elements of M are represented by functions f in $L(\mathbb{R})[S, T, \mu]$, with domain $\mathcal{P}_{\aleph_1}(\mathbb{R})$, such that $f(\sigma)$ is in $L(\sigma, S, T)$, for each $\sigma \in \mathcal{P}_{\aleph_1}(\mathbb{R})$ (let $\mathcal{F}$ be the class of such functions),

- the members of M are the equivalence classes of functions in $\mathcal{F}$, under the relation of mod-μ equivalence, and
- given two such equivalence classes $[f]_\mu$ and $[g]_\mu$, $[f]_\mu E[g]_\mu$ if and only if

$$\{\sigma \in \mathcal{P}_{\aleph_1}(\mathbb{R}) : f(\sigma) \in g(\sigma)\} \in \mu.$$

For each set $x \in L(\mathbb{R})[S,T,\mu]$, let c_x be the constant function from $\mathcal{P}_{\aleph_1}(\mathbb{R})$ to $\{x\}$. Since μ normal (and thus countably complete), and $L(\mathbb{R})[S,T,\mu]$ satisfies DC (as $\mathrm{AD}_{\mathbb{R}}$ implies $\mathrm{DC}_{\mathbb{R}}$), (M,E) is wellfounded. A standard argument by induction on subformulas, using the normality of μ for the step corresponding to existential quantifiers, shows that for any finite set of functions $f_1,\ldots,f_n$ from $\mathcal{F}$, and any n-ary formula ϕ,

$$(M,E) \models \phi([f_1]_\mu,\ldots,[f_n]_\mu)$$

if and only if

$$L(\sigma,S,T) \models \phi(f_1(\sigma),\ldots,f_n(\sigma))$$

for μ-many σ. Let us call this fact the *elementarity* of the ultraproduct. One consequence of this fact (and the wellfoundedness of (M,E)) is that there is an isomorphism π from (M,E) to an inner model of the form $L(\mathbb{R},S^*,T^*)$, where $S^* = \pi([c_S]_\mu)$ and $T^* = \pi([c_T]_\mu)$. Then S^* and T^* are sets of ordinals.

By the elementarity of the ultraproduct, $A_{S,\phi} \subseteq A_{S^*,\phi}$ and $A_{T,\psi} \subseteq A_{T^*,\psi}$. Since

$$A_{S,\phi} = (\omega^\omega)^{<\omega} \setminus A_{T,\psi},$$

it follows again by elementarity that $A_{S,\phi} = A_{S^*,\phi}$ and $A_{T,\psi} = A_{T^*,\psi}$. By the normality of μ (and elementarity once again), it follows that for μ-many σ, $\sigma = \mathbb{R} \cap L(\sigma,S,T)$ and $A_{S,\phi}^{L(\sigma,S,T)}$ (which is $(\mathcal{C} \times \Sigma) \cap L(\sigma,S,T)$) is the product of a tall strategically selective coideal and a strategy witnessing that it is strategically selective, in $L(\sigma,S,T)$. Fixing one such σ, there is an $L(\sigma,S,T)$-generic filter H for $P_{\mathcal{C}}^{L(\sigma,S,T)}$ which is generated by a run of $\mathcal{G}_{\mathcal{C}}$ according to Σ. One can build such a run of $\mathcal{G}_{\mathcal{C}}$ by letting II play according to Σ and having I play to meet each dense set in $L(\sigma,S,T)$ from $P_{\mathcal{C}}^{L(\sigma,S,T)}$. Note that $L(\sigma,S,T)$ is closed under Σ. Furthermore, since σ is a countable, $L(\sigma,S,T)$ is contained in a model of Choice, which implies that $\mathcal{P}(\mathcal{P}(\mathbb{R})) \cap L(\sigma,S,T)$ is countable, so there exists (in V) an enumeration of the dense subsets of $P_{\mathcal{C}}^{L(\sigma,S,T)}$ in $L(\sigma,S,T)$ in ordertype ω.

Let U be the selective ultrafilter in $L(\sigma,S,T)[H]$ given by H. By Theorem 1.7, since Σ is a winning strategy for player II, there is an $x \in \mathcal{C}$ which is $L(\sigma, S, T)[H]$-generic for the Mathias forcing Q_U. Some condition (s, B) in the corresponding generic filter then forces that the generic real will be the left coordinate of a pair in $A_{S,\phi}$. However, this is a contradiction, as some infinite subset of x containing s is not in $\mathcal{C}$, and any such set is still generic below (s, B), by Theorem 1.7. $\square$

Remark 1.8: A *one-point diagonalization* of a sequence $\langle A_i : i < \omega \rangle \in \mathcal{P}(\omega)^\omega$ is a set $E = \{e_i : i \in \omega\}$ (listed in increasing order) such that $e_i \in A_i$ for all $i \in \omega$. In an earlier version of this paper, we used one-point diagonalizations instead of fast diagonalizations in the definitions of selective, semiselective and strategically selective. Example 1.9 below shows that these definitions are not equivalent. We note that [1] uses one-point diagonalizations in the definition of Ramsey (i.e., selective) ultrafilters; Example 1.9 shows that there is a gap in the proof of Theorem 4.5.2 there claiming to show that this definition is equivalent to the standard one. Similarly, Example 1.9 shows that the corresponding version of Theorem 1.3 using one-point diagonalizations in place of fast diagonalizations is false. The paper [7] cites (the earlier version of) this paper for proving this false version of Theorem 1.3. Modifying that paper to use the correct definitions requires making minor changes.

Example 1.9: Let $\bar{F} = \{F_n : n \in \omega\}$ be a partition of ω into finite sets, such that $\{|F_n| : n \in \omega\}$ is infinite and $F_n \cap n = \emptyset$ for all $n \in \omega$. Let $\mathcal{I}_{\bar{F}}$ be the ideal of sets $x \subseteq \omega$ for which there exists an $m \in \omega$ such that $|x \cap F_n| < m$ for all $n \in \omega$. Then $\mathcal{I}_{\bar{F}}$ is F_σ, so Borel. Let $\mathcal{C}_{\bar{F}}$ be the corresponding coideal. If, for each $n \in \omega$, $A_n = \omega \setminus \bigcup_{m<n} F_m$, then each A_n is in $\mathcal{C}_{\bar{F}}$ and each fast diagonalization of the sequence $\langle A_n : n \in \omega \rangle$ intersects each F_n at most 2 points, so is not in $\mathcal{C}_{\bar{F}}$. On the other hand, if $\bar{B} = \langle B_n : n \in \omega \rangle$ is any $\subseteq$-decreasing sequence of members of $\mathcal{C}_{\bar{F}}$, $\bar{B}$ has a one-point diagonalization in $\mathcal{C}_{\bar{F}}$. This shows that changing "fast" to "one-point" in the definition of selective coideal gives a weaker notion (and similarly for semiselective and strategically selective). Assuming that the Continuum Hypothesis holds, one can easily construct an ultrafilter contained in $\mathcal{C}_{\bar{F}}$ with the property that each $\subseteq$-descending ω-sequence from U has a one-point diagonalization in U.

2. Semiselective Coideals

In this section we discuss some issues related to the question of whether $\mathsf{AD}_\mathbb{R}$ implies the nonexistence of tall semiselective coideals. First we note two alternate characterizations of semiselectivity shown by Farah in [3]. Theorem 2.1 can be proved in ZF.

Theorem 2.1 (Farah [3]). *The following statements are equivalent, for a coideal $\mathcal{C}$ on ω.*

(1) $\mathcal{C}$ is semiselective.
(2) The generic filter added by forcing with $P_{\mathcal{C}}$ is a selective ultrafilter.
(3) Forcing with $P_{\mathcal{C}}$ does not add reals, and whenever $\{E\} \cup \{B_i : i \in \omega\} \subseteq \mathcal{C}$ and for all $i \in \omega$, $E \setminus B_i \notin \mathcal{C}$, there exists an $E' \subseteq E$ in $\mathcal{C}$ such that for each $i \in \omega$, $E' \setminus (i+1) \subseteq B_i$.

Given any partial order P, let $\mathcal{G}_{ds}(P)$ be the game where players I and II alternately choose the members of a descending sequence of conditions in P, with I winning if the sequence does not have a lower bound in P. For any coideal $\mathcal{C}$, a winning strategy for I in $\mathcal{G}_{ds}(P_{\mathcal{C}})$ is a winning strategy for I in $\mathcal{G}_{\mathcal{C}}$. By the second part of statement (3) of Theorem 2.1, if $\mathcal{C}$ is semiselective, then a winning strategy for II in $\mathcal{G}_{ds}(P_{\mathcal{C}})$ is a winning strategy for II in $\mathcal{G}_{\mathcal{C}}$. Theorem 1.3 shows that, assuming $\mathrm{AD}_{\mathbb{R}}$, player I has a winning strategy in $\mathcal{G}_{\mathcal{C}}$ for each tall coideal $\mathcal{C}$ on ω. It follows that if $\mathcal{C}$ is a tall semiselective coideal on ω, and $\mathrm{AD}_{\mathbb{R}}$ holds, then I has a winning strategy in $\mathcal{G}_{ds}(P_{\mathcal{C}})$. The first part of the following proposition then shows that forcing with $P_{\mathcal{C}}$ must make $\mathrm{DC}_{\mathbb{R}}$ fail.

Proposition 2.2 (ZF). *Suppose that $\mathcal{C}$ is a coideal on ω such that forcing with $P_{\mathcal{C}}$ does not add subsets of ω.*

(1) If I has a winning strategy in $\mathcal{G}_{ds}(P_{\mathcal{C}})$, then $\mathrm{DC}_{\mathbb{R}}$ fails after forcing with with $P_{\mathcal{C}}$.
(2) If II has a winning strategy in $\mathcal{G}_{ds}(P_{\mathcal{C}})$, and $\mathrm{DC}_{\mathbb{R}}$ holds, then it holds after forcing with $P_{\mathcal{C}}$.

Proof: For the first part of the proposition, let τ be a winning strategy for I in $\mathcal{G}_{ds}(P_{\mathcal{C}})$, and let $G \subseteq \mathcal{C}$ be generic for $P_{\mathcal{C}}$. In $V[G]$, consider the set of finite sequences from G of odd length which are partial plays of $\mathcal{G}_{ds}(P_{\mathcal{C}})$ according to τ. Every such sequence is extended by a longer one, but no infinite play of $\mathcal{G}_{ds}(P_{\mathcal{C}})$ according to τ can be forced by any condition in $P_{\mathcal{C}}$ to be a subset of the generic filter.

For the second part, fix $A \in \mathcal{C}$, a winning strategy τ for II in $\mathcal{G}_{ds}(P_{\mathcal{C}})$ and a $P_{\mathcal{C}}$-name σ for a set of finite sequences of reals with the property that every sequence in the set is extended by another sequence in the set.

Consider the set of finite sequences

$$A_0, x_0, A_1, A_2, x_2, A_3, \ldots, A_{2n}, x_{2n}, A_{2n+1}$$

for which $A_0 \subseteq A$, $\langle A_0, \ldots, A_{2n+1}\rangle$ is a partial play of $\mathcal{G}_{ds}(P_{\mathcal{C}})$ according to τ, and each A_{2i} forces that $\langle x_0, x_2, \ldots, x_{2i}\rangle$ is a member of the realization of σ. Then $\mathrm{DC}_{\mathbb{R}}$, plus the fact that τ is a winning strategy for II, gives a condition below A forcing some infinite sequence to be a path through the realization of σ. □

A positive answer to any part of the following question would show that no tall semiselective coideals exist, assuming $\mathrm{AD}_{\mathbb{R}}$.

Question 2.3: Suppose that $\mathrm{AD}_{\mathbb{R}}$ holds, and let $\mathcal{C}$ be a coideal on ω. Must $P_{\mathcal{C}}$ preserve $\mathrm{DC}_{\mathbb{R}}$? What if $\mathcal{C}$ is tall, or if $P_{\mathcal{C}}$ is assumed not to add subsets of ω, or to be semiselective, or tall and semiselective?

Proposition 2.4 shows that if $\mathcal{C}$ is a semiselective coideal, and there exist suitable choice functions, then $\mathcal{C}$ is strategically selective (from which it follows that under AC the two notions are equivalent). Given a function a as in the statement of Proposition 2.4, the existence of a function b follows from Uniformization.

Proposition 2.4 (ZF). *Let $\mathcal{C}$ be a coideal, and let Σ be a winning strategy for I in $\mathcal{G}_{\mathcal{C}}$. Let S be the set of finite partial runs τ in $\mathcal{G}_{\mathcal{C}}$ according to Σ for which it is II's turn to move. Suppose that there exist functions a on S and b on $S \times \mathcal{C}$ such that*

- *for each $\tau \in S$, $a(\tau)$ is a maximal antichain in $(\mathcal{C}, \subseteq)$ below the last member of τ, contained in the set of members of $\mathcal{C}$ which are responses by Σ to a move for II following τ;*
- *for each $\tau \in S$ and $B \in a(\tau)$, $b(\tau, B)$ is an element of $\mathcal{C}$ such that $\tau^\frown\langle b(\tau, B), B\rangle$ is a partial run of $\mathcal{G}_{\mathcal{C}}$ according to Σ.*

Then $\mathcal{C}$ is not semiselective.

Proof: Let A_0 be the first move made by Σ. Let H be a generic filter for $P_{\mathcal{C}}$, with $A_0 \in H$. In $V[H]$, consider the collection T consisting of those sequences of the form $\langle B_0, \ldots, B_{2n}\rangle$ contained in H, where for each even $i < 2n$, B_{i+2} is in $a(\langle B_0, \ldots, B_i\rangle)$ and $B_{i+1} = b(\langle B_0, \ldots, B_i\rangle, B_{i+2})$. By genericity, each sequence in T has an extension in T. Since each $a(\tau)$ ($\tau \in S$) is an antichain, the members of T extend one another, and there exists a unique sequence $\bar{B} = \langle B_i : i < \omega \rangle$ whose finite initial segments are all in T.

Suppose now toward a contradiction that $\mathcal{C}$ is semiselective. Then forcing with $P_{\mathcal{C}}$ does not add reals, so $\bar{B}$ is in V. Since Σ is a winning strategy for I in $\mathcal{G}_{\mathcal{C}}$, $\bar{B}$ does not have a fast diagonalization in $\mathcal{C}$. By part (3) of Theorem 2.1, then, $\bar{B}$ does not have a lower bound in $\mathcal{C}$, so no element of $\mathcal{C}$ could force all of the elements of $\bar{B}$ to be in H, giving a contradiction. □

Proposition 2.4 gives the following.

Theorem 2.5 (ZF). *If there exists a wellordering of* $\mathbb{R}$*, and* M *is a model of* ZF + $\mathrm{AD}_{\mathbb{R}}$ *containing* $\mathbb{R}$*, then there is no tall coideal in* M *which is semiselective in* V.

Finally, we note that the following theorem of Todorcevic (a version of which appears in [3]; the form given here is proved in [5]) allows one to argue from the point of view of a model of the Axiom of Choice that certain inner models do not contain tall coideals which are semiselective in V. In the presence of suitably large cardinals (for instance, a measurable cardinal above infinitely many Woodin cardinals), typical inner models of determinacy (such as $L(\mathbb{R})$) have the property that all of their sets of reals are at least $\mathfrak{c}$-universally Baire in V (see, for instance, Theorems 3.3.9 and 3.3.13 of [6]).

Theorem 2.6 (Todorcevic). *If* U *is a selective ultrafilter and* I *is a tall ideal on* ω *containing* Fin *which is* $\mathfrak{c}$*-universally Baire, then* $U \cap I \neq \emptyset$.

Corollary 2.7: *If* $\mathcal{C}$ *is a tall coideal in an inner model* M *containing the reals, and every set of reals in* M *is* $\mathfrak{c}$*-universally Baire, then* $\mathcal{C}$ *is not semiselective in* V

Proof: If $\mathcal{C}$ were semiselective in V, then forcing with $P_{\mathcal{C}}$ would produce a selective ultrafilter disjoint from $\mathcal{P}(\omega) \setminus \mathcal{C}$. □

As with Theorem 2.5, the corollary to Todorcevic's result leaves open the possibility that there is a tall coideal which is semiselective in the model in question, but no longer semiselective in any outer model of Choice.

Acknowledgments

The first author's research was supported in part by NSF Grant DMS-1201494. The second author was partially supported by National University of Singapore research grant number R-146-000-211-112.

References

1. Tomek Bartoszyński, Tomek, Haim Judah. **Set theory, on the structure of the real line**. A K Peters, Ltd., Wellesley, MA, 1995
2. Carlos Di Prisco, José G. Mijares and Carlos Uzcátegui. Ideal games and Ramsey sets, Proc. Amer. Math. Soc. 140 (2012) 7, 2255–2265
3. Ilijas Farah. Semiselective coideals. Mathematika 45 (1998) 1, 79-103
4. Richard Ketchersid. More structural consequences of AD. **Set theory and its applications**, 71-105. Contemp. Math., 533, Amer. Math. Soc., Providence, RI, 2011
5. Richard Ketchersid, Paul Larson and Jindrich Zapletal. Ramsey ultrafilters and countable-to-one uniformization. Topology and Its Applications 213 (2016), 190-198
6. Paul B. Larson. **The Stationary Tower**. Notes on a course by W. Hugh Woodin. American Mathematical Society University Lecture Series vol. 32, 2004
7. Paul B. Larson. A Choice function on countable sets, from determinacy. Proc. Amer. Math. Soc. 43 (2015) 4, 1763-1770
8. Adrian Mathias. Happy families. *Annals of Mathematical Logic* 12 (1977), 59-111
9. Robert M. Solovay. The independence of DC from AD. *Cabal Seminar 76-77*. Lecture Notes in Mathematics Volume 689, 1978, 171-183
10. J. R. Steel. The Derived Model Theorem. **Logic Colloquium 2006**, 280-327. Lect. Notes Log., Assoc. Symbol. Logic, Chicago, IL, 2009
11. W. Hugh Woodin. **The Axiom of Determinacy, forcing axioms and the nonstationary ideal**. Walter de Gruyter, Berlin, 2010

COBHAM RECURSIVE SET FUNCTIONS AND WEAK SET THEORIES

Arnold Beckmann
Department of Computer Science
Swansea University
Swansea SA2 8PP, UK
a.beckmann@swansea.ac.uk

Sam Buss[a]
Department of Mathematics
University of California, San Diego
La Jolla, California 92130-0112, USA
sbuss@ucsd.edu

Sy-David Friedman[b]
Kurt Gödel Research Center
Universität Wien
A-1090 Vienna, Austria
sdf@logic.univie.ac.at

Moritz Müller
Kurt Gödel Research Center
Universität Wien
A-1090 Vienna, Austria
moritz.mueller@univie.ac.at

Neil Thapen[c]
Institute of Mathematics
Czech Academy of Sciences
115 67 Prague, Czech Republic
thapen@math.cas.cz

---

[a]Supported in part by NSF grants DMS-1101228 and CCR-1213151, and by the Simons Foundation, award 306202.
[b]Supported in part by FWF (Austrian Science Fund) through Project P24654.
[c]Research leading to these results has received funding from the European Research Council under the European Union's Seventh Framework Programme (FP7/2007-2013), ERC grant agreement 339691. The Institute of Mathematics of the Czech Academy of Sciences is supported by RVO:67985840.

The Cobham recursive set functions (CRSF) provide a notion of polynomial time computation over general sets. In this paper, we determine a subtheory $\mathrm{KP}_1^{\mathrm{u}}$ of Kripke-Platek set theory whose Σ_1-definable functions are precisely CRSF. The theory $\mathrm{KP}_1^{\mathrm{u}}$ is based on the $\in$-induction scheme for Σ_1-formulas whose leading existential quantifier satisfies certain boundedness and uniqueness conditions. Dropping the uniqueness condition and adding the axiom of global choice results in a theory $\mathrm{KPC}_1^{\preccurlyeq}$ whose Σ_1-definable functions are CRSF^C, that is, CRSF relative to a global choice function C. We further show that the addition of global choice is conservative over certain local choice principles.

1. Introduction

Barwise begins his chapter on admissible set recursion theory with: "There are many equivalent definitions of the class of recursive functions on the natural numbers. [...] As the various definitions are lifted to domains other than the integers (e.g., admissible sets) some of the equivalences break down. This break-down provides us with a laboratory for the study of recursion theory." ([5, p. 153])

Let us informally distinguish two types of characterization of the computable functions or subsets thereof, namely, *recursion theoretic* and *definability theoretic* ones. Recursion theoretically, the computable functions on ω are those obtainable from certain simple initial functions by means of composition, primitive recursion and the μ-operator. As a second example, the primitive recursive functions are similarly defined but without the μ-operator. A third example is the recursion theoretic definition of the polynomial time functions by Cobham recursion [13] or by Bellantoni-Cook safe-normal recursion [9]. Definability theoretically, the computable functions are those that are Σ_1-definable in the true theory of arithmetic. A more relevant example of a definability theoretic definition is the classic theorem of Parsons and Takeuti (see [12]) that the primitive recursive functions are those that are Σ_1-definable in the theory $\mathrm{I}\Sigma_1$; namely, one additionally requires that this theory proves the totality and functionality of the defining Σ_1-formula. Analogously, the polynomial time functions have a definability theoretic definition as the Σ_1^b-definable functions of S_2^1 [11]. For more definability theoretic definitions of weak subrecursive classes, see Cook-Nguyen [15].

Admissible set recursion theory provides a definability theoretic generalization of computability: one considers functions which are Σ_1-definable (in the language of set theory) in an admissible set, that is, a transitive standard model of Kripke-Platek set theory KP. Recall that KP consists of the axioms for Extensionality, Union, Pair, Δ_0-Separation, Δ_0-Collection and $\in$-Induction for all formulas $\varphi(x, \vec{w})$:

$$\forall y\,(\forall u{\in}y\,\varphi(u, \vec{w}) \to \varphi(y, \vec{w})) \to \varphi(x, \vec{w}).$$

To some extent this generalization of computability extends to the recursion theoretic view. By the Σ-Recursion Theorem ([5, Chapter I, Theorem 6.4]) the Σ_1-definable functions of KP are closed under $\in$-recursion. This implies that the *primitive recursive set functions* (PRSF) of [20] are all Σ_1-definable in KP. By definition, a function on the universe of sets is in PRSF if it is obtained from certain simple initial functions by means of composition and $\in$-recursion. Hence, PRSF is a recursion theoretic generalization of the primitive recursive functions. Paralleling Parson's theorem, Rathjen [22] showed that this generalization extends to the definability theoretic view: PRSF contains precisely those functions that are Σ_1-definable in KP_1, the fragment of KP where $\in$-Induction is adopted for Σ_1-formulas only. One can thus view PRSF as a reasonable generalization of primitive recursive computation to the universe of sets.

It is natural to wonder whether one can find a similarly good analogue of polynomial time computation on the universe of sets. In [7] we proposed such an analogue, following Cobham's [13] characterization of the polynomial time computable functions on ω as those obtained from certain simple initial functions, including the smash function #, by means of composition and *limited recursion on notation.* Limited recursion on notation restricts both the depth of the recursion and the size of values. Namely, a recursion on notation on x has depth roughly $\log x$; being limited means that all values are required to be bounded by some smash term $x\#\cdots\#x$.

In [7] a smash function for sets is introduced. The role of recursion on notation is taken by $\in$-recursion, and being limited is taken to mean being in a certain sense embeddable into some #-term. In this way, [7] defines the class of *Cobham recursive set functions* (CRSF), a recursion theoretic generalization of polynomial time computation from ω to the universe of sets. This paper extends the analogy to the definability theoretic view.

A definability theoretic characterization of polynomial time on ω has been given by Buss (cf. [12]). It is analogous to Parsons' theorem, with $\mathrm{I}\Sigma_1$ replaced by S^1_2 and Σ_1 replaced by a class Σ^b_1 of "bounded" Σ_1-formulas. The theory S^1_2 has a language including the smash function # and is based on a restricted form of induction scheme for Σ^b_1-formulas, in which the depth of an induction is similar to the depth of a recursion on notation.

Both directions in Buss' characterization hold in a strong form. First, S^1_2

defines polynomial time functions in the sense that one can conservatively add Σ_1^b-defined symbols and prove Cook's PV_1 [14], a theory based on the equations that can arise from derivations of functions in Cobham's calculus. Second, polynomial time functions *witness* simple theorems of S_2^1. More precisely, if S_2^1 proves $\exists y\, \varphi(y, \vec{x})$ with φ in Δ_0^b, that is, a "bounded" Δ_0-formula, then $\forall \vec{x}\, \varphi(f(\vec{x}), \vec{x})$ is true for some polynomial time computable f, even provably in PV_1 and S_2^1.

In the present paper, we analogously replace Rathjen's theory KP_1 and Σ_1-definability with a theory $\mathrm{KP}_1^{\preccurlyeq}$ and $\Sigma_1^{\preccurlyeq}$-definability; here $\Sigma_1^{\preccurlyeq}$-formulas are "bounded" Σ_1-formulas, defined using set smash and the embeddability notion $\preccurlyeq$ of [7]. The theory $\mathrm{KP}_1^{\preccurlyeq}$ has a finite language containing, along with $\in$, some basic CRSF functions including the set smash and has $\in$-Induction restricted to $\Sigma_1^{\preccurlyeq}$-formulas.

As we shall see, $\mathrm{KP}_1^{\preccurlyeq}$ defines CRSF analogously to the first part of Buss' characterization. An analogy of the second part, the witnessing theorem, would state that whenever $\mathrm{KP}_1^{\preccurlyeq}$ proves $\exists y\, \varphi(y, \vec{x})$ for a Δ_0-formula φ, then $\varphi(f(\vec{x}), \vec{x})$ is provable in ZFC (or ideally in a much weaker theory) for some f in CRSF. But this fails: a function witnessing $\exists y\, (x \neq 0 \to y \in x)$ would be a global choice function, and this is not available in $\mathrm{KP}_1^{\preccurlyeq}$.

We discuss two ways around this obstacle. If we add the axiom of global choice we get a theory $\mathrm{KPC}_1^{\preccurlyeq}$ and indeed can prove a witnessing theorem as desired (Theorem 6.13). The functions Σ_1-definable in $\mathrm{KPC}_1^{\preccurlyeq}$ are precisely those that are CRSF with a global choice function as an additional initial function (Corollary 6.2). Thus, Buss' theorems for S_2^1 and polynomial time on ω have full analogues on universes of sets equipped with global choice, if we consider the global choice function as a feasible function in such a universe.

Our second way around the obstacle is to further weaken the induction scheme, the crucial restriction being that the witness to the existential quantifier in a $\Sigma_1^{\preccurlyeq}$-formula is required to be unique. The resulting theory $\mathrm{KP}_1^{\mathrm{u}}$ still defines CRSF in the strong sense that one can conservatively add $\Sigma_1^{\preccurlyeq}$-defined function symbols and prove $\mathrm{T}_{\mathrm{crsf}}$, an analogue of PV_1 containing the equations coming from derivations in the CRSF calculus (Theorem 5.2). We prove a weak form of witnessing (Theorem 6.10): if $\mathrm{KP}_1^{\mathrm{u}}$ proves $\exists y\, \varphi(y, \vec{x})$ for φ a Δ_0-formula, then $\mathrm{T}_{\mathrm{crsf}}$ proves $\exists y \in f(\vec{x})\, \varphi(y, \vec{x})$ for some f in CRSF. This suffices to infer a definability theoretic characterization of CRSF on an arbitrary universe of sets: the Σ_1-definable functions of $\mathrm{KP}_1^{\mathrm{u}}$ are precisely those in CRSF (Corollary 6.1). We do not know whether this holds for $\mathrm{KP}_1^{\preccurlyeq}$.

To conclude the paper, we address the question on how much stronger $\mathrm{KPC}_1^{\preccurlyeq}$ is compared to $\mathrm{KP}_1^{\preccurlyeq}$. We show that the difference can be encapsulated in certain local choice principles.

The outline of the paper is as follows. Section 2 recalls the development of CRSF from [7]. For the formalizations in this paper we will use a slightly different, but equivalent, definition of CRSF which we describe in Proposition 2.9. Section 3 defines the three theories $\mathrm{KP}_1^{\preccurlyeq}$, $\mathrm{KP}_1^{\mathrm{u}}$ and $\mathrm{T}_{\mathrm{crsf}}$ mentioned above. They extend a base theory T_0 that we "bootstrap" in Section 4, in particular deriving various lemmas which allow us to manipulate embedding bounds. Section 5 proves the Definability Theorem 5.2 for $\mathrm{KP}_1^{\mathrm{u}}$. A technical difficulty is that $\mathrm{KP}_1^{\mathrm{u}}$ is too weak to eliminate $\Sigma_1^{\preccurlyeq}$-defined function symbols in the way this is usually done in developments of KP or S_2^1 (see Section 5.2). Section 6 proves the Witnessing Theorems 6.10 and 6.13 for $\mathrm{KP}_1^{\mathrm{u}}$ and $\mathrm{KPC}_1^{\preccurlyeq}$, and the Corollaries 6.1 and 6.2 on definability theoretic characterizations of CRSF. This is done via a modified version of Avigad's model-theoretic approach to witnessing [4] (see Section 6.1). Our proof gives some insight about CRSF: roughly, its definition can be given using only a certain simple form of embedding (see Section 6.4). Section 7 proves the conservativity of global choice over certain local choice principles (Theorem 7.2). Here, we use a class forcing as in [17] to construct a generic expansion of any (possibly non-standard) model of our set theory. Some extra care is needed since our set theory is rather weak.

Several related recursion theoretic notions of polynomial time set functions have been described earlier by other authors. The characterization of polynomial time by Turing machines has been generalized in Hamkins and Lewis [18] to allow binary input strings of length ω. We refer to [8] for some comparison with CRSF. Yet another characterization of polynomial time comes from the Immerman-Vardi Theorem from descriptive complexity theory (cf. [16]). Following this, Sazonov [23] gives a theory operating with terms allowing for least fixed-point constructs to capture polynomial time computations on (binary encodings of) Mostowski graphs of hereditarily finite sets. Not all of Sazonov's set functions are CRSF [7]. But under a suitable encoding of binary strings by hereditarily finite sets, CRSF does capture polynomial time [7, Theorems 30, 31].

Arai [2] gives a different such class of functions. His *Predicatively Computable Set Functions* (PCSF) form a subclass of the *Safe Recursive Set Functions* (SRSF) from [6]. SRSF is defined in analogy to Bellantoni and Cook's recursion theoretic characterization of polynomial time [9], different from Cobham's. Bellantoni and Cook's functions have two sorts of arguments, called "normal" and "safe", and recursion on notation is allowed to recurse only on normal arguments, while values obtained by such recursions are safe. Similarly, SRSF and PCSF contain two-sorted functions. It is shown in [7] that CRSF coincides with the functions having only normal arguments in PCSF$^+$ (from [2]), a slight strengthening of PCSF. In a recent manuscript [1], Arai gives a definability theoretic characterization of PCSF$^\iota$, a class of set functions intermediate between PCSF and PCSF$^+$. He proves a weak form of witnessing akin to ours. He uses two-sorted set-theoretic proof systems whose normal sort ranges over a transitive substructure of the universe, and which contains an inference rule ensuring closure of this substructure under certain definable functions. Like $\mathrm{KP}^{\mathrm{u}}_1$, these systems contain a form of "unique" Σ_1-Induction. As in our setting, eliminating defined function symbols is problematic; the final system in [1] is a union of a hierarchy of systems, each level introducing infinitely many function symbols. Thus, dealing with similar problems, Arai's solution is quite different from the one presented here; as is his proof, which is based on cut-elimination.

2. Cobham Recursive Set Functions

In this section we review some definitions and results from [7]. In later sections, many of these results will be formalized in suitable fragments of KP.

As mentioned in the introduction, [7] generalizes Cobham's recursion theoretic characterization of polynomial time to arbitrary sets. We recall Cobham's characterization. On ω the smash $x\#y$ is defined as $2^{|x|\cdot|y|}$ where $|x| := \lceil \log(x+1) \rceil$ is the *length* (of the binary representation) of x. We have successor functions $s_0(x) := 2x$ and $s_1(x) = 2x+1$ which add respectively 0 and 1 to the end of the binary representation of x.

Theorem 2.1: (Cobham 1965) *The* polynomial time *functions on ω are obtained from* initial functions, *namely, projections $\pi^r_j(x_1,\dots,x_r) := x_j$, constant 0, successors s_0, s_1 and the smash $\#$, by composition and* limited recursion on notation*: if $h(\vec{x})$, $g_0(y,z,\vec{x})$, $g_1(y,z,\vec{x})$ and $t(y,\vec{x})$ are polynomial time, then so is the function $f(y,\vec{x})$ given by*

$$f(0,\vec{x}) = h(\vec{x}),$$
$$f(s_b(y),\vec{x}) = g_b(y, f(y,\vec{x}), \vec{x}) \quad \text{for } b \in \{0,1\} \text{ and } s_b(y) \neq 0,$$

provided that $f(y,\vec{x}) \leqslant t(y,\vec{x})$ holds for all $y, \vec{x}$.

One can equivalently ask t to be built by composition from only projections, 1 and $\#$; or just demand $|f(y, x_1, \ldots, x_k)| \leqslant p(|y|, |x_1|, \ldots, |x_k|)$ for some polynomial p.

We move to some fixed universe of sets, that is, a model of ZFC. The analogue of smash defined in [7] is best understood in terms of Mostowski graphs. The *Mostowski graph* of a set x has as vertices the elements of the transitive closure $\mathrm{tc}^+(x) := \mathrm{tc}(\{x\})$ and has a directed edge from u to v if $u \in v$. Every such graph has a unique source and a unique sink.

The *set smash* $x\#y$ replaces each vertex of x by (a copy of the graph of) y with incoming edges now going to the source of y and outgoing edges now leaving the sink of y. It can be defined using *set composition* $x \odot y$, which places a copy of x above y and identifies the source of x with the sink of y. Writing 0 for $\emptyset$,

$$x \odot y := \begin{cases} y & \text{if } x = 0, \\ \{u \odot y : u \in x\} & \text{otherwise} \end{cases}$$

$$x\#y := y \odot \{u\#y : u \in x\}.$$

The Mostowski graph of $x\#y$ is isomorphic to the graph with vertices $\mathrm{tc}^+(x) \times \mathrm{tc}^+(y)$ and directed edges from $\langle u', v' \rangle$ to $\langle u, v \rangle$ if either $u' = u \wedge v' \in v$ or $u' \in u \wedge v' = y \wedge v = 0$ (see [7, Section 2]). An isomorphism is given by $\sigma_{x,y}(u, v) := v \odot \{u'\#y : u' \in u\}$.

A *$\#$-term* is built by composition from projections, $\#, \odot$ and the constant $1 = \{0\}$. Such terms serve as analogues of polynomial length bounds, with the bounding relation $\preccurlyeq$ defined as follows: $x \preccurlyeq y$ means that there is a *(multi-valued) embedding* that maps vertices $u \in \mathrm{tc}(x)$ to pairwise disjoint non-empty sets $V_u \subseteq \mathrm{tc}(y)$ such that whenever $u' \in u$ and $v \in V_u$, then there exists $v' \in V_{u'} \cap \mathrm{tc}(v)$. The notation $\tau(\cdot, \vec{w}) : x \preccurlyeq y$ means that $u \mapsto \tau(u, \vec{w})$ is such an embedding. Then [7] generalizes Cobham's definition as follows.

Definition 2.2: The *Cobham recursive set functions* (CRSF) are obtained from *initial functions*, namely projections, constant $0 := \emptyset$, pair $\{x, y\}$, union $\bigcup x$, set smash $x\#y$, and the conditional

$$cond_{\in}(x, y, u, v) := \begin{cases} x & \text{if } u \in v \\ y & \text{otherwise,} \end{cases}$$

by composition and *Cobham recursion*: if $g(x, z, \vec{w})$, $\tau(u, x, \vec{w})$ and $t(x, \vec{w})$ are CRSF, then so is the function $f(x, \vec{w})$ given by

$$f(x, \vec{w}) = g(x, \{f(y, \vec{w}) : y \in x\}, \vec{w})$$

provided that $\tau(\cdot, x, \vec{w}) : f(x, \vec{w}) \preccurlyeq t(x, \vec{w})$ holds for all $x, \vec{w}$.

Here the *embedding proviso* $\tau(\cdot, x, \vec{w}) : f(x, \vec{w}) \preccurlyeq t(x, \vec{w})$ ensures, intuitively, that a definition by recursion is allowed only provided that we can already bound the "structural complexity" of the defined function f. A relation is CRSF if its characteristic function is. Direct arguments show (see [7, Theorem 13]):

Proposition 2.3:

(a) (Separation) *If* $g(u, \vec{w})$ *is in* CRSF, *then so is* $f(x, \vec{w}) := \{u \in x : g(u, \vec{w}) \neq 0\}$.

(b) The CRSF *relations contain* $x \in y$ *and* $x = y$, *are closed under Boolean combinations and* $\in$*-bounded quantifications* $\exists u \in x$ *and* $\forall u \in x$.

It is then not hard to show that transitive closure $\mathrm{tc}(x)$, set composition $x \odot y$, the isomorphism $\sigma_{x,y}(u, v)$ and its inverses $\pi_{1,x,y}(z), \pi_{2,x,y}(z)$ are CRSF [7, Theorem 13]. In particular, #-terms are CRSF. Further, one can derive the following central lemma [7, Lemma 20]. It says that $\preccurlyeq$ is a pre-order and that #-terms enjoy some monotonicity properties one would expect from a reasonable analogue of "polynomial length bounds".

Lemma 2.4: *Below if* τ_0 *and* τ_1 *are in* CRSF *then* σ *can also be chosen in* CRSF.

(a) (Transitivity) *If* $\tau_0(\cdot, x, y, \vec{w}) : x \preccurlyeq y$ *and* $\tau_1(\cdot, y, z, \vec{w}) : y \preccurlyeq z$, *then there exists* $\sigma(u, x, y, z, \vec{w})$ *such that* $\sigma(\cdot, x, y, z, \vec{w}) : x \preccurlyeq z$.

(b) (Monotonocity) *Let* $t(x, \vec{w})$ *be a #-term. If* $\tau_0(\cdot, x, z, \vec{w}) : z \preccurlyeq t(x, \vec{w})$ *and* $\tau_1(\cdot, x, y, \vec{w}) : x \preccurlyeq y$, *then there exists* $\sigma(u, x, y, z, \vec{w})$ *such that* $\sigma(\cdot, x, y, z, \vec{w}) : z \preccurlyeq t(y, \vec{w})$.

Based on this lemma, a straightforward induction on the length of a derivation of a CRSF function shows [7, Theorem 17]:

Theorem 2.5: (Bounding) *For every* $f(\vec{x})$ *in* CRSF *there are a #-term* $t(\vec{x})$ *and a* CRSF *function* $\tau(u, \vec{x})$ *such that* $\tau(\cdot, \vec{x}) : f(\vec{x}) \preccurlyeq t(\vec{x})$.

In fact, in the definition of Cobham recursion one can equivalently require the function t in the embedding proviso to be a #-term [7, Theorem 21]. Using the Bounding Theorem and the Monotonicity Lemma one can obtain, similarly to Theorems 23, 29 and 30 of [7]:

Theorem 2.6:

(a) (Replacement) *If* $f(y, \vec{w})$ *is* CRSF, *then so is*

$$f"(x, \vec{w}) = \{f(y, \vec{w}) : y \in x\}.$$

(b) (Course of values recursion) *If* $g(x, z, \vec{w})$, $\tau(u, x, \vec{w})$ *and* $t(x, \vec{w})$ *are* CRSF, *then so is*

$$f(x, \vec{w}) := g(x, \{\langle u, f(u, \vec{w})\rangle : u \in \mathrm{tc}(x)\}, \vec{w})$$

provided $\tau(\cdot, x, \vec{w}) : f(x, \vec{w}) \preccurlyeq t(x, \vec{w})$ *holds for all* $x, \vec{w}$.

(c) (Impredicative Cobham recursion) *If* $g(x, z, \vec{w})$, $\tau(u, y, x, \vec{w})$ *and* $t(x, \vec{w})$ *are* CRSF, *then so is*

$$f(x, \vec{w}) = g(x, f\text{"}(x, \vec{w}), \vec{w})$$

provided $\tau(\cdot, f(x, \vec{w}), x, \vec{w}) : f(x, \vec{w}) \preccurlyeq t(x, \vec{w})$ *holds for all* $x, \vec{w}$.

Closure under replacement (a) implies that $x \times y$ is CRSF [7, Theorem 14]. Impredicative Cobham recursion (c) is, intuitively, somewhat circular in that the embedding τ may use as a parameter the set $f(x, \vec{w})$ whose existence it is supposed to justify.

We introduce a variant definition of CRSF that uses *syntactic Cobham recursion.* The name "syntactic" indicates that it does not have an embedding proviso, but rather constructs a new function from any CRSF functions g, τ and #-term t. We also allow the bound to be impredicative in the sense of (c) above.

Definition 2.7: Let $g(x, z, \vec{w})$ and $\tau(u, y, x, \vec{w})$ be functions and $t(x, \vec{w})$ a #-term. Then *syntactic Cobham recursion* gives the function $f(x, \vec{w})$ defined by

$$f(x, \vec{w}) = \begin{cases} g(x, f\text{"}(x, \vec{w}), \vec{w}) & \text{if } \tau \text{ is an embedding into } t \text{ at } x, \vec{w} \\ 0 & \text{otherwise} \end{cases} \tag{2.1}$$

where the condition "τ is an embedding into t at $x, \vec{w}$" stands for

$$\tau(\cdot, g(x, f\text{"}(x, \vec{w}), \vec{w}), x, \vec{w}) : g(x, f\text{"}(x, \vec{w}), \vec{w}) \preccurlyeq t(x, \vec{w}). \tag{2.2}$$

Note that Proposition 2.3 implies that the condition (2.2) is a CRSF relation, cf. (3.2) in Section 3.2.

Proposition 2.8: *The* CRSF *functions are precisely those obtained from the initial functions by composition and syntactic Cobham recursion.*

Proof: Since the condition (2.2) is a CRSF relation, (2.1) can be written $f(x, \vec{w}) = g'(x, f\text{"}(x, \vec{w}), \vec{w})$ for some g' in CRSF. The embedding proviso $\tau(\cdot, f(x, \vec{w}), x, \vec{w}) : f(x, \vec{w}) \preccurlyeq t(x, \vec{w})$ holds for all $x, \vec{w}$, since either (2.2) holds, in which case $f(x, \vec{w}) = g(x, f\text{"}(x, \vec{w}), \vec{w})$ so (2.2) gives us the embedding, or $f(x, \vec{w}) = 0$, in which case any function is an embedding of

$f(x, \vec{w})$ into $t(x, \vec{w})$. We thus get that f is CRSF by impredicative Cobham recursion.

Conversely, assume $f(x, \vec{w})$ is obtained from g, τ, t by Cobham recursion, and in particular that the embedding proviso is satisfied for all $x, \vec{w}$. Then f satisfies (2.1), so f can be obtained by syntactic Cobham recursion. By [7, Theorem 21], we can assume that t is a #-term. □

The next proposition describes the definition of CRSF that we will formalize with the theory $\mathrm{T_{crsf}}$ in Section 3.4. Closure under replacement and the extra initial functions are included to help with the formalization.

Proposition 2.9: *The* CRSF *functions are precisely those obtained from projection, zero, pair, union, conditional, transitive closure, cartesian product, set composition and set smash functions by composition, replacement and syntactic Cobham recursion.*

Remark 2.10: For an arbitrary function $g(\vec{x})$, let CRSF^g be defined as CRSF but with $g(\vec{x})$ as additional initial function. This class might be interpreted as a set-theoretic analogue of polynomial time computations with an oracle function $g(\vec{x})$. If there is $\tau(u, \vec{x})$ in CRSF^g such that $\tau(\cdot, \vec{x}) : g(\vec{x}) \preccurlyeq t(\vec{x})$ for some #-term $t(\vec{x})$, then all results mentioned in this section "relativize", that is, hold true with CRSF replaced by CRSF^g.

3. Theories for CRSF

3.1. *The language* $\mathrm{L_0}$ *and theory* $\mathrm{T_0}$

The language $\mathrm{L_0}$ contains the relation symbol $\in$ and symbols for the following CRSF functions:

$$0,\ 1,\ \bigcup x,\ \{x, y\},\ x \times y,\ \mathrm{tc}(x),\ x \odot y,\ x \# y.$$

The meaning of these symbols is given by their defining axioms:

symbol	*defining axiom*
0	$u \notin 0$
1	$u \in 1 \leftrightarrow u = 0$
$\bigcup x$	$u \in \bigcup x \leftrightarrow \exists y \in x\, (u \in y)$
$\{x, y\}$	$u \in \{x, y\} \leftrightarrow u = x \vee u = y$
$x \times y$	$u \in x \times y \leftrightarrow \exists x' \in x\, \exists y' \in y\, (u = \langle x', y' \rangle)$
$\mathrm{tc}(x)$	$u \in \mathrm{tc}(x) \leftrightarrow u \in x \vee \exists y \in x\, (u \in \mathrm{tc}(y))$
$x \odot y$	$0 \odot y = y \wedge (x \neq 0 \rightarrow x \odot y = \{z \odot y : z \in x\})$
$x \# y$	$\exists w \in \mathrm{tc}^+(x \# y)\ (w = \{z \# y : z \in x\} \wedge (x \# y = y \odot w))$

The table above uses some special notations: as usual, $\{x\}$ stands for the term $\{x, x\}$, $\langle x, y\rangle$ for the term $\{\{x\}, \{x, y\}\}$, $x \cup y$ for the term $\bigcup\{x, y\}$, and $x \subseteq y$ for the formula $\forall u \in x\, (u \in y)$. We write $\mathrm{tc}^+(x)$ for the term $\mathrm{tc}(\{x\})$. The final two lines of the table use "replacement terms". More generally, we use three types of comprehension terms:

Definition 3.1: The following notations are used for comprehension terms:

- *Proper comprehension terms*: for a formula $\varphi(u, \vec{x})$, we write

  $$z = \{u \in x : \varphi(u, \vec{x})\}$$

  for

  $$\forall u \in z\, (u \in x \wedge \varphi(u, \vec{x})) \wedge \forall u \in x\, (\varphi(u, \vec{x}) \to u \in z).$$

- *Collection terms*: for a formula $\varphi(u, v, \vec{x})$, we write

  $$z = \{v : \exists u \in x\, \varphi(u, v, \vec{x})\}$$

  for

  $$\forall v \in z\, \exists u \in x\, \varphi(u, v, \vec{x}) \wedge \forall v\, ((\exists u \in x\, \varphi(u, v, \vec{x})) \to v \in z). \tag{3.1}$$

- *Replacement terms*: for a term $t(u, \vec{x})$, we write

  $$z = \{t(u, \vec{x}) : u \in x\}$$

  for

  $$z = \{v : \exists u \in x\, (v = t(u, \vec{x}))\}.$$

Such terms may not be used as arguments to functions.

We use collection terms only in contexts where we have

$$\forall u \in x\, \exists! v\, \varphi(u, v, \vec{x}),$$

so that (3.1) is equivalent to

$$\forall v \in z\, \exists u \in x\, \varphi(u, v, \vec{x}) \wedge \forall u \in x\, \exists v \in z\, \varphi(u, v, \vec{x}).$$

In particular, this is the case for replacement terms. These restrictions ensure that our formulas are $\Delta_0(\mathrm{L}_0)$ whenever φ is, provided that the comprehension terms have the uniqueness property. As usual, we write $\exists! y\, \varphi(y, \vec{x})$ for $\exists^{\leqslant 1} y\, \varphi \wedge \exists y\, \varphi$, where $\exists^{\leqslant 1} y\, \varphi$ stands for

$$\forall y, y'\, (\varphi(y, \vec{x}) \wedge \varphi(y', \vec{x}) \to y = y').$$

Instead of Δ_0 etc. we use more precise notation, making the language explicit:

Definition 3.2: Let L be a language containing L_0. Then $\Delta_0(L)$ is the set of L-formulas φ in which all quantifiers are $\in$-bounded, that is, of the form $\exists x \in t$ or $\forall x \in t$ where t is an L-term not involving x. We refer to t as an *$\in$-bounding term in φ.*

The classes $\Sigma_1(L)$ and $\Pi_1(L)$ contain the formulas obtained from $\Delta_0(L)$-formulas by respectively existential and universal quantification, and $\Pi_2(L)$ contains those obtained from $\Sigma_1(L)$-formulas by universal quantification.

We define our basic theory, which the other theories we consider will extend.

Definition 3.3: The theory T_0 consists of

- the defining axioms for the symbols in L_0
- the *Extensionality* axiom: $x \neq y \to \exists u \in x\,(u \notin y) \vee \exists u \in y\,(u \notin x)$
- the *Set Foundation* axiom: $x \neq 0 \to \exists y \in x\, \forall u \in y\,(u \notin x)$
- the tc-*Transitivity* axiom: $y \in \mathrm{tc}(x) \to y \subseteq \mathrm{tc}(x)$
- the $\Delta_0(\mathrm{L}_0)$-*Separation* scheme: $\exists z\,(z = \{u \in x : \varphi(u, \vec{w})\})$ for $\varphi \in \Delta_0(\mathrm{L}_0)$.

Lemma 3.4: *The theory* T_0 *proves the* $\Delta_0(\mathrm{L}_0)$*-Induction scheme*

$$\forall y\,(\forall u \in y\, \varphi(u, \vec{w}) \to \varphi(y, \vec{w})) \to \varphi(x, \vec{w}) \quad \textit{for } \varphi \in \Delta_0(\mathrm{L}_0).$$

Proof: $\Delta_0(\mathrm{L}_0)$-Induction is logically equivalent to $\Delta_0(\mathrm{L}_0)$-*Foundation*

$$\varphi(x, \vec{w}) \to \exists y\,(\varphi(y, \vec{w}) \wedge \forall u \in y\, \neg\varphi(u, \vec{w}))$$

for φ in $\Delta_0(\mathrm{L}_0)$ which is derived in T_0 as follows. Assume $\varphi(x, \vec{w})$ and use $\Delta_0(\mathrm{L}_0)$-Separation to get the set $z = \{y \in \mathrm{tc}^+(x) : \varphi(y, \vec{w})\}$. Then $x \in \mathrm{tc}^+(x)$ by the defining axiom for tc, so $x \in z \neq 0$. Choose y as the $\in$-minimal element in z according to Set Foundation. Then $\varphi(y, \vec{w})$ and, if $u \in y$, then $u \notin z$, and thus $\neg\varphi(u, \vec{w})$ because $u \in y \subseteq \mathrm{tc}^+(x)$ by tc-Transitivity. □

Remark 3.5: It is for the sake of the previous lemma that the tc-Transitivity axiom is included in T_0. In fact, this axiom is equivalent to $\Delta_0(\mathrm{L}_0)$-Induction with respect to the remaining axioms of T_0.

3.2. *Embeddings*

An *embedding* of a set x into a set y is an injective multifunction τ from $\mathrm{tc}(x)$ to $\mathrm{tc}(y)$ which respects the $\in$-ordering on $\mathrm{tc}(x)$ in a certain sense. There are several variants of embeddings, depending on how τ is defined.

Definition 3.6: A function symbol $\tau(u, \vec{w})$ is a *strongly uniform embedding* (with parameters $\vec{w}$) of x in y if the following $\Delta_0(\mathrm{L}_0 \cup \{\tau\})$-formula holds (where for the sake of readability we suppress the parameters $\vec{w}$):

$$\begin{aligned}
&\forall u{\in}\mathrm{tc}(x)\,(\tau(u) \subseteq \mathrm{tc}(y))\\
&\quad\wedge\ \forall u{\in}\mathrm{tc}(x)\,(\tau(u) \neq 0)\\
&\quad\wedge\ \forall u, u'{\in}\mathrm{tc}(x)\,(u \neq u' \rightarrow (\tau(u) \text{ and } \tau(u') \text{ are disjoint}))\\
&\quad\wedge\ \forall u{\in}\mathrm{tc}(x)\,\forall u'{\in}u\,\forall v{\in}\tau(u)\,\exists v'{\in}\tau(u')\,(v' \in \mathrm{tc}(v)).
\end{aligned} \tag{3.2}$$

The last conjunct is read as "for all $u, u' \in \mathrm{tc}(x)$, if $u' \in u$ then for every v in the image of u there is some v' in the image of u' with $v' \in \mathrm{tc}(v)$." Note that the "identity" multifunction $u \mapsto \{u\}$ is an embedding of x in x; we will call an embedding of this form *the identity embedding.*

We abbreviate (3.2) by $\tau(\cdot, \vec{w}) : x \preccurlyeq y$. We next introduce terminology for embeddings whose graphs are given by formulas and embeddings whose graphs are given by sets.

Definition 3.7: Given a formula $\varepsilon(u, v, \vec{w})$, we define $\varepsilon(\cdot, \cdot, \vec{w}) : x \preccurlyeq y$ to be condition (3.2) with $v \in \tau(u, \vec{w})$ replaced by $v \in \mathrm{tc}(y) \wedge \varepsilon(u, v, \vec{w})$. More precisely, $\varepsilon(\cdot, \cdot, \vec{w}) : x \preccurlyeq y$ means:

$$\begin{aligned}
&\forall u{\in}\mathrm{tc}(x)\,\exists v \in \mathrm{tc}(y)\,\varepsilon(u, v, \vec{w})\\
&\quad\wedge\ \forall u, u'{\in}\mathrm{tc}(x)\,\forall v{\in}\mathrm{tc}(y)\,(u \neq u' \rightarrow \neg\varepsilon(u, v, \vec{w}) \vee \neg\varepsilon(u', v, \vec{w}))\\
&\quad\wedge\ \forall u{\in}\mathrm{tc}(x)\,\forall u'{\in}u\,\forall v{\in}\mathrm{tc}(y)\,(\varepsilon(u, v, \vec{w}) \rightarrow \exists v' \in \mathrm{tc}(v)\,\varepsilon(u', v', \vec{w})).
\end{aligned} \tag{3.3}$$

This kind of embedding is called a *weakly uniform embedding.*

Definition 3.8: For a set e, we write $e : x \preccurlyeq y$ if $e \subseteq \mathrm{tc}(x) \times \mathrm{tc}(y)$ and $\varepsilon(\cdot, \cdot, e) : x \preccurlyeq y$ holds when $\varepsilon(u, v, e)$ is the formula $\langle u, v\rangle \in e$. We write simply $x \preccurlyeq y$ to abbreviate $\exists e\,(e : x \preccurlyeq y)$. This is called a *nonuniform embedding.*

Note that $e : x \preccurlyeq y$ is $\Delta_0(\mathrm{L}_0)$. More generally, if ε is a $\Delta_0(L)$-formula in a language $L \supseteq \mathrm{L}_0$ then $\varepsilon(\cdot, \cdot, \vec{w}) : x \preccurlyeq y$ is also $\Delta_0(L)$.

Definition 3.9: We say that a theory *defines a* $\Delta_0(\mathrm{L}_0)$*-embedding* $x \preccurlyeq y$ if there is a $\Delta_0(\mathrm{L}_0)$-formula $\varepsilon(u, v, x, y)$ such that the theory proves $\varepsilon(\cdot, \cdot, x, y) : x \preccurlyeq y$.

The next lemma is useful for constructing embeddings. We state it for nonuniform embeddings, but there are analogous versions for strongly and weakly uniform embeddings. Say that two embeddings $e : x \preccurlyeq z$ and $f : y \preccurlyeq z$ are *compatible* if their union is still an injective multifunction, that is, if it satisfies the disjointness condition of (3.2). In particular, embeddings with disjoint ranges are automatically compatible.

Lemma 3.10: *Provably in* T_0, *if two embeddings* $e : x \preccurlyeq z$ *and* $f : y \preccurlyeq z$ *are compatible, then* $e \cup f : x \cup y \preccurlyeq z$.

Proof: It follows from the axioms that $u \in \mathrm{tc}(x \cup y)$ if and only if $u \in \mathrm{tc}(x)$ or $u \in \mathrm{tc}(y)$. The proof is then immediate. □

3.3. *The theories* $\mathbf{KP}_1^{\preccurlyeq}$ *and* $\mathbf{KP}_1^{\mathbf{u}}$

This section defines theories $\mathrm{KP}_1^{\preccurlyeq}$ and $\mathrm{KP}_1^{\mathrm{u}}$ that, intuitively, are to Rathjen's KP_1 as S_2^1 is to $\mathrm{I}\Sigma_1$. The role of "sharply bounded" quantification in S_2^1 is now played by $\in$-bounded quantification. The analogue of a "bounded" quantifier in our context is one where the quantified variable is embeddable in a #-term:

Definition 3.11: A *#-term* is a $\{1, \odot, \#\}$-term.

Saying that a set is embeddable in a #-term $t(x)$ is analogous to saying that a number/string has length at most $p(|x|)$ for some polynomial p. When we write a #-term, we will use the convention that the # operation takes precedence over $\odot$, and otherwise we omit right-associative parentheses. So for example $1 \odot x\#y \odot z$ is read as $1 \odot ((x\#y) \odot z)$.

Definition 3.12: Let L be a language containing L_0. The class $\Sigma_1^{\preccurlyeq}(L)$ consists of L-formulas of the form

$$\exists x{\preccurlyeq}t(\vec{x})\, \varphi(x, \vec{x})$$

where t is a #-term not involving x and φ is $\Delta_0(L)$. Here $\exists x{\preccurlyeq}t\, \varphi$ stands for $\exists x\, (x{\preccurlyeq}t \wedge \varphi)$. Recall that $x \preccurlyeq y$ denotes a nonuniform embedding, i.e., it stands for $\exists e\, (e : x \preccurlyeq y)$. Hence a $\Sigma_1^{\preccurlyeq}(L)$-formula is also a $\Sigma_1(L)$-formula. (See also Lemma 4.13.)

Note that the term $\preccurlyeq$-bounding the leading existential quantifier in a $\Sigma_1^{\preccurlyeq}(L)$-formula is required to be a #-term while the $\in$-bounding terms in the $\Delta_0(L)$-part can be arbitrary L-terms.

Definition 3.13: The theory $\mathrm{KP}_1^{\preccurlyeq}$ consists of T_0 without tc-Transitivity, that is, the defining axioms for the symbols in L_0, Extensionality, Set Foundation and $\Delta_0(\mathrm{L}_0)$-Separation, together with the two schemes:

- $\Delta_0(\mathrm{L}_0)$-*Collection:*

$$\forall u{\in}x\, \exists v\, \varphi(u,v,\vec{w}) \rightarrow \exists y\, \forall u{\in}x\, \exists v \in y\, \varphi(u,v,\vec{w}) \quad \text{for } \varphi \in \Delta_0(\mathrm{L}_0),$$

- $\Sigma_1^{\preccurlyeq}(\mathrm{L}_0)$-*Induction:*

$$\forall y\, (\forall u{\in}y\, \varphi(u,\vec{w}) \rightarrow \varphi(y,\vec{w})) \rightarrow \varphi(x,\vec{w}) \quad \text{for } \varphi \in \Sigma_1^{\preccurlyeq}(\mathrm{L}_0).$$

We omitted tc-Transitivity from the definition of $\mathrm{KP}_1^{\preccurlyeq}$ because it is not one of the usual axioms for Kripke-Platek set theories. However, tc-Transitivity can be proven by $\Delta_0(\mathrm{L}_0)$-Induction from the rest of the axioms of T_0. Since $\Delta_0(\mathrm{L}_0)$-Induction is contained in $\mathrm{KP}_1^{\preccurlyeq}$, it follows that tc-Transitivity is a consequence of $\mathrm{KP}_1^{\preccurlyeq}$. Thus $\mathrm{KP}_1^{\preccurlyeq}$ contains T_0. The same holds for the theory $\mathrm{KP}_1^{\mathrm{u}}$ defined next.

Our goal is to $\Sigma_1^{\preccurlyeq}(\mathrm{L}_0)$-define all CRSF functions in $\mathrm{KP}_1^{\preccurlyeq}$ in the following sense. Fix a universe of sets V (a model of ZFC); of course, we may view V as interpreting L_0. Let T be a theory and Φ a set of formulas. A function $f(\vec{x})$ over V is Φ-*definable in* T if there is $\varphi(y,\vec{x}) \in \Phi$ such that $\mathrm{V} \models \forall \vec{x}\, \varphi(f(\vec{x}),\vec{x})$ and T proves $\exists! y\, \varphi(y,\vec{x})$.

In fact, we will show that an apparently weaker theory $\mathrm{KP}_1^{\mathrm{u}}$ is sufficient for this purpose. $\mathrm{KP}_1^{\mathrm{u}}$ is defined in the same way as $\mathrm{KP}_1^{\preccurlyeq}$, except that the induction scheme is restricted to $\Sigma_1^{\preccurlyeq}(\mathrm{L}_0)$-formulas of a special form, where the witness to the leading existential quantifier is required to be *u*nique and *u*niformly embeddable into a #-term (hence the superscript *u*). We will see later (in Lemma 4.13) that it is only the uniqueness requirement that distinguishes this from $\mathrm{KP}_1^{\preccurlyeq}$.

Definition 3.14: The theory $\mathrm{KP}_1^{\mathrm{u}}$ consists of T_0 without tc-Transitivity, that is, the defining axioms for the symbols in L_0, Extensionality, Set Foundation and $\Delta_0(\mathrm{L}_0)$-Separation, together with $\Delta_0(\mathrm{L}_0)$-Collection and the scheme:

- *Uniformly Bounded Unique* $\Sigma_1^{\preccurlyeq}(\mathrm{L}_0)$-*Induction*

$$\forall u\, \exists^{\leqslant 1} v\, \varphi(u,v,\vec{w}) \wedge \forall y\, (\forall u{\in}y\, \exists v\, \varphi^{\varepsilon,t}(u,v,\vec{w}) \rightarrow \exists v\, \varphi^{\varepsilon,t}(y,v,\vec{w}))$$
$$\rightarrow \exists v\, \varphi^{\varepsilon,t}(x,v,\vec{w})$$

where $\varphi^{\varepsilon,t}(u,v,\vec{w})$ abbreviates the formula

$$\varphi(u,v,\vec{w}) \wedge \varepsilon(\cdot,\cdot,v,u,\vec{w}) : v \preccurlyeq t(u,\vec{w})$$

and the scheme ranges over $\Delta_0(\mathrm{L}_0)$-formulas φ, ε and #-terms t.

3.4. *The language* $\mathrm{L_{crsf}}$ *and theory* $\mathrm{T_{crsf}}$

Our final main theory, $\mathrm{T_{crsf}}$, is an analogue of the bounded arithmetic theory PV_1. $\mathrm{T_{crsf}}$ has a function symbol for every CRSF function, and Π_1 axioms describing how the CRSF functions are defined from each other. By comparison, $\mathrm{KP}_1^{\preccurlyeq}$ and $\mathrm{KP}_1^{\mathrm{u}}$ are analogues of S_2^1. One of our main results is Theorem 6.9, which states that a definitional expansion of $\mathrm{KP}_1^{\mathrm{u}}$ is $\Pi_2(\mathrm{L_{crsf}})$-*conservative* over $\mathrm{T_{crsf}}$, that is, every $\Pi_2(\mathrm{L_{crsf}})$-sentence provable in the former theory is also provable in the latter. Theorem 6.12 states an analogous result for $\mathrm{KP}_1^{\preccurlyeq}$, but only with the addition of a global choice function to $\mathrm{T_{crsf}}$.

Definition 3.15: The language $\mathrm{L_{crsf}}$ consists of $\in$ and the function symbols listed below. The theory $\mathrm{T_{crsf}}$ contains the axioms of Extensionality, Set Foundation and tc-Transitivity, together with a defining axiom for each function symbol of $\mathrm{L_{crsf}}$, as follows.

- $\mathrm{L_{crsf}}$ contains the function symbols from L_0, and $\mathrm{T_{crsf}}$ contains their defining axioms.
- $\mathrm{L_{crsf}}$ contains the function symbols $proj_i^n$ for $1 \leqslant i \leqslant n$ and $cond_\in(x,y,u,v)$ with defining axioms $proj_i^n(x_1,\ldots,x_n) = x_i$ and

$$cond_\in(x,y,u,v) = \begin{cases} x & \text{if } u \in v \\ y & \text{otherwise.} \end{cases}$$

- (Closure under composition) For all function symbols $h, g_1, \ldots, g_k \in \mathrm{L_{crsf}}$ of suitable arities, $\mathrm{L_{crsf}}$ contains the function symbol $f_{h,\vec{g}}$ with defining axiom

$$f_{h,\vec{g}}(\vec{x}) = h(g_1(\vec{x}),\ldots,g_k(\vec{x})).$$

- (Closure under replacement) For all function symbols $f \in \mathrm{L_{crsf}}$, $\mathrm{L_{crsf}}$ contains the function symbol f" with defining axiom

$$f\text{"}(x,\vec{z}) = \{f(y,\vec{z}) : y \in x\}.$$

- (Closure under syntactic Cobham recursion) Suppose g, τ are function symbols in $\mathrm{L_{crsf}}$ and t is a #-term. Let us write "τ is an embedding into t at $x, \vec{w}$" for the $\Delta_0(\mathrm{L_{crsf}})$-formula

$$\tau(\cdot, g(x, f"(x, \vec{w}), \vec{w}), x, \vec{w}) : g(x, f"(x, \vec{w}), \vec{w}) \preccurlyeq t(x, \vec{w}).$$

Then $\mathrm{L_{crsf}}$ contains the function symbol $f = f_{g,\tau,t}$ with defining axiom

$$f(x, \vec{w}) = \begin{cases} g(x, f"(x, \vec{w}), \vec{w}) & \text{if } \tau \text{ is an embedding into } t \text{ at } x, \vec{w} \\ 0 & \text{otherwise.} \end{cases}$$

Proposition 3.16: *The universe* V *of sets can be expanded uniquely to a model of* $\mathrm{T_{crsf}}$. *The* $\mathrm{L_{crsf}}$*-function symbols then name exactly the* CRSF *functions as defined in Proposition 2.9.*

Because of closure under composition, every $\mathrm{L_{crsf}}$-term is equivalent to an $\mathrm{L_{crsf}}$-function symbol, provably in $\mathrm{T_{crsf}}$. Hence we will not always be careful to distinguish between terms and function symbols in $\mathrm{L_{crsf}}$.

Lemma 3.17: *For every function symbol* $f \in \mathrm{L_{crsf}}$, *there is a function symbol* $g \in \mathrm{L_{crsf}}$ *such that* $\mathrm{T_{crsf}}$ *proves* $g(x, \vec{w}) = \{y \in x : f(y, \vec{w}) \neq 0\}$.

Proof: Using $cond_{\in}$ we may construct a function symbol $h(y, \vec{w})$ which takes the value $\{y\}$ if $f(y, \vec{w}) \notin \{0\}$ and the value 0 otherwise. We put $g(x, \vec{w}) = \bigcup h"(x, \vec{w})$. □

The next lemma is proved as in the development of CRSF in [7, Theorem 13]. Note that we do not need recursion to prove either Lemma 3.17 or Lemma 3.18.

Lemma 3.18: *Every* $\Delta_0(\mathrm{L_{crsf}})$*-formula is provably equivalent in* $\mathrm{T_{crsf}}$ *to a formula of the form* $f(\vec{x}) \neq 0$ *for some* $\mathrm{L_{crsf}}$*-function symbol* f. *It follows that the* $\mathrm{L_{crsf}}$ *functions are closed under* $\Delta_0(\mathrm{L_{crsf}})$*-Separation provably in* $\mathrm{T_{crsf}}$*; that is, for every* $\Delta_0(\mathrm{L_{crsf}})$*-formula* $\varphi(y, \vec{w})$ *there is an* $\mathrm{L_{crsf}}$*-function symbol* f *such that* $\mathrm{T_{crsf}}$ *proves* $f(x, \vec{w}) = \{y \in x : \varphi(y, \vec{w})\}$.

Corollary 3.19: *The theory* $\mathrm{T_{crsf}}$ *extends* $\mathrm{T_0}$.

4. Bootstrapping

4.1. *Bootstrapping the defining axioms*

We first derive some simple consequences of the defining axioms, namely basic properties of tc, a description of the Mostowski graph of $x \odot y$, injectivity of $\odot$ in its first argument, and associativity of $\odot$.

Lemma 4.1: *The theory* T_0 *proves*

(*a*) $x \subseteq y \to \mathrm{tc}(x) \subseteq \mathrm{tc}(y),\ \ \mathrm{tc}(x) = \mathrm{tc}(\mathrm{tc}(x))$,
(*b*) $u \in \mathrm{tc}(x \odot y) \leftrightarrow (u \in \mathrm{tc}(y) \vee \exists u' \in \mathrm{tc}(x)\,(u = u' \odot y))$,
(*c*) $x \neq x' \to x \odot y \neq x' \odot y$,
(*d*) $x \odot (y \odot z) = (x \odot y) \odot z$.

Proof: We omit the proof of (a).

For (b) argue in T_0 as follows. If $x = 0$, the claim follows from the $\odot$-axiom, so assume $x \neq 0$. We prove ($\to$) by $\Delta_0(\mathrm{L}_0)$-Induction (recall Lemma 3.4), so assume it to hold for all $x' \in x$. Let $u \in \mathrm{tc}(x \odot y)$. By the tc,$\odot$-axioms, either $u \in \mathrm{tc}(x' \odot y)$ for some $x' \in x$ or $u \in x \odot y$. In the first case our claim follows by induction noting $\mathrm{tc}(x') \subseteq \mathrm{tc}(x)$ by (a). In the second case, $u = u' \odot y$ for some $u' \in x$ by the $\odot$-axiom.

Conversely, we first show $u \in \mathrm{tc}(y) \to u \in \mathrm{tc}(x \odot y)$. By $\Delta_0(\mathrm{L}_0)$-Induction we can assume this holds for all $z \in x$. Assume $u \in \mathrm{tc}(y)$ and let $z \in x$ be arbitrary. Then $u \in \mathrm{tc}(z \odot y)$ by induction. But $z \odot y \in x \odot y$, so $\mathrm{tc}(z \odot y) \subseteq \mathrm{tc}(x \odot y)$ by (a).

Finally, we show $u \in \mathrm{tc}(x) \to u \odot y \in \mathrm{tc}(x \odot y)$. We assume this for all $z \in x$. Let $u \in \mathrm{tc}(x)$. If $u \in x$, then $u \odot y \in x \odot y \subseteq \mathrm{tc}(x \odot y)$. Otherwise $u \in \mathrm{tc}(z)$ for some $z \in x$. Then $u \odot y \in \mathrm{tc}(z \odot y)$ by induction; but $\mathrm{tc}(z \odot y) \subseteq \mathrm{tc}(x \odot y)$ by $z \odot y \in x \odot y$ and (a).

For (c) argue in T_0 as follows. Suppose there are y, x_0, x_0' such that $x_0 \neq x_0'$ and $x_0 \odot y = x_0' \odot y$. It is easy to derive $\forall x\,(x \in \mathrm{tc}^+(x))$, so the set

$$z := \{x \in \mathrm{tc}^+(x_0) : \exists x' \in \mathrm{tc}^+(x_0')\,(x \neq x' \wedge x \odot y = x' \odot y)\}$$

is non-empty because it contains x_0. The set exists by $\Delta_0(\mathrm{L}_0)$-Separation. By Foundation, it contains an $\in$-minimal element x_1. Choose $x_1' \in \mathrm{tc}^+(x_0')$ with $x_1 \neq x_1'$ and $x_1 \odot y = x_1' \odot y$.

We claim that x_1, x_1' are both non-empty. Assume otherwise, say, $x_1' = 0$ and hence $x_1 \neq 0$. An easy $\Delta_0(\mathrm{L}_0)$-Induction shows $x \notin \mathrm{tc}(x)$ and $(x \neq 0 \to 0 \in \mathrm{tc}(x))$ for all x. Then $y \notin y = x_1' \odot y$ and $y = 0 \odot y \in \mathrm{tc}(x_1 \odot y)$ by (b). This contradicts $x_1 \odot y = x_1' \odot y$.

Choose x_2 such that either $x_2 \in x_1 \wedge x_2 \notin x_1'$ or $x_2 \in x_1' \wedge x_2 \notin x_1$. Assume the former (the latter case is similar). By the $\odot$-axiom, $x_2 \odot y \in x_1 \odot y = x_1' \odot y$. Since $x_1' \neq 0$ the $\odot$-axiom gives $x_2' \in x_1'$ such that $x_2 \odot y = x_2' \odot y$. As $x_2 \notin x_1'$ we have $x_2 \neq x_2'$. By (a), $x_2 \in x_1 \subseteq \mathrm{tc}^+(x_0)$ and $x_2' \in x_1' \subseteq \mathrm{tc}^+(x_0')$. Thus $x_2 \in z$, contradicting the minimality of x_1.

For (d), an easy induction shows that $u \odot 0 = u$ for all u. Item (d) is then true immediately if any of x, y or z is 0. Otherwise it follows by induction on x, using the $\odot$-axiom. □

We write $2^{\odot} := 1 \odot 1$, $3^{\odot} := 1 \odot 1 \odot 1$, etc. Notice that, in T_0, $1 \odot x = \{x\}$. We give an example of how we can now begin to build useful embeddings.

Example 4.2: There is a #-term $t_{\mathrm{pair}}(x, y)$ such that T_0 defines a $\Delta_0(\mathrm{L}_0)$-embedding from $\langle x, y \rangle$ into $t_{\mathrm{pair}}(x, y)$.

Proof: We put $t_{\mathrm{pair}}(x, y) := 4^{\odot} \odot x \odot 1 \odot y$. Consider the relations

$$\begin{aligned} e &:= \{\langle u, u \rangle : u \in \mathrm{tc}^+(y)\} \\ f &:= \{\langle u, u \odot 1 \odot y \rangle : u \in \mathrm{tc}^+(x)\} \\ g &:= \{\langle \{x\}, 2^{\odot} \odot x \odot 1 \odot y \rangle\} \\ h &:= \{\langle \{x, y\}, 3^{\odot} \odot x \odot 1 \odot y \rangle\}. \end{aligned}$$

Then $e : \{y\} \preccurlyeq t_{\mathrm{pair}}(x, y)$ and $f : \{x\} \preccurlyeq t_{\mathrm{pair}}(x, y)$, and these two embeddings are compatible since they have disjoint ranges. So $e \cup f : \{x, y\} \preccurlyeq t_{\mathrm{pair}}(x, y)$ (appealing to Lemma 3.10), hence $e \cup f \cup h : \{\{x, y\}\} \preccurlyeq t_{\mathrm{pair}}(x, y)$. On the other hand $f \cup g : \{\{x\}\} \preccurlyeq t_{\mathrm{pair}}(x, y)$. These are compatible, so $e \cup f \cup g \cup h : \{\{x\}, \{x, y\}\} \preccurlyeq t_{\mathrm{pair}}(x, y)$, as required. All these embeddings can be expressed straightforwardly in T_0 by $\Delta_0(\mathrm{L}_0)$-formulas. □

4.2. *Adding* $\in$*-bounded functions*

We give a small expansion T_0^+ of T_0.

Definition 4.3: Let L_0^+ be the language obtained from L_0 by adding a relation symbol $R(\vec{x})$ for every $\Delta_0(\mathrm{L}_0)$-formula $\varphi(\vec{x})$, and a function symbol $f(\vec{x})$ for every $\Delta_0(\mathrm{L}_0)$-formula $\psi(y, \vec{x})$ such that T_0 proves $\exists! y \in t(\vec{x})\, \psi(y, \vec{x})$ for some L_0-term $t(\vec{x})$.

The theory T_0^+ has language L_0^+ and is obtained from T_0 by adding for every relation symbol $R(\vec{x})$ in L_0^+ as above the defining axiom $R(\vec{x}) \leftrightarrow \varphi(\vec{x})$, and for every function symbol $f(\vec{x})$ in L_0^+ as above the defining axiom $\psi(f(\vec{x}), \vec{x})$.

Proposition 4.4: T_0^+ *is a conservative extension of* T_0. *Every* $\Delta_0(\mathrm{L}_0^+)$-*formula is* T_0^+-*provably equivalent to a* $\Delta_0(\mathrm{L}_0)$-*formula. In particular,* T_0^+ *proves* $\Delta_0(\mathrm{L}_0^+)$-*Induction and* $\Delta_0(\mathrm{L}_0^+)$-*Separation.*

We omit the proof. The language L_0^+ and the theory T_0^+ are introduced mainly for notational convenience. Interesting functions often do not have $\in$-bounded values.

Lemma 4.5: *Every function symbol introduced in* L_0^+ *has a copy in* $\mathrm{L}_{\mathrm{crsf}}$ *for which* $\mathrm{T}_{\mathrm{crsf}}$ *proves the defining axiom.*

Proof: Suppose T_0 proves $\exists! y \in t(\vec{x})\, \psi(y, \vec{x})$. Then using Lemma 3.18 we can compute y in $\mathrm{T}_{\mathrm{crsf}}$ as $\bigcup\{y \in t(\vec{x}) : \psi(y, \vec{x})\}$. □

Example 4.6: The language L_0^+ contains the relation symbol $\mathit{IsPair}(x)$ with defining axiom $\exists u, v \in \mathrm{tc}(x)\, (x = \langle u, v\rangle)$, the function symbol $\mathit{cond}_\in(x, y, u, v)$ with defining axiom as in Definition 3.15, and function symbols $\pi_1(x)$, $\pi_2(x)$ and $w\text{'}x$ such that T_0^+ proves $\pi_1(\langle x_1, x_2\rangle) = x_1$, $\pi_2(\langle x_1, x_2\rangle) = x_2$ and

$$w\text{'}x = \begin{cases} y \text{ if } y \text{ is unique with } \langle x, y\rangle \in w \\ 0 \text{ otherwise.} \end{cases}$$

We now formalize the graph isomorphism for $\#$ mentioned in Section 2. We introduce $\#\text{"}(u, y)$ below as an auxiliary function to formulate the defining axiom for $\sigma_{x,y}(u, v)$.

Lemma 4.7: *There are function symbols* $\#\text{"}(u, y)$, $\sigma_{x,y}(u, v)$, $\pi_{1,x,y}(w)$ *and* $\pi_{2,x,y}(w)$ *in* L_0^+ *such that* T_0^+ *proves*

$$\#\text{"}(u, y) = \{u'\#y : u' \in u\},$$

$$\sigma_{x,y}(u, v) = \begin{cases} v \odot \#\text{"}(u, y) \textit{ if } u \in \mathrm{tc}^+(x) \textit{ and } v \in \mathrm{tc}^+(y) \\ 0 \qquad\qquad\qquad \textit{otherwise.} \end{cases}$$

Moreover, T_0^+ *proves that*

(a) $\sigma_{x,y}$ *is injective on arguments* $u \in \mathrm{tc}^+(x), v \in \mathrm{tc}^+(y)$.
(b) Every $w \in \mathrm{tc}^+(x\#y)$ *has a* $\sigma_{x,y}$*-preimage* $(\pi_{1,x,y}(w), \pi_{2,x,y}(w))$.
(c) For all $u, u' \in \mathrm{tc}^+(x)$ *and* $v, v' \in \mathrm{tc}^+(y)$,

$$\sigma_{x,y}(u', v') \in \sigma_{x,y}(u, v) \leftrightarrow (u' = u \wedge v' \in v) \vee (u' \in u \wedge v' = y \wedge v = 0).$$

Proof: The functions $\#\text{"}(u, y), \sigma_{x,y}(u, v)$ have obvious defining axioms. Concerning bounding terms, from the $\#$-axiom we get that $\#\text{"}(u, y) \in \mathrm{tc}^+(u\#y)$ and

$$x\#y = y \odot \#\text{"}(x, y). \tag{4.1}$$

By induction on x, using Lemma 4.1(b), we get that if $u \in \mathrm{tc}^+(x)$ then $u\#y \in \mathrm{tc}^+(x\#y)$. Another use of Lemma 4.1(b) shows that if $v \in \mathrm{tc}^+(y)$ then $v \odot \#\text{"}(u, y) \in \mathrm{tc}^+(u\#y)$. Hence $\sigma_{x,y}(u, v) \in \mathrm{tc}^+(x\#y)$.

Observe that for $u \in \mathrm{tc}^+(x), v \in \mathrm{tc}^+(y)$,

$$\begin{aligned} v \neq 0 \to \sigma_{x,y}(u,v) &= \{\sigma_{x,y}(u,v') : v' \in v\}, \\ \sigma_{x,y}(u,0) &= \{\sigma_{x,y}(u',y) : u' \in u\}. \end{aligned} \tag{4.2}$$

The first line is from the definition. For the second, note $z \in \sigma_{x,y}(u,0)$ is equivalent to $z \in \#\text{"}(u,y)$, and hence to $z = u'\#y$ for some $u' \in u$; but $u'\#y = \sigma_{x,y}(u',y)$ by (4.1).

For (a), let $u, \tilde{u}, \ldots$ range over $\mathrm{tc}^+(x)$ and $v, \tilde{v}, \ldots$ range over $\mathrm{tc}^+(y)$. We claim that $\sigma_{x,y}(u,v) = \sigma_{x,y}(\tilde{u},\tilde{v})$ implies $u = \tilde{u}$ and $v = \tilde{v}$. By Lemma 4.1(c) it suffices to show it implies $u = \tilde{u}$. Assume otherwise. By $\Delta_0(\mathrm{L}_0^+)$-Foundation choose u $\in$-minimal such that there exist $\tilde{u}, v, \tilde{v}$ with $\sigma_{x,y}(u,v) = \sigma_{x,y}(\tilde{u},\tilde{v})$ and $u \neq \tilde{u}$; then choose $\tilde{u}$ $\in$-minimal such that there are $v, \tilde{v}$ with this property, and so on for $v, \tilde{v}$. We distinguish two cases, as in (4.2).

First suppose $v \neq 0$. Then there is $v' \in v$ such that $\sigma_{x,y}(u,v') \in \sigma_{x,y}(\tilde{u},\tilde{v})$. If $\tilde{v} \neq 0$, then $\sigma_{x,y}(u,v') = \sigma_{x,y}(\tilde{u},v'')$ for some $v'' \in \tilde{v}$, and this contradicts the choice of v. If $\tilde{v} = 0$, then $\sigma_{x,y}(u,v') = \sigma_{x,y}(u',y)$ for some $u' \in \tilde{u}$, and this contradicts the choice of $\tilde{u}$.

Now suppose $v = 0$. If $\tilde{v} = 0$, then $\{\sigma_{x,y}(u',y) : u' \in u\} = \{\sigma_{x,y}(u'',y) : u'' \in \tilde{u}\}$, so for each $u' \in u$ there is $u'' \in \tilde{u}$ such that $\sigma_{x,y}(u',y) = \sigma_{x,y}(u'',y)$, so then $u' = u''$ by choice of u; thus $u \subseteq \tilde{u}$. Similarly $\tilde{u} \subseteq u$, contradicting our assumption $u \neq \tilde{u}$. If $\tilde{v} \neq 0$, then for each $u' \in u$ there is $v' \in \tilde{v}$ such that $\sigma_{x,y}(u',y) = \sigma_{x,y}(\tilde{u},v')$, so $u' = \tilde{u}$ by choice of u. Thus $u = 0$ or $u = \{\tilde{u}\}$; the latter is impossible by choice of u, so $u = 0$; then $\sigma_{x,y}(\tilde{u},\tilde{v}) = \sigma_{x,y}(u,v) = \sigma_{x,y}(0,0) = 0$, so $\tilde{v} = 0$, a contradiction.

For (b) we will show surjectivity; $\pi_{1,x,y}(w)$ and $\pi_{2,x,y}(w)$ can then easily be constructed. So let $w \in \mathrm{tc}^+(x\#y)$. If $w = x\#y$, put $u := x, v := y$. Otherwise $w \in \mathrm{tc}(x\#y) = \mathrm{tc}(y \odot \#\text{"}(x,y))$ by (4.1). By Lemma 4.1(c) we have two cases. If $w = v' \odot \#\text{"}(x,y)$ for some $v' \in y$, put $u := x, v := v'$. If $w \in \mathrm{tc}(\#\text{"}(x,y))$, then $w \in \mathrm{tc}^+(x'\#y)$ for some $x' \in x$ and, using $\Delta_0(\mathrm{L}_0^+)$-Induction on x, we find $u \in \mathrm{tc}^+(x') \subseteq \mathrm{tc}^+(x)$ and $v \in \mathrm{tc}^+(y)$ with $w = \sigma_{x',y}(u,v)$. Since $\sigma_{x,y}(u,v)$ does not depend on x, we have $w = \sigma_{x',y}(u,v) = \sigma_{x,y}(u,v)$.

Claim (c) follows by (4.2). □

4.3. *Monotonicity lemma*

We can now formally derive non-uniform and weakly uniform versions of the Monotonicity Lemma 2.4, meaning "monotonicity of #-terms with respect to embeddings." Note that Lemma 4.8 includes as a special case the transitivity of embeddings,

$$z \preccurlyeq x \wedge x \preccurlyeq y \rightarrow z \preccurlyeq y. \tag{4.3}$$

Lemma 4.8: (Monotonicity) *For all #-terms $t(x, \vec{w})$ the theory T_0 proves*

$$z \preccurlyeq t(x, \vec{w}) \wedge x \preccurlyeq y \rightarrow z \preccurlyeq t(y, \vec{w}). \tag{4.4}$$

Moreover, for all $\Delta_0(\mathrm{L}_0)$-formulas ε_0, ε_1 there is a $\Delta_0(\mathrm{L}_0)$-formula ε_2 such that T_0 proves

$$\begin{aligned} &\varepsilon_0(\cdot,\cdot,x,z,\vec{w}) : z \preccurlyeq t(x,\vec{w}) \ \wedge \ \varepsilon_1(\cdot,\cdot,x,y,\vec{w}) : x \preccurlyeq y \\ &\rightarrow \varepsilon_2(\cdot,\cdot,x,y,z,\vec{w}) : z \preccurlyeq t(y,\vec{w}). \end{aligned}$$

Proof: We only verify the first statement; the second follows by inspection of the proof. We proceed by induction on t. We work in T_0^+, which is sufficient by Proposition 5.5.

If $t(x, \vec{w})$ is 1 or a variable distinct from x, then there is nothing to show. If $t(x, \vec{w})$ equals x then we have to show (4.3). So assume $e : z \preccurlyeq x$ and $f : x \preccurlyeq y$. Then

$$g := \{\langle u, w\rangle \in \mathrm{tc}(z) \times \mathrm{tc}(y) : \exists v \in \mathrm{tc}(x)\, (\langle u, v\rangle \in e \wedge \langle v, w\rangle \in f)\}$$

exists by $\Delta_0(\mathrm{L}_0)$-Separation. We claim $g : z \preccurlyeq y$. It is easy to see that $\langle u, w\rangle, \langle u', w\rangle \in g$ implies $u = u'$. Assume $u' \in u \in \mathrm{tc}(x)$ and $\langle u, w\rangle \in g$. Choose v such that $\langle u, v\rangle \in e$ and $\langle v, w\rangle \in f$. Then there is $v' \in \mathrm{tc}(v)$ such that $\langle u', v'\rangle \in e$. It now suffices to show that, generally, for all v, v', w we have

$$v' \in \mathrm{tc}(v) \wedge \langle v, w\rangle \in f \rightarrow \exists w' \in \mathrm{tc}(w)\, \langle v', w'\rangle \in f.$$

This is clear if $v' \in v$. Otherwise, $v' \in \mathrm{tc}(v'')$ for some $v'' \in v$. Then choose $w'' \in \mathrm{tc}(w)$ such that $\langle v'', w''\rangle \in f$. Appealing to $\Delta_0(\mathrm{L}_0)$-Induction, we can find $w' \in \mathrm{tc}(w'')$ such that $\langle v', w'\rangle \in f$. Then $w' \in \mathrm{tc}(w)$ by Lemma 4.1(a), as claimed.

As preparation for the induction step in our induction on t, we show

$$x \preccurlyeq x' \wedge y \preccurlyeq y' \rightarrow x \odot y \preccurlyeq x' \odot y' \wedge x\#y \preccurlyeq x'\#y'. \tag{4.5}$$

Assume $e : x \preccurlyeq x'$ and $f : y \preccurlyeq y'$. By $\Delta_0(\mathrm{L}_0)$-Separation the set

$$\{\langle u, v\rangle \in \mathrm{tc}(x \odot y) \times \mathrm{tc}(x' \odot y') :\\ \exists u' \in \mathrm{tc}(x)\, \exists v' \in \mathrm{tc}(x')\, (u = u' \odot y \wedge v = v' \odot y' \wedge \langle u', v'\rangle \in e)\}$$

exists. We leave it to the reader to check that its union with f witnesses $x \odot y \preccurlyeq x' \odot y'$. For #, observe $e_+ := e \cup \{\langle x, x'\rangle\} : \{x\} \preccurlyeq \{x'\}$ and $f_+ := f \cup \{\langle y, y'\rangle\} : \{y\} \preccurlyeq \{y'\}$. Let g be the set containing the pairs $\langle \sigma_{x,y}(u,v), \sigma_{x',y'}(u',v')\rangle$ such that $\langle u, u'\rangle \in e_+$ and $\langle v, v'\rangle \in f_+$. This set g exists by $\Delta_0(\mathrm{L}_0^+)$-Separation. Using Lemma 4.7 it is straightforward to check that $g : x\#y \preccurlyeq x'\#y'$.

Now the induction step is easy. We are given embeddings $z \preccurlyeq t(x, \vec{w})$ and $x \preccurlyeq y$. Assume first that $t(x, \vec{w}) = t_1(x, \vec{w}) \odot t_2(x, \vec{w})$. By the identity embedding $t_1(x, \vec{w}) \preccurlyeq t_1(x, \vec{w})$, and applying the inductive hypothesis gives $t_1(x, \vec{w}) \preccurlyeq t_1(y, \vec{w})$. Similarly $t_2(x, \vec{w}) \preccurlyeq t_2(y, \vec{w})$. Applying (4.5) we get

$$t_1(x, \vec{w}) \odot t_2(x, \vec{w}) \preccurlyeq t_1(y, \vec{w}) \odot t_2(y, \vec{w}),$$

that is, $t(x, \vec{w}) \preccurlyeq t(y, \vec{w})$. This, together with (4.3), implies (4.4). The case of $t_1(x, \vec{w})\#t_2(x, \vec{w})$ is analogous. □

4.4. *Some useful embeddings*

We first show that we can embed all L_0^+-terms into #-terms.

Lemma 4.9: *The theory* T_0 *defines a* $\Delta_0(\mathrm{L}_0)$*-embedding of* $x \times y$ *into a #-term* $t_\times(x, y)$.

Proof: By Proposition 4.4 it suffices to prove this for T_0^+ and a $\Delta_0(\mathrm{L}_0^+)$-embedding. We set

$$t_\times(x, y) = (x\#y) \odot (x\#y) \odot x \odot y \odot x.$$

The formula $\varepsilon_\times(z, z', x, y)$ implements the following informal procedure on input z, z', x, y. In the description of this procedure we understand that whenever a "check" is carried out then the computation halts, and the procedure rejects or accepts depending on whether the check failed or not. For example, line 2 is reached only if $z \notin \mathrm{tc}(x)$.

It is easy to check that the condition that z, z', x, y is accepted is expressible as a $\Delta_0(\mathrm{L}_0)$-formula.

Input: z, z', x, y

1. **if** $z \in \mathrm{tc}(x)$ **then** check $z' = z$
2. **if** $z \in \mathrm{tc}(y)$ **then** check $z' = z \odot x$
3. **guess** $u \in x$
4. **if** $z = \{u\}$ **then** check $z' = u \odot y \odot x$
5. **guess** $v \in y$
6. **if** $z = \{u, v\}$ **then** check $z' = \sigma_{x,y}(u, v) \odot x \odot y \odot x$
7. **if** $z = \{\{u\}, \{u, v\}\}$ **then** check $z' = \sigma_{x,y}(u, v) \odot (x \# y) \odot x \odot y \odot x$
8. reject

It is clear that any $z \in \mathrm{tc}(x \times y)$ is mapped to at least one z'. Further, distinct $z \neq \tilde{z}$ cannot be mapped to the same $z' = \tilde{z}'$: any z' satisfies the check of at most one line and this line determines the pre-image z (Lemmas 4.7 and 4.1(c)).

Assume $z \in \tilde{z}$ and $\tilde{z}$ is mapped to $\tilde{z}'$. We have to find $z' \in \mathrm{tc}(\tilde{z}')$ such that z is mapped to z'. This is easy if $z \subset \mathrm{tc}(x) \cup \mathrm{tc}(y)$, so assume this is not the case. Then $\tilde{z}$ cannot satisfy any "if" condition before line 7. Hence $\tilde{z} = \{\{u\}, \{u, v\}\}$ for some $u \in x, v \in y$ and $\tilde{z}'$ satisfies the check in line 7. As $z \in \tilde{z}$ we have $z = \{u\}$ or $z = \{u, v\}$ and for suitable guesses in lines 3 and 5, z satisfies the "if" condition of line 4 or 6. Then choose z' satisfying the (first) corresponding check. □

Lemma 4.10: *For each* L_0^+*-term* $s(\vec{x})$ *the theory* T_0 *defines a* $\Delta_0(\mathrm{L}_0)$*-embedding of* $s(\vec{x})$ *into a* $\#$*-term* $s^{\#}(\vec{x})$.

Proof: This follows by an induction on $s(\vec{x})$ using Lemma 4.8 once we verify it for the base case that $s(\vec{x})$ is a function symbol in L_0^+.

For any such $s(\vec{x})$, there is an L_0-term $r(\vec{x})$ such that T_0^+ proves $s(\vec{x}) \in r(\vec{x})$. By Lemma 4.1 $\mathrm{tc}(s(\vec{x})) \subseteq \mathrm{tc}(r(\vec{x}))$, so the identity embedding (expressed by the formula $u = v$) embeds $s(\vec{x})$ in $r(\vec{x})$. By transitivity of $\preccurlyeq$ (Lemma 4.8), it thus suffices to verify the lemma for L_0-terms $r(\vec{x})$. As for L_0-terms, this follows by an induction on $r(\vec{x})$ using Lemma 4.8 once we verify it for the base case that $r(\vec{x})$ is a function symbol in L_0. The only non-trivial case now is crossproduct $\times$, and this is handled by the previous lemma. □

For tuples $\vec{u} = u_1, \ldots, u_k$ let us abbreviate $\bigwedge_i u_i \in z$ as $\vec{u} \in z$. We show that given a family of sets parametrized by tuples $\vec{u} \in z$, where each set is uniformly embeddable in s, we can embed the whole family (if it exists as a set) in a #-term $t(z, s)$. Note that the existence of V in the lemma is automatic in the presence of the Collection scheme.

Lemma 4.11: *Let $\varphi(v, \vec{u}, \vec{w})$ and $\varepsilon(z, z', v, \vec{u}, \vec{w})$ be $\Delta_0(\mathrm{L}_0^+)$-formulas and $s(\vec{w})$ a #-term. There is a $\Delta_0(\mathrm{L}_0^+)$-formula $\delta(z, z', V, z, \vec{w})$ and a #-term $t(z, x)$ such that if*

$$\begin{aligned} &\forall \vec{u} {\in} z\, \exists^{\leqslant 1} v\, \varphi(v, \vec{u}, \vec{w}) \\ &\quad \wedge\ \forall \vec{u} {\in} z\, \exists v\, (\varphi(v, \vec{u}, \vec{w}) \wedge \varepsilon(\cdot, \cdot, v, \vec{u}, \vec{w}) : v \preccurlyeq s(\vec{w})) \\ &\quad \wedge\ V = \{v : \exists \vec{u} \in z\, \varphi(v, \vec{u}, \vec{w})\} \end{aligned} \tag{4.6}$$

then $\delta(\cdot, \cdot, V, z, \vec{w}) : V \preccurlyeq t(z, s(\vec{w}))$, provably in T_0^+.

Proof: For notational simplicity we suppress the side variables $\vec{w}$. We first consider the case in which $\vec{u}$ is a single variable u. Note that in the second line of the assumption (4.6) we may assume without loss of generality that we actually have $\varepsilon(\cdot, \cdot, v, u) : \{v\} \preccurlyeq s$, since otherwise we could modify ε so that $\varepsilon(v, s, v, u)$ holds and replace the bound s with $1 \odot s$. Now put $t(z, s) := z\#s$ and define

$$\varepsilon'(y, \tilde{y}', V, u) := \exists v \in V\, \exists y' \in \mathrm{tc}(s)\, (\varphi(v, u) \wedge \varepsilon(y, y', v, u) \wedge \tilde{y}' = \sigma_{z,s}(u, y')).$$

For each $u \in z$, if $\varphi(u, v)$ then the formula $\varepsilon'(\cdot, \cdot, V, u)$ describes an embedding of $\{v\}$ into $z\#s$ which is a copy of the embedding $\varepsilon(\cdot, \cdot, v, u)$, but with its range moved to lie entirely within the uth copy of s inside $z\#s$. These embeddings have disjoint ranges for distinct u, so as in Lemma 3.10 their union $\delta(y, \tilde{y}', V, z) := \exists u \in z\, \varepsilon'(y, \tilde{y}', V, u)$ describes an embedding of V into $z\#s$, since $y \in \mathrm{tc}(V)$ implies $y \in \mathrm{tc}(\{v\})$ for some $v \in V$.

When $\vec{u}$ is a tuple of k variables, we reduce to the first case by coding $\vec{u}$ as an ordered k-tuple in the usual way. So the quantifier $\forall \vec{u} {\in} z$ becomes $\forall u {\in} (z \times \cdots \times z)$ and we replace φ and ε with formulas accessing the values of $\vec{u}$ from u using projection functions. The first case then gives an embedding of V into $t(z \times \cdots \times z, s)$, and by Lemma 4.10 and the Monotonicity Lemma 4.8 we get an embedding of V into some #-term $t'(z, s)$. □

We finish this section by showing, in Lemma 4.13, that the non-uniform embedding bounding the existential quantifier in a $\Sigma_1^{\preccurlyeq}$-formula (over any language) can be replaced with a weakly uniform embedding. This will be useful when we want to show that structures satisfy $\Sigma_1^{\preccurlyeq}$-Induction. We first show a suitable embedding exists.

Lemma 4.12: *There is a* $\Delta_0(\mathrm{L_0})$*-formula* $\varepsilon_{\mathrm{emb}}(u,v,e,x,y)$ *and a* #*-term* $t_{\mathrm{emb}}(y)$ *such that* $\mathrm{T_0}$ *proves* $e : x \preccurlyeq y \to \varepsilon_{\mathrm{emb}}(\cdot,\cdot,e,x,y) : \langle e,x\rangle \preccurlyeq t_{\mathrm{emb}}(y)$.

Proof: Let $\varepsilon_\times(u,v,x,y) \in \Delta_0(\mathrm{L_0})$ describe (in $\mathrm{T_0}$) an embedding of $x \times y$ into $t_\times(x,y)$. Then $\varepsilon_\times(u,v,\mathrm{tc}(x),\mathrm{tc}(y))$ describes an embedding of $\mathrm{tc}(x) \times \mathrm{tc}(y)$ into $t_\times(\mathrm{tc}(x),\mathrm{tc}(y))$. The identity embedding embeds $\mathrm{tc}(x)$ into x. Combining these, Lemma 4.8 gives a $\Delta_0(\mathrm{L_0})$-formula describing an embedding $\mathrm{tc}(x) \times \mathrm{tc}(y) \preccurlyeq t_\times(x,y)$. But $e : x \preccurlyeq y$ implies $e \subseteq \mathrm{tc}(x) \times \mathrm{tc}(y)$, so this formula also describes an embedding $e \preccurlyeq t_\times(x,y)$.

Using Example 4.2 there is a $\Delta_0(\mathrm{L_0})$-formula describing an embedding $\langle e,x\rangle \preccurlyeq t_{\mathrm{pair}}(e,x)$. By Lemma 4.8 and the previous paragraph, there is a $\Delta_0(\mathrm{L_0})$-formula describing an embedding $\langle e,x\rangle \preccurlyeq t_{\mathrm{pair}}(t_\times(x,y),x)$. Since $e : x \preccurlyeq y$ it is easy to write a $\Delta_0(\mathrm{L_0})$-formula with parameter e describing an embedding $x \preccurlyeq y$, so using Lemma 4.8 again we can replace x by y, that is, construct a $\Delta_0(\mathrm{L_0})$-embedding $\langle e,x\rangle \preccurlyeq t_{\mathrm{emb}}(y) := t_{\mathrm{pair}}(t_\times(y,y),y)$. □

Lemma 4.13: *Let* L *be a language extending* $\mathrm{L_0}$. *Provably in* $\mathrm{T_0}$, *every* $\Sigma_1^{\preccurlyeq}(L)$*-formula* $\theta(\vec{x})$ *is equivalent to a formula of the form*

$$\exists v\,(\varphi(v,\vec{x}) \wedge \varepsilon(\cdot,\cdot,v,\vec{x}) : v \preccurlyeq t(\vec{x})) \tag{4.7}$$

where φ *is a* $\Delta_0(L)$*-formula,* ε *is a* $\Delta_0(\mathrm{L_0})$*-formula and* t *is a* #*-term.*

Proof: Expanding the existential quantifier implicit in the nonuniform embedding bound, there is a $\Delta_0(L)$-formula ψ and a #-term s such that $\theta(\vec{x})$ has the form

$$\exists w \exists e\,(e : w \preccurlyeq s(\vec{x}) \wedge \psi(w,\vec{x})).$$

By Lemma 4.12, $\mathrm{T_0}$ proves

$$e : w \preccurlyeq s(\vec{x}) \to \varepsilon_{\mathrm{emb}}(\cdot,\cdot,e,w,s(\vec{x})) : \langle e,w\rangle \preccurlyeq t_{\mathrm{emb}}(s(\vec{x})).$$

Hence $\theta(\vec{x})$ is equivalent to

$$\exists \langle e,w\rangle\,(e : w \preccurlyeq s(\vec{x}) \wedge \psi(w,\vec{x}) \wedge \varepsilon_{\mathrm{emb}}(\cdot,\cdot,e,w,s(\vec{x})) : \langle e,w\rangle \preccurlyeq t_{\mathrm{emb}}(s(\vec{x}))).$$

For clarity we have written this rather informally. Strictly speaking, $\langle e,w\rangle$ should be a single variable v, and e and w should be respectively $\pi_1(v)$ and $\pi_2(v)$; then apply Proposition 4.4. □

5. Definability

This section develops $\mathrm{KP}_1^{\mathrm{u}}$ with the goal of proving that it $\Sigma_1^{\preccurlyeq}(\mathrm{L}_0)$-defines all CRSF functions. The Definability Theorem 5.2 below states this in a syntactic manner, without reference to the universe of sets.

Definition 5.1: A $\Sigma_1^{\preccurlyeq}(\mathrm{L}_0)$-*expansion of* $\mathrm{KP}_1^{\mathrm{u}}$ is obtained from $\mathrm{KP}_1^{\mathrm{u}}$ by adding a set of formulas of the following forms:

- $\varphi(f(\vec{x}), \vec{x})$ where $f(\vec{x})$ is a function symbol outside L_0 and $\varphi(y, \vec{x})$ is $\Sigma_1^{\preccurlyeq}(\mathrm{L}_0)$ such that $\mathrm{KP}_1^{\mathrm{u}}$ proves $\exists! y\, \varphi(y, \vec{x})$
- $R(\vec{x}) \leftrightarrow \varphi(\vec{x})$ where $R(\vec{x})$ is a relation symbol outside L_0 and $\varphi(\vec{x})$ is $\Delta_0(\mathrm{L}_0)$.

For example, it is not hard to give a $\Sigma_1^{\preccurlyeq}(\mathrm{L}_0)$-definition of a proper comprehension term for a formula $\varphi \in \Delta_0(\mathrm{L}_0)$ as a new function symbol.

Theorem 5.2: *(Definability) There is a* $\Sigma_1^{\preccurlyeq}(\mathrm{L}_0)$-*expansion of* $\mathrm{KP}_1^{\mathrm{u}}$ *which contains all function symbols in* $\mathrm{L}_{\mathrm{crsf}}$ *in its expanded language, and proves all axioms of* $\mathrm{T}_{\mathrm{crsf}}$.

A $\Sigma_1^{\preccurlyeq}(\mathrm{L}_0)$-expansion of $\mathrm{KP}_1^{\mathrm{u}}$ is an expansion by definitions, and hence is conservative over $\mathrm{KP}_1^{\mathrm{u}}$. Thus Theorem 5.2 immediately implies that every CRSF function is denoted by a symbol in the language, and hence is $\Sigma_1^{\preccurlyeq}(\mathrm{L}_0)$-definable in $\mathrm{KP}_1^{\mathrm{u}}$. Note that such an expansion does not include the axiom schemes of $\mathrm{KP}_1^{\mathrm{u}}$ for formulas in the expanded language. (Definition 3.14 describes the axiom schemes of $\mathrm{KP}_1^{\mathrm{u}}$.) The lack of these schemes is the main technical difficulty in proving the Theorem 5.2. Some further comments can be found in Section 5.2, where this difficulty is tackled. In Section 5.5 we describe a particular well-behaved expansion which proves all these axiom schemes in the expanded language.

We will prove Theorem 5.2 indirectly. We first define an expansion $\mathrm{KP}_1^{\mathrm{u}} + \mathrm{L}_{\mathrm{def}}$ which includes a function symbol for every function definable in $\mathrm{KP}_1^{\mathrm{u}}$ by a particular kind of $\Sigma_1^{\preccurlyeq}(\mathrm{L}_0)$-formula. We will then show that these function symbols contain the basic functions from $\mathrm{L}_{\mathrm{crsf}}$ and satisfy the right closure properties.

Remark 5.3: For the results in this section about $\mathrm{KP}_1^{\mathrm{u}}$ and its expansions, we do not need the full strength of the collection scheme in $\mathrm{KP}_1^{\mathrm{u}}$. Every instance of $\Delta_0(\mathrm{L}_0)$-Collection we use is an instance of the apparently weaker scheme

- *Uniformly Bounded $\Delta_0(\mathrm{L_O})$-Replacement*

$$\forall u{\in}x\, \exists^{\leqslant 1} v\, \varphi(u, v, \vec{x}) \wedge \forall u{\in}x\, \exists v\, \varphi^{\varepsilon,t}(u, v, \vec{x}) \\ \to \exists y\, (y = \{v : \exists u \in x\, \varphi(u, v, \vec{x})\})$$

where $\varphi^{\varepsilon,t}(u, v, \vec{x})$ abbreviates the formula

$$\varphi(u, v, \vec{x}) \wedge \varepsilon(\cdot, \cdot, v, u, \vec{x}) : v \preccurlyeq t(u, \vec{x})$$

and the scheme ranges over $\Delta_0(\mathrm{L_O})$-formulas φ, ε and #-terms t.

5.1. *The definitional expansion* $\mathbf{KP}_1^{\mathbf{u}} + \mathbf{L_{def}}$

It will be convenient to have the L_0^+ relation and function symbols available, so the first step in the expansion is a small one, allowing this.

Definition 5.4: $\mathrm{KP}_1^{\mathrm{u}+}$ is the $\Sigma_1^{\preccurlyeq}(\mathrm{L_O})$-expansion of $\mathrm{KP}_1^{\mathrm{u}}$ which adds the defining axioms for all symbols in $\mathrm{L}_0^+ \setminus \mathrm{L_O}$.

We will use the following proposition without comment. (Cf. Proposition 4.4.)

Proposition 5.5: $\mathrm{KP}_1^{\mathrm{u}+}$ *is a conservative extension of* $\mathrm{KP}_1^{\mathrm{u}}$. *Furthermore* $\mathrm{KP}_1^{\mathrm{u}+}$ *proves the axiom schemes of* $\mathrm{KP}_1^{\mathrm{u}}$ *with* L_0^+ *replacing* $\mathrm{L_O}$.

We now expand $\mathrm{KP}_1^{\mathrm{u}+}$ with functions symbols $f(\vec{x})$ with $\Sigma_1^{\preccurlyeq}(\mathrm{L}_0^+)$-definitions of a special kind. The existentially quantified witness v in such a definition is not only bounded by a #-term $t(\vec{x})$, but weakly uniformly bounded. Moreover, the witness v is uniquely described by a $\Delta_0(\mathrm{L}_0^+)$-formula $\varphi(v, \vec{x})$. Intuitively, this formula says "v is a computation of the value of f on input $\vec{x}$". The "output" function e is a very simple function that extracts the value $e(v) = f(\vec{x})$ from the computation v.

Definition 5.6: A *good definition* is a tuple

$$(\varphi(v, \vec{x}), \varepsilon(z, z', v, \vec{x}), e(v), t(\vec{x}))$$

where φ, ε are $\Delta_0(\mathrm{L}_0^+)$-formulas, $e(v)$ is an L_0^+-term and $t(\vec{x})$ is a #-term such that $\mathrm{KP}_1^{\mathrm{u}+}$ proves

(Witness Existence) $\exists v\, \varphi(v, \vec{x})$
(Witness Uniqueness) $\exists^{\leqslant 1} v\, \varphi(v, \vec{x})$
(Witness Embedding) $\varphi(v, \vec{x}) \to \varepsilon(\cdot, \cdot, v, \vec{x}) : v \preccurlyeq t(\vec{x})$.

Definition 5.7: The theory $\mathrm{KP}_1^{\mathrm{u}} + \mathrm{L}_{\mathrm{def}}$ is obtained from $\mathrm{KP}_1^{\mathrm{u}+}$ by adding for every such good definition a function symbol $f(\vec{x})$ along with the defining axiom

$$\exists v\,(\varphi(v,\vec{x}) \wedge f(\vec{x}) = e(v)). \tag{5.1}$$

We then speak of a *good definition of* f. The language $\mathrm{L}_{\mathrm{def}}$ of $\mathrm{KP}_1^{\mathrm{u}} + \mathrm{L}_{\mathrm{def}}$ consists of L_0^+ together with all such function symbols.

It is obvious that $\mathrm{KP}_1^{\mathrm{u}} + \mathrm{L}_{\mathrm{def}}$ is a conservative extension of $\mathrm{KP}_1^{\mathrm{u}+}$. Again, we stress that we do not adopt the axiom schemes of $\mathrm{KP}_1^{\mathrm{u}}$ for the language $\mathrm{L}_{\mathrm{def}}$. For example, by definition, $\mathrm{KP}_1^{\mathrm{u}} + \mathrm{L}_{\mathrm{def}}$ has just $\Delta_0(\mathrm{L}_0)$-Separation, not $\Delta_0(\mathrm{L}_{\mathrm{def}})$-Separation.

Theorem 5.8: $\mathrm{KP}_1^{\mathrm{u}} + \mathrm{L}_{\mathrm{def}}$ *proves that* $\mathrm{L}_{\mathrm{def}}$ *satisfies the closure properties of* $\mathrm{L}_{\mathrm{crsf}}$ *from Definition 3.15. That is, (a) closure under composition, (b) closure under replacement and (c) closure under syntactic Cobham recursion.*

Statement (a) is proved in Theorem 5.10, statement (b) in Theorem 5.15 (see Example 5.16) and statement (c) in Theorem 5.17.

The Definability Theorem 5.2 follows easily from Theorem 5.8. We first observe that all function symbols in $\mathrm{L}_{\mathrm{def}}$ are $\Sigma_1^{\preccurlyeq}(\mathrm{L}_0)$-definable in $\mathrm{KP}_1^{\mathrm{u}}$, since we can replace all L_0^+ symbols in (5.1) by their $\Delta_0(\mathrm{L}_0)$ definitions and appeal to the conservativity of $\mathrm{KP}_1^{\mathrm{u}+}$ over $\mathrm{KP}_1^{\mathrm{u}}$. We can then go through the function symbols in $\mathrm{L}_{\mathrm{crsf}}$ one-by-one and show that each one has a corresponding symbol in $\mathrm{L}_{\mathrm{def}}$ (see also Section 5.5 below). Notice that this gives us more than just that every CRSF function f is $\Sigma_1^{\preccurlyeq}(\mathrm{L}_0)$-definable in $\mathrm{KP}_1^{\mathrm{u}}$. In particular, we have that the witness v is unique, which we will use later in Corollary 6.17. Put differently, the value $f(\vec{x})$ is $\Delta_0(\mathrm{L}_0)$-definable from a set v which is $\Delta_0(\mathrm{L}_0)$-definable from the arguments $\vec{x}$.

A first step in the proof of Theorem 5.8 is to show that we can treat the language $\mathrm{L}_{\mathrm{def}}$ uniformly, in that every function symbol in it has a good definition.

Lemma 5.9: *For every* $f(\vec{x})$ *in* L_0^+ *there exists a good definition* $(\varphi(v,\vec{x}), \varepsilon(u,v,\vec{x}), e(v), t(\vec{x}))$ *such that* $\mathrm{KP}_1^{\mathrm{u}+}$ *proves (5.1).*

Proof: For $\varphi(v,\vec{x})$ choose $v = f(\vec{x})$, for $e(v)$ choose v, and for ε and t use Lemma 4.10. □

It is straightforward to show part (a) of Theorem 5.8, that $\mathrm{L}_{\mathrm{def}}$ is closed under composition. In this proof, and in the rest of the section, we will make frequent appeals to the Monotonocity Lemma 4.8 and will simply say "by monotonicity".

Theorem 5.10: *For all n-ary function symbols $h(x_1, \ldots, x_n)$ in $\mathrm{L_{def}}$ and m-ary function symbols $g_i(y_1, \ldots, y_m)$ for $i = 1, \ldots, n$ in $\mathrm{L_{def}}$ there is an m-ary function symbol $f(\vec{y})$ in $\mathrm{L_{def}}$ such that $\mathrm{KP}_1^{\mathrm{u}} + \mathrm{L_{def}}$ proves*

$$f(\vec{y}) = h(g_1(\vec{y}), \ldots, g_n(\vec{y})).$$

Proof: For notational simplicity, assume $n = 1$. Let $h(x)$ and $g(\vec{y})$ be function symbols in $\mathrm{L_{def}}$ with good definitions $(\varphi_h, \varepsilon_h, e_h, t_h)$ and $(\varphi_g, \varepsilon_g, e_g, t_g)$. Set

$$\begin{aligned}
\psi(v, v_g, v_h, \vec{y}) &:= (v = \langle v_h, v_g\rangle \wedge \varphi_g(v_g, \vec{y}) \wedge \varphi_h(v_h, e_g(v_g))), \\
\varphi_f(v, \vec{y}) &:= \exists v_h, v_g \in \mathrm{tc}(v)\, \psi(v, v_g, v_h, \vec{y}), \\
e_f(v) &:= e_h(\pi_1(v)).
\end{aligned}$$

We claim that there are ε_f, t_f such that $(\varphi_f, \varepsilon_f, e_f, t_f)$ is a good definition, i.e., such that $\mathrm{KP}_1^{\mathrm{u+}}$ proves

$$\psi(v, v_g, v_h, \vec{y}) \to \varepsilon_f(\cdot, \cdot, v, \vec{y}) : \langle v_h, v_g\rangle \preccurlyeq t_f(\vec{y}).$$

Argue in $\mathrm{KP}_1^{\mathrm{u+}}$. Assume $\psi(v, v_g, v_h, \vec{y})$. By Lemma 4.10 and monotonicity, we have $e_g(v_g) \preccurlyeq e_g^{\#}(t_g(\vec{y}))$ for some $\#$-term $e_g^{\#}$. By monotonicity, $v_h \preccurlyeq t_h(e_g(v_g))$ implies that $v_h \preccurlyeq t_h(e_g^{\#}(t_g(\vec{y})))$. Using the term t_{pair} from Example 4.2, $t_f(\vec{y}) := t_{\mathrm{pair}}(t_h(e_g^{\#}(t_g(\vec{y}))), t_g(\vec{y}))$ is as desired. It is easy to find a formula ε_f as desired. □

Before proving parts (b) and (c) of Theorem 5.8 we need a technical lemma. We return to the proof of part (b) in Section 5.3.

5.2. *Elimination lemma*

Recall that the axioms of $\mathrm{KP}_1^{\mathrm{u}} + \mathrm{L_{def}}$ do not include the axiom schemes of $\mathrm{KP}_1^{\mathrm{u}}$ in the language $\mathrm{L_{def}}$ but only in the language $\mathrm{L_0}$. However, in order to prove closure under syntactic Cobham recursion, Theorem 5.8 (c), we will need some version of these schemes.

In the usual development of full Kripke Platek set theory KP (e.g., [5, Chapter I]), one shows that Σ_1-expansions prove each scheme for formulas mentioning new symbols from their bigger language L, for example $\Delta_0(L)$-Separation. This is done in two steps. First, one shows that occurrences of new Σ_1-defined symbols can be eliminated in a way that transforms $\Delta_0(L)$-formulas into Δ_1-formulas. Second, one proves Δ_1-Separation in KP. An analogous procedure is employed in bounded arithmetic when developing S^1_2 (cf. [12]).

For our weak theory $\mathrm{KP}^{\mathrm{u}+}_1$ the situation is more subtle. The following lemma gives a version of the elimination step, just good enough for our purposes: it eliminates new function symbols by $\in$-bounding quantifiers, with the help of an auxiliary parameter V. Intuitively, this V is a set collecting enough computations of new functions to evaluate the given formula; it is uniquely determined by a simple formula and weakly uniformly bounded. The precise statement needs the following auxiliary notion.

Definition 5.11: We write $\Delta^+_0(\mathrm{L_{def}})$ for the class of $\Delta_0(\mathrm{L_{def}})$-formulas all of whose $\in$-bounding terms are L^+_0-terms.

Lemma 5.12: (Elimination) *For every* $\varphi(\vec{x}) \in \Delta^+_0(\mathrm{L_{def}})$ *there are* $\Delta_0(\mathrm{L}^+_0)$-*formulas* $\varphi_{\mathrm{equ}}(\vec{x}, V)$, $\varphi_{\mathrm{aux}}(\vec{x}, V)$, $\varphi_{\mathrm{emb}}(z, z', \vec{x}, V)$ *and a* $\#$-*term* $t_\varphi(\vec{x})$ *such that* $\mathrm{KP}^{\mathrm{u}}_1 + \mathrm{L_{def}}$ *proves*

$$\begin{aligned}
&\exists^{\leqslant 1} V\, \varphi_{\mathrm{aux}}(\vec{x}, V)\\
&\exists V\, (\varphi_{\mathrm{aux}}(\vec{x}, V) \wedge \varphi_{\mathrm{emb}}(\cdot, \cdot, \vec{x}, V) : V \preccurlyeq t_\varphi(\vec{x})) \qquad (5.2)\\
&\varphi_{\mathrm{aux}}(\vec{x}, V) \rightarrow (\varphi(\vec{x}) \leftrightarrow \varphi_{\mathrm{equ}}(\vec{x}, V)).
\end{aligned}$$

Proof: This is proved by induction on $\varphi(\vec{x})$. The base case for atomic $\varphi(\vec{x})$ is the most involved and is proved by induction on the number of occurrences of symbols in φ from $\mathrm{L_{def}} \setminus \mathrm{L}^+_0$. If this number is 0, there is not much to be shown. Otherwise one can write

$$\varphi(\vec{x}) = \psi(\vec{x}, f(\vec{s}(\vec{x}))),$$

where $\psi(\vec{x}, y)$ has one fewer occurrence of symbols from $\mathrm{L_{def}} \setminus \mathrm{L}^+_0$, the symbol $f(\vec{z})$ is from $\mathrm{L_{def}} \setminus \mathrm{L}^+_0$, and $\vec{s}(\vec{x})$ is a tuple of L^+_0-terms.

Let $(\varphi_f, \varepsilon_f, e_f, t_f)$ be a good definition of $f(\vec{z})$. By Lemma 4.10 and monotonicity, we have $\#$-terms $e^\#_f(v), \vec{s}^\#(\vec{x})$ and $\Delta_0(\mathrm{L}_0)$-formulas $\varepsilon_0, \varepsilon_1$ such that $\mathrm{KP}^{\mathrm{u}}_1 + \mathrm{L_{def}}$ proves

$$\begin{aligned}
&\varphi_f(v, \vec{s}(\vec{x})) \rightarrow\\
&\varepsilon_0(\cdot, \cdot, v, \vec{x}) : v \preccurlyeq t_f(\vec{s}^\#(\vec{x})) \wedge \varepsilon_1(\cdot, \cdot, v, \vec{x}) : e_f(v) \preccurlyeq e^\#_f(t_f(\vec{s}^\#(\vec{x}))).
\end{aligned}$$

By induction, there are $\psi_{\mathrm{equ}}, \psi_{\mathrm{aux}}, \psi_{\mathrm{emb}}, t_\psi$ such that $\mathrm{KP}^{\mathrm{u}}_1 + \mathrm{L_{def}}$ proves

$$\begin{aligned}
&\exists^{\leqslant 1} W\, \psi_{\mathrm{aux}}(\vec{x}, e_f(v), W)\\
&\exists W\, (\psi_{\mathrm{aux}}(\vec{x}, e_f(v), W) \wedge \psi_{\mathrm{emb}}(\cdot, \cdot, \vec{x}, e_f(v), W) : W \preccurlyeq t_\psi(\vec{x}, e_f(v)))\\
&\psi_{\mathrm{aux}}(\vec{x}, e_f(v), W) \rightarrow (\psi(\vec{x}, e_f(v)) \leftrightarrow \psi_{\mathrm{equ}}(\vec{x}, e_f(v), W)).
\end{aligned}$$

Define $\varphi_{\mathrm{aux}}(\vec{x}, V) := \exists W, v \in \mathrm{tc}(V)\,(\chi(\vec{x}, V, W, v))$ where

$$\chi(\vec{x}, V, W, v) := (V = \langle W, v\rangle \wedge \psi_{\mathrm{aux}}(\vec{x}, e_f(v), W) \wedge \varphi_f(v, \vec{s}(\vec{x}))).$$

Monotonicity lets us construct from ψ_{emb} a $\Delta_0(\mathrm{L}_0^+)$-formula ε_2 such that $\mathrm{KP}_1^{\mathrm{u}} + \mathrm{L}_{\mathrm{def}}$ proves

$$\begin{aligned}&\chi(\vec{x}, V, W, v) \to \\ &\varepsilon_2(\cdot,\cdot,\vec{x}, V) : W \preccurlyeq t_\psi(\vec{x}, e_f^{\#}(v)) \wedge \varepsilon_0(\cdot,\cdot,\pi_2(V), \vec{x}) : v \preccurlyeq t_f(\vec{s}^{\#}(\vec{x})).\end{aligned}$$

Using the term t_{pair} from Example 4.2 we define

$$t_\varphi(\vec{x}) := t_{\mathrm{pair}}(t_\psi(\vec{x}, e_f^{\#}(v)), t_f(\vec{s}^{\#}(\vec{x})))$$

and get a $\Delta_0(\mathrm{L}_0^+)$-formula φ_{emb} such that $\mathrm{KP}_1^{\mathrm{u}} + \mathrm{L}_{\mathrm{def}}$ proves

$$\chi(\vec{x}, V, W, v) \to \varphi_{\mathrm{emb}}(\cdot,\cdot,\vec{x}, V) : V \preccurlyeq t_\varphi(\vec{x}).$$

Finally, we set

$$\varphi_{\mathrm{equ}}(\vec{x}, V) := \psi_{\mathrm{equ}}(\vec{x}, e_f(\pi_2(V)), \pi_1(V)).$$

It is easy to verify (5.2). This completes the proof for the case that $\varphi(\vec{x})$ is atomic.

The induction step is easy if $\varphi(\vec{x})$ is a negation or a conjunction. We consider the case that $\varphi(\vec{x}) = \forall u{\in}s(\vec{x})\,\psi(u, \vec{x})$ for some L_0^+-term $s(\vec{x})$. By induction, there are $\psi_{\mathrm{equ}}, \psi_{\mathrm{aux}}, \psi_{\mathrm{emb}}, t_\psi$ such that $\mathrm{KP}_1^{\mathrm{u}} + \mathrm{L}_{\mathrm{def}}$ proves

$$\begin{aligned}&\forall u{\in}s(\vec{x})\,\exists^{\leqslant 1} W\,\psi_{\mathrm{aux}}(u, \vec{x}, W)\\ &\forall u{\in}s(\vec{x})\,\exists W\,(\psi_{\mathrm{aux}}(u, \vec{x}, W) \wedge \psi_{\mathrm{emb}}(\cdot,\cdot,u,\vec{x}, W) : W \preccurlyeq t_\psi(u, \vec{x}))\\ &\forall u{\in}s(\vec{x})\,\forall W\,(\psi_{\mathrm{aux}}(u, \vec{x}, W) \to (\psi(u, \vec{x}) \leftrightarrow \psi_{\mathrm{equ}}(u, \vec{x}, W))).\end{aligned}$$

By monotonicity and Lemma 4.10 there is a #-term $s^{\#}(\vec{x})$ such that T_0^+ defines a $\Delta_0(\mathrm{L}_0)$-embedding of u into $s^{\#}(\vec{x})$ when $u \in s(\vec{x})$, and hence without loss of generality we can replace the bound $t_\psi(u, \vec{x})$ above with $t_\psi(s^{\#}(\vec{x}), \vec{x})$. By $\Delta_0(\mathrm{L}_0)$-Collection, $\mathrm{KP}_1^{\mathrm{u}} + \mathrm{L}_{\mathrm{def}}$ proves that the set

$$V = \{W : \exists u \in s(\vec{x})\,\psi_{\mathrm{aux}}(u, \vec{x}, W)\} \tag{5.3}$$

exists, and by Lemma 4.11 it also $\Delta_0(\mathrm{L}_0)$-defines an embedding of V into some #-term $t(\vec{x})$. For $\varphi_{\mathrm{emb}}(z, z', \vec{x}, V)$ we choose a formula describing this embedding and we set $t_\varphi(\vec{x}) = t(\vec{x})$. Define $\varphi_{\mathrm{aux}}(\vec{x}, V)$ to be a $\Delta_0(\mathrm{L}_0^+)$-formula expressing (5.3); this is $\Delta_0(\mathrm{L}_0^+)$ because witnesses in ψ_{aux} are unique – recall the discussion of "collection terms" in Section 3.3. Define $\varphi_{\mathrm{equ}}(\vec{x}, V)$ to be the $\Delta_0(\mathrm{L}_0^+)$-formula

$$\forall u{\in}s(\vec{x})\,\exists W \in V\,(\psi_{\mathrm{equ}}(u, \vec{x}, W) \wedge \psi_{\mathrm{aux}}(u, \vec{x}, W)).$$

It is straightforward to verify (5.2) in $\mathrm{KP}_1^{\mathrm{u}} + \mathrm{L}_{\mathrm{def}}$. □

One can bootstrap the Elimination Lemma to yield a bigger auxiliary set V such that $\varphi_{\mathrm{equ}}(\vec{x}, V)$ is equivalent to $\varphi(\vec{x})$ simultaneously for all tuples $\vec{x}$ taken from a given set. As we shall use this stronger version too, we give details. Recall $\vec{x} \in z$ stands for $\bigwedge_i x_i \in z$.

Lemma 5.13: *For every $\varphi(\vec{x}, \vec{y}) \in \Delta_0^+(\mathrm{L}_{\mathrm{def}})$ and every L_0^+-term $s(\vec{y})$ there are $\Delta_0(\mathrm{L}_0^+)$-formulas $\varphi^s_{\mathrm{equ}}(\vec{x}, \vec{y}, U)$, $\varphi^s_{\mathrm{aux}}(\vec{y}, U)$, $\varphi^s_{\mathrm{emb}}(z, z', \vec{y}, U)$ and a #-term $t^s_\varphi(\vec{y})$ such that $\mathrm{KP}_1^{\mathrm{u}} + \mathrm{L}_{\mathrm{def}}$ proves*

$$\exists^{\leqslant 1} U\, \varphi^s_{\mathrm{aux}}(\vec{y}, U)$$
$$\exists U\, (\varphi^s_{\mathrm{aux}}(\vec{y}, U) \wedge \varphi^s_{\mathrm{emb}}(\cdot, \cdot, \vec{y}, U) : U \preccurlyeq t^s_\varphi(\vec{y}))$$
$$\varphi^s_{\mathrm{aux}}(\vec{y}, U) \to \forall \vec{x} {\in} s(\vec{y})\, (\varphi(\vec{x}, \vec{y}) \leftrightarrow \varphi^s_{\mathrm{equ}}(\vec{x}, \vec{y}, U)).$$

Proof: Using Lemma 5.12, choose φ_{equ}, φ_{aux}, φ_{emb} and t_φ for which $\mathrm{KP}_1^{\mathrm{u}} + \mathrm{L}_{\mathrm{def}}$ proves

$$\forall \vec{x} {\in} s(\vec{y})\, \exists^{\leqslant 1} V\, \varphi_{\mathrm{aux}}(\vec{x}, \vec{y}, V)$$
$$\forall \vec{x} {\in} s(\vec{y})\, \exists V\, (\varphi_{\mathrm{aux}}(\vec{x}, \vec{y}, V) \wedge \varphi_{\mathrm{emb}}(\cdot, \cdot, \vec{x}, \vec{y}, V) : V \preccurlyeq t_\varphi(\vec{x}, \vec{y}))$$
$$\forall \vec{x} {\in} s(\vec{y})\, \big(\varphi_{\mathrm{aux}}(\vec{x}, \vec{y}, V) \to (\varphi(\vec{x}, \vec{y}) \leftrightarrow \varphi_{\mathrm{equ}}(\vec{x}, \vec{y}, V))\big).$$

From the first and second lines, exactly as in the universal quantification step in the proof of Lemma 5.12, we get a $\Delta_0(\mathrm{L}_0^+)$-formula $\varphi^s_{\mathrm{emb}}(z, z', \vec{y}, U)$ and a #-term $t_\varphi(\vec{y})$ such that $\mathrm{KP}_1^{\mathrm{u}} + \mathrm{L}_{\mathrm{def}}$ proves the existence of a set U with

$$U = \{V : \exists \vec{x} \in s(\vec{y})\, \varphi_{\mathrm{aux}}(\vec{x}, \vec{y}, V)\} \ \text{ and } \ \varphi^s_{\mathrm{emb}}(\cdot, \cdot, \vec{y}, U) : U \preccurlyeq t_\varphi(\vec{y}). \quad (5.4)$$

For φ^s_{aux} take a $\Delta_0(\mathrm{L}_0^+)$-formula expressing the first conjunct of (5.4), and for φ^s_{equ} take the $\Delta_0(\mathrm{L}_0^+)$-formula $\exists V \in U\, (\varphi_{\mathrm{aux}}(\vec{x}, \vec{y}, V) \wedge \varphi_{\mathrm{equ}}(\vec{x}, \vec{y}, V))$. □

As a simple application of Lemma 5.12 we derive a separation scheme.

Corollary 5.14: *The theory $\mathrm{KP}_1^{\mathrm{u}} + \mathrm{L}_{\mathrm{def}}$ proves $\Delta_0^+(\mathrm{L}_{\mathrm{def}})$-Separation.*

Proof: Let $\varphi(u, x, \vec{w})$ be a $\Delta_0^+(\mathrm{L}_{\mathrm{def}})$-formula. We want to show that $\{u \in x : \varphi(u, x, \vec{w})\}$ exists. Choose $\varphi^x_{\mathrm{aux}}, \varphi^x_{\mathrm{equ}}$ according to the previous lemma, substituting $x, \vec{w}$ for $\vec{y}$, substituting u for $\vec{x}$, and substituting x for $s(\vec{y})$. Choose U such that $\varphi^x_{\mathrm{aux}}(x, \vec{w}, U)$. Then the set $\{u \in x : \varphi(u, x, \vec{w})\}$ equals $\{u \in x : \varphi^x_{\mathrm{equ}}(u, x, \vec{w}, U)\}$, so exists by $\Delta_0(\mathrm{L}_0^+)$-Separation. □

5.3. *Closure under replacement*

The following theorem is crucial. It provides a formalized version of Theorem 2.6(a) showing, more generally, that $\mathrm{KP}_1^{\mathrm{u}} + \mathrm{L}_{\mathrm{def}}$ can handle comprehension terms coming from Replacement. Similar terms are basic computation steps in Sazonov's term calculus [23] and in the logic of Blass et al. [10]. Recall that $\vec{x} \in u$ stands for $\bigwedge_i x_i \in u$.

Theorem 5.15: *Let $\theta(u, \vec{y}, \vec{x})$ be a $\Delta_0^+(\mathrm{L}_{\mathrm{def}})$-formula and $g(\vec{y}, \vec{x})$ a function symbol in $\mathrm{L}_{\mathrm{def}}$. Then there exists a function symbol $f(u, \vec{y})$ in $\mathrm{L}_{\mathrm{def}}$ such that $\mathrm{KP}_1^{\mathrm{u}} + \mathrm{L}_{\mathrm{def}}$ proves*

$$f(u, \vec{y}) = \{g(\vec{y}, \vec{x}) : \theta(u, \vec{y}, \vec{x}) \wedge \vec{x} \in u\}.$$

Proof: For notational simplicity we assume $\vec{y}$ is the empty tuple. It is sufficient to prove the theorem for g such that $\mathrm{KP}_1^{\mathrm{u}} + \mathrm{L}_{\mathrm{def}}$ proves $g(\vec{x}) \neq 0$. We first show that $\mathrm{KP}_1^{\mathrm{u}} + \mathrm{L}_{\mathrm{def}}$ proves the existence of

$$z := \{g(\vec{x}) : \theta(u, \vec{x}) \wedge \vec{x} \in u\}$$

and furthermore describes an embedding of z into $t_1(u)$ for a suitable #-term t_1.

Let $(\varphi_g, \varepsilon_g, e_g, t_g)$ be a good definition of g and choose $\theta_{\mathrm{equ}}, \theta_{\mathrm{aux}}, \theta_{\mathrm{emb}}, t_\theta$ for θ according to the Elimination Lemma 5.12. Argue in $\mathrm{KP}_1^{\mathrm{u}} + \mathrm{L}_{\mathrm{def}}$. For every $\vec{x} \in u$ there exists a unique w such that

$$\exists y, v_g, V \in \mathrm{tc}(w)\ \psi(w, y, v_g, V, u, \vec{x}),$$

where $\psi(w, y, v_g, V, u, \vec{x})$ expresses that $w = \langle\langle y, v_g\rangle, V\rangle$ where either $(y = g(\vec{x}) \wedge \theta(u, \vec{x}))$ or $(y = 0 \wedge \neg\theta(u, \vec{x}))$, and the computations of g and θ are witnessed by v_g and V. Formally, $\psi(w, y, v_g, V, u, \vec{x})$ is the following $\Delta_0(\mathrm{L}_0^+)$-formula:

$$\begin{aligned} w = \langle\langle y, v_g\rangle, V\rangle \wedge \theta_{\mathrm{aux}}(u, \vec{x}, V) \wedge \varphi_g(v_g, \vec{x}) \\ \wedge\ \big((y = e_g(v_g) \wedge \theta_{\mathrm{equ}}(u, \vec{x}, V)) \vee (y = 0 \wedge \neg\theta_{\mathrm{equ}}(u, \vec{x}, V))\big). \end{aligned}$$

As in the proof of Lemma 5.12, from θ_{emb} and ε_g we can construct a $\Delta_0(\mathrm{L}_0)$-formula ε and #-term $t_2(u, \vec{x})$ such that $\varepsilon(\cdot, \cdot, w, u, \vec{x}) : w \preccurlyeq t_2(u, \vec{x})$ for this w. By Collection the set

$$W = \{w : \exists \vec{x} \in u\, \exists y, v_g, V \in \mathrm{tc}(w)\ \psi(w, y, v_g, V, u, \vec{x})\},$$

exists, and by Lemma 4.11 we have $\varepsilon'(\cdot, \cdot, W, u) : W \preccurlyeq t_3(u)$ for a suitable $\Delta_0(\mathrm{L}_0)$-formula ε' and #-term $t_3(u)$. The definition of W above can be expressed by the $\Delta_0(\mathrm{L}_0^+)$-formula (as discussed in Section 3.3)

$$\forall w{\in}W\, \exists \vec{x} \in u\, \exists y, v_g, V \in \mathrm{tc}(w)\ \psi \ \wedge\ \forall \vec{x}{\in}u\, \exists w \in W\, \exists y, v_g, V \in \mathrm{tc}(w)\ \psi. \tag{5.5}$$

We see that z exists by $\Delta_0(\mathrm{L_O})$-Separation:

$$z = \{y \in \mathrm{tc}(W) : \exists w \in W\, (y = \pi_1(\pi_1(w)) \wedge y \neq 0)\}. \tag{5.6}$$

Note $\varepsilon'(\cdot,\cdot,W,u) : z \preccurlyeq t_3(u)$ since z is a subset of $\mathrm{tc}(W)$. Recalling t_{pair} from Example 4.2, we construct a good definition $(\theta_f, \varepsilon_f, e_f, t_f)$ of $f(u)$:

$$\begin{aligned}\theta_f(v,u) &:= \exists W, z \in \mathrm{tc}(v)\, (v = \langle W, z\rangle \wedge (5.5) \text{ and } (5.6) \text{ hold})\\ e_f(v) &:= \pi_2(v)\\ t_f(u) &:= t_{\mathrm{pair}}(t_3(u), t_3(u)),\end{aligned}$$

and ε_f such that $\varepsilon_f(\cdot,\cdot,v,u) : v \preccurlyeq t_f(u)$ for the unique v with $\theta_f(v,u)$. □

We can now show that, in $\mathrm{KP}_1^{\mathrm{u}} + \mathrm{L_{def}}$, weakly uniform embeddings (given by $\Delta_0(\mathrm{L_O})$-formulas) and strongly uniform embeddings (given by function symbols) are closely related. For suppose we are given a $\Delta_0(\mathrm{L_O})$-embedding $\varepsilon(\cdot,\cdot,\vec{x}) : s(\vec{x}) \preccurlyeq t(\vec{x})$. Then a function τ satisfying $\tau(z,\vec{x}) = \{z' \in \mathrm{tc}(t(\vec{x})) : \varepsilon(z,z',\vec{x})\}$ is in $\mathrm{L_{def}}$ by Theorem 5.15, and we have $\tau(\cdot,\vec{x}) : s(\vec{x}) \preccurlyeq t(\vec{x})$. On the other hand, suppose $\tau \in \mathrm{L_{def}}$ and $\tau(\cdot,\vec{x}) : s(\vec{x}) \preccurlyeq t(\vec{x})$. If we define $\varepsilon(z,z',\vec{x})$ as $z' \in \tau(z,\vec{x})$, then $\varepsilon(\cdot,\cdot,\vec{x}) : s(\vec{x}) \preccurlyeq t(\vec{x})$. The embedding ε is $\Delta_0^+(\mathrm{L_{def}})$ rather than $\Delta_0(\mathrm{L_O})$, but using the Elimination Lemma 5.12 we can find an equivalent $\Delta_0(\mathrm{L_O})$-embedding, at the cost of involving a unique, bounded parameter V.

We will use constructions like this in the next subsection, where we need to show, using induction with bounds given by weakly uniform embeddings, that $\mathrm{L_{def}}$ is closed under Cobham recursion where the bound is given by a strongly uniform embedding.

Example 5.16: Let $f(x,\vec{w})$ be a function symbol in $\mathrm{L_{def}}$. Then $\mathrm{L_{def}}$ contains a function symbol $f"(x,\vec{w})$ such that $\mathrm{KP}_1^{\mathrm{u}} + \mathrm{L_{def}}$ proves

$$f"(x,\vec{w}) = \{f(u,\vec{w}) : u \in x\}.$$

Furthermore, $\mathrm{L_{def}}$ contains the function symbols $x \cap y$ and $x \setminus y$ and $\mathrm{KP}_1^{\mathrm{u}} + \mathrm{L_{def}}$ proves the usual defining axioms for them.

5.4. *Closure under syntactic Cobham recursion*

We are ready to verify statement (c) of Theorem 5.8, that $\mathrm{L_{def}}$ is closed under syntactic Cobham recursion.

Theorem 5.17: *For all function symbols $g(x, z, \vec{w})$ and $\tau(u, v, x, \vec{w})$ in* $\mathrm{L_{def}}$ *and all #-terms $t(x, \vec{w})$ there is a function symbol $f(x, \vec{w})$ in* $\mathrm{L_{def}}$ *such that* $\mathrm{KP}_1^{\mathrm{u}} + \mathrm{L_{def}}$ *proves*

$$f(x, \vec{w}) = \begin{cases} g(x, f\text{''}(x, \vec{w}), \vec{w}) & \text{if } \tau \text{ is an embedding into } t \text{ at } x, \vec{w} \\ 0 & \text{otherwise,} \end{cases} \tag{5.7}$$

where "τ is an embedding into t at $x, \vec{w}$" stands for the $\Delta_0(\mathrm{L_{def}})$*-formula*

$$\tau(\cdot, g(x, f\text{''}(x, \vec{w}), \vec{w}), x, \vec{w}) \,:\, g(x, f\text{''}(x, \vec{w}), \vec{w}) \preccurlyeq t(x, \vec{w}).$$

Proof: Let g, τ, t be as stated. For notational simplicity we assume $\vec{w}$ is the empty tuple. We are looking for a good definition $(\varphi_f, \varepsilon_f, e_f, t_f)$ of the function $f(x)$, that is, for a good definition for which $\mathrm{KP}_1^{\mathrm{u}} + \mathrm{L_{def}}$ proves (5.7) for the associated function symbol $f(x)$ in $\mathrm{L_{def}}$.

We intend to let $\varphi_f(v, x)$ say that v encodes the course of values of f, namely the set of all pairs $\langle u, f(u)\rangle, u \in \mathrm{tc}^+(x)$. More precisely, we will express this by writing a $\Delta_0^+(\mathrm{L_{def}})$-formula $\psi(w, x)$ which asserts that the values in a sequence w are recursively computed by g, and then applying the Elimination Lemma 5.12 to get the required $\Delta_0(\mathrm{L}_0^+)$-formula $\varphi_f(v, x)$. Hence the witness v will consist of w plus some parameters needed for the elimination of $\mathrm{L_{def}}$-symbols.

By Theorem 5.15 there is a binary function symbol $w\text{''}y$ in $\mathrm{L_{def}}$ such that $\mathrm{KP}_1^{\mathrm{u}} + \mathrm{L_{def}}$ proves $w\text{''}y = \{w\text{'}z : z \in y\}$. We define an auxiliary formula

$$\xi(w, y) := \tau(\cdot, g(y, w\text{''}y), y) : g(y, w\text{''}y) \preccurlyeq t(y).$$

We then let $\psi(w, x)$ express that w is a function with domain $\mathrm{tc}^+(x)$ such that

$$\forall y \in \mathrm{tc}^+(x) \big((\xi(w, y) \wedge w\text{'}y = g(y, w\text{''}y)) \ \vee \ (\neg\xi(w, y) \wedge w\text{'}y = 0) \big).$$

Claim 1. There is a $\Delta_0^+(\mathrm{L_{def}})$-formula δ and a #-term s such that $\mathrm{KP}_1^{\mathrm{u}} + \mathrm{L_{def}}$ proves

$$\psi(w, x) \to \delta(\cdot, \cdot, w, x) : w \preccurlyeq s(x).$$

Proof of Claim 1. Argue in $\mathrm{KP}_1^{\mathrm{u}} + \mathrm{L_{def}}$. Suppose that $\psi(w, x)$ holds. Then for all $y \in \mathrm{tc}^+(x)$ we have $\tau(\cdot, w\text{'}y, y) : w\text{'}y \preccurlyeq t(y)$. Let $\delta_0(z, z', w, y)$ be the formula $z' \in \tau(z, w\text{'}y, y)$. Then we have a weakly uniform $\Delta_0^+(\mathrm{L_{def}})$-embedding $\delta_0(\cdot, \cdot, w, y) : w\text{'}y \preccurlyeq t(y)$ for all $y \in \mathrm{tc}^+(x)$, and by the proof of the Monotonicity Lemma 4.8, adapted for $\Delta_0^+(\mathrm{L_{def}})$-formulas, we can construct a $\Delta_0^+(\mathrm{L_{def}})$-embedding $\delta_1(\cdot, \cdot, w, y, x) : w\text{'}y \preccurlyeq t(x)$.

Using t_{pair} from Example 4.2 we can find $\delta_2 \in \Delta_0^+(\mathrm{L_{def}})$ and a #-term t' such that

$$\delta_2(\cdot,\cdot,w,y,x) : \langle y, w\text{'}y\rangle \preccurlyeq t'(x)$$

for $y \in \mathrm{tc}^+(x)$. Writing $\varphi(v,y,w)$ for the formula $v = \langle y, w\text{'}y\rangle$, we have $\exists! v\, \varphi(v,y,w)$ and an embedding of v for each y, so can apply Lemma 4.11, adapted for $\Delta_0^+(\mathrm{L_{def}})$-embeddings, to combine these into a single embedding of w into $t''(\mathrm{tc}^+(x), t'(x))$ for a #-term t''. As usual by Lemma 4.10 and monotonicity we can replace this bound with a #-term $s(x)$. ⊣

We can begin to construct a good definition of f. Since ψ is a $\Delta_0^+(\mathrm{L_{def}})$-formula there exist $\Delta_0(\mathrm{L}_0^+)$-formulas $\psi_{\mathrm{equ}}(w,x,V)$, $\psi_{\mathrm{aux}}(w,x,V)$, $\psi_{\mathrm{emb}}(z,z',w,x,V)$ and a #-term $t_\psi(w,x)$ satisfying the Elimination Lemma 5.12 for ψ. Choose δ and s to satisfy Claim 1. By Lemma 5.13, since δ is a $\Delta_0^+(\mathrm{L_{def}})$-formula there exist $\Delta_0(\mathrm{L}_0^+)$-formulas $\delta^r_{\mathrm{equ}}(z,z',w,x,U)$, $\delta^r_{\mathrm{aux}}(w,x,U)$, $\delta^r_{\mathrm{emb}}(z,z',w,x,U)$ and a #-term $t^r_\delta(w,x)$ such that $\mathrm{KP}_1^{\mathrm{u}} + \mathrm{L_{def}}$ proves

$$\begin{aligned}
&\exists^{\leqslant 1} U\, \delta^r_{\mathrm{aux}}(w,x,U)\\
&\exists U\, (\delta^r_{\mathrm{aux}}(w,x,U) \wedge \delta^r_{\mathrm{emb}}(\cdot,\cdot,w,x,U) : U \preccurlyeq t^r_\delta(w,x))\\
&\delta^r_{\mathrm{aux}}(w,x,U) \to \forall z,z' \in r(w,x)\, (\delta(z,z',w,x) \leftrightarrow \delta^r_{\mathrm{equ}}(z,z',w,x,U)),
\end{aligned}$$

where $r(w,x)$ is the term $\mathrm{tc}(\{w, s(x)\})$. From the third line it follows that

$$\delta^r_{\mathrm{aux}}(w,x,U) \to (\delta(\cdot,\cdot,w,x) : w \preccurlyeq s(x) \leftrightarrow \delta^r_{\mathrm{equ}}(\cdot,\cdot,w,x,U) : w \preccurlyeq s(x)).$$

Now define

$$\begin{aligned}
\varphi_f(v,x) :=\ &\exists w, V, U \in \mathrm{tc}(v)\\
&\quad (v = \langle w, \langle V, U\rangle\rangle \wedge \psi_{\mathrm{aux}}(w,x,V) \wedge \psi_{\mathrm{equ}}(w,x,V) \wedge \delta^r_{\mathrm{aux}}(w,x,U))\\
e_f(v) :=\ &\pi_1(v)\text{'}top(\pi_1(v)),
\end{aligned}$$

where $top(w)$ is an L_0^+-term that recovers x from w, as the unique member x' of the domain of w such that $\mathrm{tc}(x')$ contains all other members of the domain of w.

We obtain ε_f and t_f from the following claim. Recall that the properties of a good definition of an $\mathrm{L_{def}}$ symbol must be provable in $\mathrm{KP}_1^{\mathrm{u}+}$.

Claim 2. There is a $\Delta_0(\mathrm{L}_0^+)$-formula ε_f and a #-term t_f such that $\mathrm{KP}_1^{\mathrm{u}+}$ proves

$$\varphi_f(v,x) \to \varepsilon_f(\cdot,\cdot,v,x) : v \preccurlyeq t_f(x).$$

Proof of Claim 2. By conservativity we can argue in $\mathrm{KP}_1^{\mathrm{u}} + \mathrm{L}_{\mathrm{def}}$. Assume $\varphi_f(v,x)$ and write $v = \langle w, \langle V, U\rangle\rangle$. We have $\Delta_0(\mathrm{L}_0^+)$-formulas ψ_{emb}, δ^r_{emb} and δ^r_{equ} such that

- $\psi_{\mathrm{emb}}(\cdot,\cdot,w,x,V) : V \preccurlyeq t_\psi(w,x)$,
- $\delta^r_{\mathrm{emb}}(\cdot,\cdot,w,x,U) : U \preccurlyeq t^r_\delta(w,x)$,
- $\delta^r_{\mathrm{equ}}(\cdot,\cdot,w,x,U) : w \preccurlyeq s(x)$,

where $t_\psi(w,x)$, $t^r_\delta(w,x)$ and $s(x)$ are #-terms. Define

$$t_f(x) := t_{\mathrm{pair}}\Big(s(x),\ t_{\mathrm{pair}}\big(t_\psi(s(x),x), t^r_\delta(s(x),x)\big)\Big)$$

and use monotonicity to get a $\Delta_0(\mathrm{L}_0^+)$-formula $\varepsilon_f(z,z',v,x)$ describing an embedding of v into $t_f(x)$. ⊣

We must show that $(\varphi_f, \varepsilon_f, e_f, t_f)$ is a good definition. Claim 2 gives (Witness Embedding) and the next two claims show (Witness Uniqueness) and (Witness Existence).

Claim 3. The tuple $(\varphi_f, \varepsilon_f, e_f, t_f)$ satisfies (Witness Uniqueness).

Proof of Claim 3. It suffices to prove in $\mathrm{KP}_1^{\mathrm{u}} + \mathrm{L}_{\mathrm{def}}$ that

$$\psi(w,x) \wedge \psi(\tilde{w},x) \to w = \tilde{w}$$

since uniqueness of V and U is then guaranteed by ψ_{aux} and δ^r_{aux}. So suppose $\psi(w,x)$, $\psi(\tilde{w},x)$ and $w \neq \tilde{w}$. Then the set $\{y \in \mathrm{tc}^+(x) : w\text{'}y \neq \tilde{w}\text{'}y\}$ is nonempty. By set foundation it contains an $\in$-minimal element y_0. Then $w\text{"}y_0 = \tilde{w}\text{"}y_0$ since $w, \tilde{w}$ both have domain $\mathrm{tc}^+(x) \supseteq y_0$. It follows that $g(y_0, w\text{"}y_0) = g(y_0, \tilde{w}\text{"}y_0)$ and $\xi(w,y_0) \leftrightarrow \xi(\tilde{w},y_0)$. Since $\psi(w,x)$ and $\psi(\tilde{w},x)$, we get $w\text{'}y_0 = \tilde{w}\text{'}y_0$, a contradiction. ⊣

The proof of the next claim is the only place where we use the full strength of induction available in $\mathrm{KP}_1^{\mathrm{u}}$.

Claim 4. The tuple $(\varphi_f, \varepsilon_f, e_f, t_f)$ satisfies (Witness Existence).

Proof of Claim 4. Again we will work in $\mathrm{KP}_1^{\mathrm{u}} + \mathrm{L}_{\mathrm{def}}$ and appeal to conservativity. We will use uniformly bounded unique $\Sigma_1^{\preccurlyeq}(\mathrm{L}_0)$-Induction to prove $\exists v\, \varphi_f(v,x)$. We already know by (Witness Uniqueness) that $\exists^{\leqslant 1} v\, \varphi_f(v,x)$. Furthermore by Claim 2, the witness v is automatically uniformly bounded by the embedding ε_f. It thus suffices to show

$$\forall y {\in} x\, \exists v\, \varphi_f(v,y) \to \exists v\, \varphi_f(v,x).$$

Suppose the antecedent holds. By $\Delta_0(\mathrm{L}_0)$-Collection and $\Delta_0(\mathrm{L}_0^+)$-Separation the set

$$W := \{\pi_1(v) : \exists y \in x\, \varphi_f(v, y)\}$$

exists. For each $y \in x$ this contains exactly one w_y such that $\psi(w_y, y)$, that is, such that w_y is a function with domain $\mathrm{tc}^+(y)$ which recursively applies g. By the same argument as in the proof of Claim 3, any two such functions agree on arguments where they are both defined. Hence, $w := \bigcup W$ is a function with domain $\mathrm{tc}(x)$, and we put $w' := w \cup \{\langle x, y\rangle\}$ where $y = g(x, w\text{"}x)$ if $\xi(x, w)$, and $y = 0$ otherwise. Then $\psi(w', x)$ holds. Furthermore $\mathrm{KP}_1^{\mathrm{u}} + \mathrm{L}_{\mathrm{def}}$ proves that there exist U and V such that $\psi_{\mathrm{aux}}(w', x, V)$ and $\delta^{r}_{\mathrm{aux}}(w', x, U)$. This yields $\varphi_f(v, x)$ for $v = \langle w', \langle V, U\rangle\rangle$. ⊣

We have shown $(\varphi_f, \varepsilon_f, e_f, t_f)$ is a good definition. Let f be the symbol in $\mathrm{L}_{\mathrm{def}}$ associated to this definition. To conclude the proof we verify the conclusion of the theorem, that is, that $\mathrm{KP}_1^{\mathrm{u}} + \mathrm{L}_{\mathrm{def}}$ proves

$$f(x) = \begin{cases} g(x, f\text{"}(x)) & \text{if } \tau(\cdot, g(x, f\text{"}(x)), x) : g(x, f\text{"}(x)) \preccurlyeq t(x) \\ 0 & \text{otherwise.} \end{cases}$$

Argue in $\mathrm{KP}_1^{\mathrm{u}} + \mathrm{L}_{\mathrm{def}}$. The witness v for $f(x)$ has the form $\langle w, \langle V, U\rangle\rangle$ such that $\psi(w, x)$ and $f(x) = e_f(v) = w\text{'}x$. From $\psi(w, x)$ we get

$$w\text{'}x = \begin{cases} g(x, w\text{"}x) & \text{if } \tau(\cdot, g(x, w\text{"}x), x) : g(x, w\text{"}x) \preccurlyeq t(x) \\ 0 & \text{otherwise.} \end{cases}$$

It now suffices to verify $f\text{"}(x) = w\text{"}x$. This follows from $f(y) = w\text{'}y$ for every $y \in \mathrm{tc}^+(x)$ which is seen similarly as in the proofs of Claims 3 and 4. □

This completes the proof of Theorem 5.8 and thus of Theorem 5.2.

5.5. *The expanded theories*

We fix a fragment of $\mathrm{KP}_1^{\mathrm{u}} + \mathrm{L}_{\mathrm{def}}$ whose language is exactly $\mathrm{L}_{\mathrm{crsf}}$. We show it proves the schemes in the language $\mathrm{L}_{\mathrm{crsf}}$.

Definition 5.18: $\mathrm{KP}_1^{\mathrm{u}}(\mathrm{L}_{\mathrm{crsf}})$ is the theory in the language $\mathrm{L}_{\mathrm{crsf}}$ which consists of $\mathrm{KP}_1^{\mathrm{u}}$ together with, for each function symbol f in $\mathrm{L}_{\mathrm{crsf}} \setminus \mathrm{L}_0$, a defining axiom for f from $\mathrm{KP}_1^{\mathrm{u}} + \mathrm{L}_{\mathrm{def}}$ chosen in such a way that $\mathrm{KP}_1^{\mathrm{u}}(\mathrm{L}_{\mathrm{crsf}})$ proves the defining axiom for f from $\mathrm{T}_{\mathrm{crsf}}$.

Since $\mathrm{KP}_1^{\mathrm{u}}(\mathrm{L}_{\mathrm{crsf}})$ is an expansion by definitions of $\mathrm{KP}_1^{\mathrm{u}}$, we have:

Proposition 5.19: $\mathrm{KP}_1^{\mathrm{u}}(\mathrm{L}_{\mathrm{crsf}})$ *is conservative over* $\mathrm{KP}_1^{\mathrm{u}}$ *and proves* $\mathrm{T}_{\mathrm{crsf}}$.

We will later show that $\mathrm{KP}_1^{\mathrm{u}}(\mathrm{L}_{\mathrm{crsf}})$ is $\Pi_2(\mathrm{L}_{\mathrm{crsf}})$-conservative over $\mathrm{T}_{\mathrm{crsf}}$ (Theorem 6.9). Observe that, by Lemma 3.18, $\mathrm{KP}_1^{\mathrm{u}}(\mathrm{L}_{\mathrm{crsf}})$ proves $\Delta_0(\mathrm{L}_{\mathrm{crsf}})$-Separation.

Lemma 5.20: *Let the theory T consist of $\mathrm{T}_{\mathrm{crsf}}$ together with the axiom schemes of $\mathrm{KP}_1^{\mathrm{u}}$ expanded to the language $\mathrm{L}_{\mathrm{crsf}}$, that is, the $\Delta_0(\mathrm{L}_{\mathrm{crsf}})$-Collection and uniformly bounded unique $\Sigma_1^{\preccurlyeq}(\mathrm{L}_{\mathrm{crsf}})$-Induction schemes, where the uniform embeddings may be given by $\Delta_0(\mathrm{L}_{\mathrm{crsf}})$-formulas. Then $\mathrm{KP}_1^{\mathrm{u}}(\mathrm{L}_{\mathrm{crsf}})$ is equivalent to T.*

Proof: Consider a model M of T; we must show $M \vDash \mathrm{KP}_1^{\mathrm{u}}(\mathrm{L}_{\mathrm{crsf}})$. The reduct of M to L_0 is a model of $\mathrm{KP}_1^{\mathrm{u}}$, and Definition 5.18 is an extension by definitions. Thus M can be expanded to a model $\tilde{M}$ of $\mathrm{KP}_1^{\mathrm{u}}(\tilde{\mathrm{L}}_{\mathrm{crsf}})$, where $\tilde{\mathrm{L}}_{\mathrm{crsf}}$ is a disjoint copy of $\mathrm{L}_{\mathrm{crsf}}$ containing a function symbol $\tilde{f}$ for every function symbol f in $\mathrm{L}_{\mathrm{crsf}}$. The $\tilde{\mathrm{L}}_{\mathrm{crsf}}$ functions satisfy the defining axioms from $\mathrm{T}_{\mathrm{crsf}}$; we claim that this implies that the $\tilde{\mathrm{L}}_{\mathrm{crsf}}$ functions are identical to the original functions from $\mathrm{L}_{\mathrm{crsf}} \setminus \mathrm{L}_0$ in M, and thus $M \vDash \mathrm{KP}_1^{\mathrm{u}}(\mathrm{L}_{\mathrm{crsf}})$. The claim is immediate in the case of functions defined by composition or replacement, but for recursion we need to appeal to induction in M. Suppose $f = f_{g,\tau,s}$ is defined by syntactic Cobham recursion, where $g = \tilde{g}$ and $\tau = \tilde{\tau}$ in $\tilde{M}$ and s is a #-term. We have in $\tilde{M}$, where for clarity we suppress the side variables $\vec{w}$, both

$$\forall x\, \tilde{f}(x) = \begin{cases} g(x, \tilde{f}\text{"}(x)) & \text{if } \tau \text{ is an embedding into } s \text{ at } x \\ 0 & \text{otherwise} \end{cases}$$

and the same formula with f and f" in place of $\tilde{f}$ and $\tilde{f}$". The function $\tilde{f}$ has a good definition in the sense of Definition 5.6, so there are $\Delta_0(\mathrm{L}_0^+)$-formulas φ, ε, an L_0^+-term e and a #-term t such that for all $x \in \tilde{M}$,

$$\begin{aligned} \tilde{M} \vDash \quad & \exists! v\, \varphi(x, v) \\ & \wedge\ \forall v\, (\varphi(x,v) \to \varepsilon(\cdot,\cdot,x,v) : v \preccurlyeq t(x)) \\ & \wedge\ \big(\tilde{f}(x) = f(x) \leftrightarrow \exists v\, (\varphi(x,v) \wedge e(v) = f(x))\big). \end{aligned}$$

Hence $\tilde{f}(x) = f(x)$ can be expressed as a uniformly bounded $\Sigma_1^{\preccurlyeq}(\mathrm{L}_{\mathrm{crsf}})$-formula, for which witnesses are unique. Therefore we can prove it holds for all x by induction in M, as the induction step follows immediately from the recursive equations for $\tilde{f}$ and f.

For the other direction, suppose $M \vDash \mathrm{KP}_1^{\mathrm{u}}(\mathrm{L}_{\mathrm{crsf}})$. We must show that M satisfies the induction and collection schemes of T.

Suppose $\varphi(x,y)$ and $\varepsilon(z,z',x,y)$ are $\Delta_0(\mathrm{L_{crsf}})$-formulas and $t(x)$ is a $\#$-term, all with parameters from M, and that $M \models \forall x\, \exists^{\le 1} y\, \varphi(x,y)$. Let $\varphi^{\varepsilon,t}(x,y)$ abbreviate the formula $\varphi(x,y) \wedge \varepsilon(\cdot,\cdot,x,y) : y \preccurlyeq t(x)$. We will find $\Delta_0(\mathrm{L}_0^+)$-formulas $\tilde{\varphi}(x,w)$, $\tilde{\varepsilon}'(z,z',x,w)$ and a $\#$-term $\tilde{t}(x)$, with the same, unwritten, parameters, such that

$$M \models \forall x\, \exists^{\le 1} w\, \tilde{\varphi}(x,w) \wedge \forall x\, (\exists y\, \varphi^{\varepsilon,t}(x,y) \leftrightarrow \exists w\, \tilde{\varphi}^{\tilde{\varepsilon},\tilde{t}}(x,w)) \tag{5.8}$$

from which it follows that M satisfies uniformly bounded unique $\Sigma_1^{\preccurlyeq}(\mathrm{L_{crsf}})$-Induction.

Let $\chi_\varepsilon(u,x,y)$ express $u = \{\langle z,z'\rangle \in \mathrm{tc}(y) \times \mathrm{tc}(t(x)) : \varepsilon(z,z',x,y)\}$, which implies in M that $u : y \preccurlyeq t(x) \leftrightarrow \varepsilon(\cdot,\cdot,x,y) : y \preccurlyeq t(x)$. By Lemma 3.18 there is $f \in \mathrm{L_{crsf}}$ such that

$$M \models (\varphi^{\varepsilon,t}(x,y) \wedge \chi_\varepsilon(u,x,y)) \leftrightarrow f(x,y,u) \neq 0.$$

The function f has a good definition in the sense of Definition 5.6. Therefore there are $\Delta_0(\mathrm{L}_0^+)$-formulas ψ, δ, an L_0^+-term e and a $\#$-term s such that for all $x, y \in M$,

$$\begin{aligned} M \models\quad & \exists! v\, \psi(x,y,u,v) \\ & \wedge\ \forall v\, (\psi(x,y,u,v) \rightarrow \delta(\cdot,\cdot,x,y,u,v) : v \preccurlyeq s(x,y,u)) \\ & \wedge\ ((\varphi^{\varepsilon,t}(x,y) \wedge \chi_\varepsilon(u,x,y)) \leftrightarrow \exists v\, (\psi(x,y,u,v) \wedge e(v) \neq 0)). \end{aligned}$$

We can now define

$$\tilde{\varphi}(x,w) := \exists y,u,v \in \mathrm{tc}(w)\, (w = \langle\langle u,y\rangle,v\rangle \wedge \psi(x,y,u,v) \wedge e(v) \neq 0).$$

Then $\tilde{\varphi}$ satisfies the uniqueness condition; furthermore, the right-to-left implication in (5.8) will hold for any choice of $\tilde{\varepsilon}$ and $\tilde{t}$. For the other direction, $\Delta_0(\mathrm{L_{crsf}})$-Separation yields u satisfying $\chi_\varepsilon(u,x,y)$. To construct $\tilde{\varepsilon}$ and $\tilde{t}$ for the embedding, Lemma 4.12 gives $\varepsilon_{\mathrm{emb}} \in \Delta_0(\mathrm{L}_0)$ and a $\#$-term t_{emb} such that $M \models u : y \preccurlyeq t(x) \rightarrow \varepsilon_{\mathrm{emb}}(\cdot,\cdot,u,y,t(x)) : \langle u,y\rangle \preccurlyeq t_{\mathrm{emb}}(t(x))$. Thus, as in the proof of the Elimination Lemma 5.12, using monotonicity and the term t_{pair} we can find a $\Delta_0(\mathrm{L}_0^+)$-formula $\tilde{\varepsilon}$ and a $\#$-term $\tilde{t}$ such that for all $x,y,v \in M$,

$$\begin{aligned} M \models u : y \preccurlyeq t(x) \wedge \delta(\cdot,\cdot,x,y,v) : v \preccurlyeq s(x,y,u) \\ \rightarrow \tilde{\varepsilon}(\cdot,\cdot,x,\langle\langle u,y\rangle,v\rangle) : \langle\langle u,y\rangle,v\rangle \preccurlyeq \tilde{t}(x). \end{aligned}$$

Then $\tilde{\varphi}$, $\tilde{\varepsilon}$ and $\tilde{t}$ satisfy (5.8).

For collection, suppose $M \models \forall x \in u\, \exists y\, \varphi(x,y)$ for $\varphi \in \Delta_0(\mathrm{L_{crsf}})$. Then $\varphi(x,y)$ is equivalent to $f(x,y) \neq 0$ for some $f \in \mathrm{L_{crsf}}$. The good definition of f gives $\psi \in \Delta_0(\mathrm{L}_0^+)$ such that $\varphi(x,y) \leftrightarrow \exists v\, \psi(x,y,v)$ in M for all x, y. By $\Delta_0(\mathrm{L}_0^+)$-Collection there is $W \in M$ such that

$$M \models \forall x{\in}u\, \exists y, v \in W\, \psi(x,y,v).$$

Thus $M \models \forall x{\in}u\, \exists y \in W\, \varphi(x,y)$ as required. □

Lemma 5.21: *Let* $\mathrm{KP}_1^{\preccurlyeq}(\mathrm{L_{crsf}})$ *be the theory* $\mathrm{KP}_1^{\preccurlyeq} + \mathrm{KP}_1^{\mathrm{u}}(\mathrm{L_{crsf}})$. *Then* $\mathrm{KP}_1^{\preccurlyeq}(\mathrm{L_{crsf}})$ *is conservative over* $\mathrm{KP}_1^{\preccurlyeq}$, *and is equivalent to the theory consisting of* $\mathrm{T_{crsf}}$ *plus the* $\Delta_0(\mathrm{L_{crsf}})$*-Collection and* $\Sigma_1^{\preccurlyeq}(\mathrm{L_{crsf}})$*-Induction schemes.*

Proof: By Lemma 5.20 it is sufficient to show that any model M of $\mathrm{KP}_1^{\preccurlyeq}(\mathrm{L_{crsf}})$ satisfies the $\Sigma_1^{\preccurlyeq}(\mathrm{L_{crsf}})$-Induction scheme. By Lemma 4.13 it is enough to show that uniformly bounded $\Sigma_1^{\preccurlyeq}(\mathrm{L_{crsf}})$-Induction holds, and this follows by the same argument as in the proof of Lemma 5.20, ignoring the conditions about the witnesses y and w being unique. □

6. Witnessing

Theorem 5.2 established that every CRSF function is $\Sigma_1^{\preccurlyeq}$-definable in $\mathrm{KP}_1^{\preccurlyeq}$, and in fact already in $\mathrm{KP}_1^{\mathrm{u}}$. We would like to show that every function $\Sigma_1^{\preccurlyeq}$-definable in $\mathrm{KP}_1^{\preccurlyeq}$ is in CRSF. By analogy with bounded arithmetic, one could aim to prove that whenever $\mathrm{KP}_1^{\preccurlyeq} \vdash \exists y\, \varphi(y, \vec{x})$ with $\varphi \in \Delta_0(\mathrm{L_0})$, then $\mathrm{T_{crsf}} \vdash \varphi(f(\vec{x}), \vec{x})$ or at least $\mathrm{ZFC} \vdash \varphi(f(\vec{x}), \vec{x})$ for some "witnessing" function $f(\vec{x})$ in CRSF. As mentioned in the introduction, this fails: a witnessing function $C(x)$ for $(x \neq 0 \to \exists y\, (y \in x))$ would satisfy

$$(x \neq 0 \to C(x) \in x)$$

and not even ZFC can define such a C as a CRSF function.[d] This section shows two ways around this obstacle.

The first is to weaken the conclusion of the witnessing theorem from $\varphi(f(\vec{x}), \vec{x})$ to $\exists y \in f(\vec{x})\, \varphi(y, \vec{x})$. We prove such a witnessing theorem for $\mathrm{KP}_1^{\mathrm{u}}$ (Theorem 6.10), and this has as a corollary the following definability theoretic characterization of CRSF. We do not know whether Theorem 6.10 or Corollary 6.1 hold for $\mathrm{KP}_1^{\preccurlyeq}$ instead of $\mathrm{KP}_1^{\mathrm{u}}$.

Corollary 6.1: *A function is in* CRSF *if and only if it is* $\Sigma_1(\mathrm{L_0})$*-definable in* $\mathrm{KP}_1^{\mathrm{u}}$.

[d]This is well known: otherwise ZFC would define a global well-order and thus prove V = HOD; but V ≠ HOD is relatively consistent (see e.g., [19, p.222]).

The second is to simply add a global choice function C to CRSF as one of the initial functions, resulting in CRSF^C (Remark 2.10). In this way we are able to prove full witnessing, even for the stronger theory $\mathrm{KPC}_1^{\preccurlyeq}$ obtained by adding the axiom of global choice (Theorem 6.10). Again this has as a corollary the following definability theoretic characterization of CRSF^C where we write $\mathrm{L}_0^C := \mathrm{L}_0 \cup \{C\}$.

Corollary 6.2: *A function is in* CRSF^C *if and only if it is* $\Sigma_1(\mathrm{L}_0^C)$*-definable in* $\mathrm{KPC}_1^{\preccurlyeq}$.

We do not know whether some form of witnessing holds for $\mathrm{KP}_1^{\preccurlyeq}$ without choice. In particular, the following question is open: if $\mathrm{KP}_1^{\preccurlyeq} \vdash \exists! y\, \varphi(y, x)$, for φ a $\Delta_0(\mathrm{L}_0)$-formula, does this imply that there is a CRSF function f such that (provably in $\mathrm{KP}_1^{\preccurlyeq}$) $\forall x\, \varphi(f(x), x)$ holds?

It would also be interesting to prove a result of this type that needs only an appropriate form of local choice, rather than global choice. For example: if $\mathrm{KP}_1^{\preccurlyeq} \vdash \exists! y\, \varphi(y, x)$, for φ a $\Delta_0(\mathrm{L}_0)$-formula, does this imply that there is a CRSF function $f(x, r)$ such that (provably in $\mathrm{KP}_1^{\preccurlyeq}$) $\forall x\, \varphi(f(x, r), x)$ holds whenever r is a well-ordering of $\mathrm{tc}(x)$?

6.1. *Witnessing* $\mathrm{T}_{\mathrm{crsf}}$ *and Herbrand saturation*

We use a method introduced by Avigad in [4] as a general tool for model-theoretic proofs of witnessing theorems, in particular subsuming Zambella's witnessing proof for bounded arithmetic [24]. A structure is *Herbrand saturated* if it satisfies every $\exists\forall$ sentence, with parameters, which is consistent with its universal diagram. To get a witnessing theorem for a theory T, one uses Herbrand saturation to show that T is $\forall\exists$-conservative over a suitable universal theory S. Since S is universal, a form of witnessing for S follows directly from Herbrand's theorem; conservativity means that this carries over to T.

We want to use this approach where T is $\mathrm{KP}_1^{\mathrm{u}}$ and S is $\mathrm{T}_{\mathrm{crsf}}$. We cannot do this directly since $\mathrm{T}_{\mathrm{crsf}}$ is not universal but, as $\mathrm{T}_{\mathrm{crsf}}$ is Π_1, it turns out that something similar works. Below we prove a version of Herbrand's theorem for $\mathrm{T}_{\mathrm{crsf}}$, in which a witness to a $\Sigma_1(\mathrm{L}_{\mathrm{crsf}})$ sentence is not necessarily equal to a term, but is always contained in some term.

Theorem 6.3: *Suppose* $\mathrm{T}_{\mathrm{crsf}} \vdash \exists y\, \varphi(y, \vec{x})$ *where* φ *is* $\Delta_0(\mathrm{L}_{\mathrm{crsf}})$. *Then there is an* $\mathrm{L}_{\mathrm{crsf}}$ *function symbol* f *such that* $\mathrm{T}_{\mathrm{crsf}} \vdash \exists y \in f(\vec{x})\, \varphi(y, \vec{x})$.

Proof: Take a new tuple $\vec{c}$ of constants and let $P(\vec{c})$ be the theory

$$\mathrm{T_{crsf}} + \{\forall y \in t(\vec{c})\, \neg\varphi(y,\vec{c}) : t(\vec{x}) \text{ an } \mathrm{L_{crsf}}\text{-term}\}.$$

It suffices to show that $P(\vec{c})$ is inconsistent. Then $\mathrm{T_{crsf}}$ proves $\bigvee_i \exists y \in t_i(\vec{x})\, \varphi(y,\vec{x})$ for finitely many terms $t_1(\vec{x}),\ldots,t_k(\vec{x})$; we can choose $f(\vec{x})$ so that $\mathrm{T_{crsf}}$ proves $f(\vec{x}) = t_1(\vec{x}) \cup \cdots \cup t_k(\vec{x})$ using closure under composition.

For the sake of a contradiction assume $P(\vec{c})$ has a model M. Define

$$N := \{a \in M : M \models a \in t(\vec{c}) \text{ for some } \mathrm{L_{crsf}}\text{-term } t(\vec{x})\}.$$

Note N contains each component c_i^M of $\vec{c}^M$ via the term $\{c_i\}$. We first show that N is a substructure of M. To see this, suppose g is an r-ary function symbol in $\mathrm{L_{crsf}}$ and $\vec{a} \in N^r$. We must show $g(\vec{a}) \in N$. For each component a_i of $\vec{a}$ there is a term $t_i(\vec{x})$ such that $M \models a_i \in t_i(\vec{c})$. Choose a function symbol $G(z)$ in $\mathrm{L_{crsf}}$ such that $\mathrm{T_{crsf}}$ proves $G(z) = g(\pi_1^r(z),\ldots,\pi_r^r(z))$, where π_i^r is the standard projection function for ordered r-tuples (which is in L_0^+). Then in M we have $g(\vec{a}) \in G\text{"}(t_1(\vec{c}) \times \cdots \times t_r(\vec{c}))$.

Next we show that N is a $\Delta_0(\mathrm{L_{crsf}})$-elementary substructure of M, that is, for every $\Delta_0(\mathrm{L_{crsf}})$-formula θ and $\vec{a} \in N$, we have $N \models \theta(\vec{a}) \Leftrightarrow M \models \theta(\vec{a})$. This is proved by induction on θ, and the only non-trivial case is where $\theta(\vec{a})$ has the form $\exists u \in t(\vec{a})\, \psi(u,\vec{a})$ for some term t, and we have $M \models b \in t(\vec{a}) \wedge \psi(b,\vec{a})$ for some b in M. As N is a substructure, $t(\vec{a}) \in N$ and hence $M \models t(\vec{a}) \in s(\vec{c})$ for some term $s(\vec{x})$. Thus $M \models b \in \bigcup s(\vec{c})$, so $b \in N$. By the induction hypothesis $N \models \psi(b,\vec{a})$ which gives $N \models \theta(\vec{a})$ as required.

Thus $N \models \mathrm{T_{crsf}}$ since $\mathrm{T_{crsf}}$ is $\Pi_1(\mathrm{L_{crsf}})$. Further, $N \models \forall y\, (\neg\varphi(y,\vec{c}))$, since in M there is no witness for $\varphi(y,\vec{c})$ inside any term in $\vec{c}$. This contradicts the assumption of the theorem. □

Corollary 6.4: *If* $\mathrm{T_{crsf}} \vdash \exists! y\, \varphi(y,\vec{x})$, *where* φ *is* $\Delta_0(\mathrm{L_{crsf}})$, *then there is an* $\mathrm{L_{crsf}}$ *function symbol* g *such that* $\mathrm{T_{crsf}} \vdash \varphi(g(\vec{x}),\vec{x})$.

Proof: Appealing to Lemma 3.18, take $g(\vec{x})$ computing $\bigcup\{y \in f(\vec{x}) : \varphi(y,\vec{x})\}$ where f is given by Theorem 6.3. □

We give our version of Herbrand saturation. Let $L \supseteq \mathrm{L}_0$ be a countable language.

Definition 6.5: A structure M is $\Delta_0(L)$-*Herbrand saturated* if it satisfies every $\Sigma_2(L)$-sentence with parameters from M which is consistent with the $\Pi_1(L)$-diagram of M.

The next two lemmas do not use any special properties of the class $\Delta_0(L)$, beyond that it is closed under subformulas, negations and substitution.

Lemma 6.6: *Every consistent $\Pi_1(L)$ theory T has a $\Delta_0(L)$-Herbrand saturated model.*

Proof: Let L^+ be L together with names for countably many new constants. Enumerate all $\Delta_0(L^+)$-formulas as $\varphi_1, \varphi_2, \ldots$. Let $T_1 = T$ and define a sequence of theories $T_1 \subseteq T_2 \subseteq \ldots$ as follows: if $T_i + \exists \vec{x}\, \forall \vec{y}\, \varphi_i(\vec{x}, \vec{y})$ is consistent, let $T_{i+1} = T_i + \forall \vec{y}\, \varphi_i(\vec{c}, \vec{y})$ where $\vec{c}$ is a tuple of constant symbols that do not appear in T_i or φ_i. Otherwise let $T_{i+1} = T_i$. Let $T^* = \bigcup_i T_i$. By construction, T^* is consistent and $\Pi_1(L^+)$.

Let M be a model of T^* and let N be the substructure of M consisting of elements named by L^+-terms. We claim that $N \models T^*$. It is enough to show that for every $\Delta_0(L^+)$-formula φ and every tuple $\vec{a}$ from N, we have $N \models \theta(\vec{a}) \Leftrightarrow M \models \theta(\vec{a})$. We prove this by induction on θ. For the only interesting case, suppose $M \models \exists x\, \psi(\vec{a}, x)$ where the inductive hypothesis holds for ψ. Since the components of $\vec{a}$ are named by terms, $\psi(\vec{a}, x)$ is equivalent in M to some formula $\varphi_i(x)$ from our enumeration. But $M \models \exists x\, \varphi_i(x)$ implies that $\exists x\, \varphi_i(x)$ is consistent with T_i and hence that $\varphi_i(c)$ is in T_{i+1} for some constant c. Thus $M \models \varphi_i(c)$ and therefore $M \models \psi(\vec{a}, c)$, so $N \models \psi(\vec{a}, c)$ by the inductive hypothesis.

Finally, N is $\Delta_0(L)$-Herbrand saturated. For suppose that ψ is $\Delta_0(L)$ and $\exists \vec{x}\, \forall \vec{y}\, \psi(\vec{x}, \vec{y}, \vec{a})$ is consistent with the $\Pi_1(L)$-diagram of N, and hence with T^*. Then as above $\psi(\vec{x}, \vec{y}, \vec{a})$ is equivalent to $\varphi_i(\vec{x}, \vec{y})$ for some i, and since $\exists \vec{x}\, \forall \vec{y}\, \varphi_i(\vec{x}, \vec{y})$ is consistent with T_i it is witnessed in T_{i+1} by a tuple of constants and hence is true in N. □

Lemma 6.7: *If S, T are theories such that S is $\Pi_1(L)$ and every $\Delta_0(L)$-Herbrand saturated model of S is a model of T, then T is $\Pi_2(L)$-conservative over S.*

Proof: Suppose T proves $\forall \vec{x}\, \exists \vec{y}\, \varphi(\vec{x}, \vec{y})$ but S does not, where φ is $\Delta_0(L)$. Then, letting $\vec{c}$ be a tuple of new constants, the theory $S + \forall \vec{y}\, (\neg \varphi(\vec{c}, \vec{y}))$ has a $\Delta_0(L)$-Herbrand saturated model by Lemma 6.6. This contradicts the assumptions about S and T. □

We now describe the most useful property of $\Delta_0(\mathrm{L}_{\mathrm{crsf}})$-Herbrand saturated models.

Lemma 6.8: *Suppose that* $M \models \mathrm{T}_{\mathrm{crsf}}$ *is* $\Delta_0(\mathrm{L}_{\mathrm{crsf}})$*-Herbrand saturated and that* $\varphi(y, \vec{x}, \vec{a})$ *is a* $\Delta_0(\mathrm{L}_{\mathrm{crsf}})$*-formula with parameters* $\vec{a} \in M$ *such that* $M \models \forall \vec{x}\, \exists y\, \varphi(y, \vec{x}, \vec{a})$*. Then there exist a function* $f \in \mathrm{L}_{\mathrm{crsf}}$ *and parameters* $\vec{m} \in M$ *such that* $M \models \forall \vec{x}\, \exists y \in f(\vec{x}, \vec{m})\, \varphi(y, \vec{x}, \vec{a})$.

Proof: Let T^* be the $\Pi_1(\mathrm{L}_{\mathrm{crsf}})$-diagram of M. Then $T^* \vdash \forall \vec{x}\, \exists y\, \varphi(y, \vec{x}, \vec{a})$ since otherwise $M \models \exists \vec{x}\, \forall y\, (\neg\varphi(y, \vec{x}, \vec{a}))$ by Herbrand saturation. The rest of the argument is standard. By compactness, there are $\vec{b} \in M$ and θ in $\Delta_0(\mathrm{L}_{\mathrm{crsf}})$ such that $M \models \forall \vec{z}\, \theta(\vec{a}, \vec{b}, \vec{z})$ and

$$\mathrm{T}_{\mathrm{crsf}} + \forall \vec{z}\, \theta(\vec{a}, \vec{b}, \vec{z}) \vdash \forall \vec{x}\, \exists y\, \varphi(y, \vec{x}, \vec{a})$$

where we treat $\vec{a}, \vec{b}$ as constant symbols. Hence, replacing $\vec{a}, \vec{b}$ with variables $\vec{u}, \vec{v}$,

$$\mathrm{T}_{\mathrm{crsf}} \vdash \exists \vec{z}\, \neg\theta(\vec{u}, \vec{v}, \vec{z}) \vee \exists y\, \varphi(y, \vec{x}, \vec{u}). \tag{6.1}$$

Using pairing and projection functions to code tuples of sets as single sets, we can apply Theorem 6.3 to formulas with more than one unbounded existential quantifier. In particular from (6.1) we get an $\mathrm{L}_{\mathrm{crsf}}$ function symbol f with

$$\mathrm{T}_{\mathrm{crsf}} \vdash \exists \vec{z}\, \neg\theta(\vec{u}, \vec{v}, \vec{z}) \vee \exists y \in f(\vec{x}, \vec{u}, \vec{v})\, \varphi(y, \vec{x}, \vec{u}).$$

Since $M \models \forall \vec{z}\, \theta(\vec{a}, \vec{b}, \vec{x})$ it follows that $M \models \forall \vec{x}\, \exists y \in f(\vec{x}, \vec{a}, \vec{b})\, \varphi(y, \vec{x}, \vec{a})$. □

6.2. *Witnessing* $\mathrm{KP}^{\mathrm{u}}_1$

We prove witnessing for $\mathrm{KP}^{\mathrm{u}}_1(\mathrm{L}_{\mathrm{crsf}})$ as a consequence of witnessing for $\mathrm{T}_{\mathrm{crsf}}$, together with the following conservativity result.

Theorem 6.9: *The theory* $\mathrm{KP}^{\mathrm{u}}_1(\mathrm{L}_{\mathrm{crsf}})$ *is* $\Pi_2(\mathrm{L}_{\mathrm{crsf}})$*-conservative over* $\mathrm{T}_{\mathrm{crsf}}$.

Proof: Let M be an arbitrary $\Delta_0(\mathrm{L}_{\mathrm{crsf}})$-Herbrand saturated model of $\mathrm{T}_{\mathrm{crsf}}$. By Lemma 6.7 it is enough to show that M is a model of $\mathrm{KP}^{\mathrm{u}}_1(\mathrm{L}_{\mathrm{crsf}})$. By Lemma 5.20 it is enough to show that M satisfies $\Delta_0(\mathrm{L}_{\mathrm{crsf}})$-Collection and uniformly bounded unique $\Sigma_1^{\preccurlyeq}(\mathrm{L}_{\mathrm{crsf}})$-Induction.

For collection, suppose that for some $a \in M$ we have

$$M \models \forall u{\in}a\, \exists v\, \varphi(u, v),$$

where φ is $\Delta_0(\mathrm{L}_{\mathrm{crsf}})$ with parameters. We rewrite this as

$$M \models \forall u\, \exists v\, (u \in a \to \varphi(u, v)).$$

By Lemma 6.8, for some $\mathrm{L_{crsf}}$ function symbol f and tuple $\vec{b} \in M$,

$$M \models \forall u \, \exists v \in f(u, \vec{b}) \, (u \in a \to \varphi(u, v)).$$

Hence if we let $c = \bigcup f"(a, \vec{b})$ we have, as required for collection,

$$M \models \forall u{\in}a \, \exists v \in c \, \varphi(u, v).$$

For induction, let $\varphi(u, v), \varepsilon(z, z', v, u) \in \Delta_0(\mathrm{L_{crsf}})$ and $t(u)$ be a #-term, all possibly with parameters, and let $\varphi^{\varepsilon,t}(u, v)$ abbreviate $\varphi(u, v) \wedge \varepsilon(\cdot, \cdot, v, u) : v \preccurlyeq t(u)$. Working in M, suppose

$$\forall u \, \exists^{\leqslant 1} v \, \varphi(u, v) \wedge \forall x \, \big(\forall u{\in}x \, \exists v \, \varphi^{\varepsilon,t}(u, v) \to \exists v' \, \varphi^{\varepsilon,t}(x, v')\big).$$

Then in particular

$$\forall x, w \, \big(\forall u{\in}x \, \exists v \in w \, \varphi^{\varepsilon,t}(u, v) \to \exists v' \, \varphi^{\varepsilon,t}(x, v')\big).$$

By Lemma 6.8 there is an $\mathrm{L_{crsf}}$-function symbol h and a tuple $\vec{a} \in M$ such that we can bound the witness v' as a member of $h(x, w, \vec{a})$. Then, since witnesses v to φ are unique, if we let $g(x, w, \vec{a})$ compute

$$\bigcup \{v' \in h(x, w, \vec{a}) : \varphi(x, v')\}$$

we have

$$\forall x, w \, \big(\forall u{\in}x \, \exists v \in w \, \varphi^{\varepsilon,t}(u, v) \to \varphi^{\varepsilon,t}(x, g(x, w, \vec{a}))\big). \tag{6.2}$$

We must be careful here with our parameters. We may assume without loss of generality that the so-far unwritten parameters in φ, ε and t are contained in the tuple $\vec{a}$, and further that $\varphi(u, v)$ is really $\varphi(u, v, \vec{a})$, $\varepsilon(z, z', v, u)$ is $\varepsilon(z, z', v, u, \vec{a})$ and $t(u)$ is $t(u, \vec{a})$.

We now use syntactic Cobham recursion to iterate g. To use the recursion available in $\mathrm{T_{crsf}}$ we need to turn the weakly uniform embedding given by ε into a strongly uniform embedding. So let τ be an $\mathrm{L_{crsf}}$-function symbol for which $\mathrm{T_{crsf}}$ proves

$$\tau(z, v, u, \vec{w}) = \{z' \in t(u, \vec{w}) : \varepsilon(z, z', v, u, \vec{w})\}.$$

Let f be the $\mathrm{L_{crsf}}$ function symbol $f_{g,\tau,t}$ with defining axiom

$$f(u, \vec{w}) = \begin{cases} g(u, f"(u, \vec{w}), \vec{w}) & \text{if } \tau \text{ is an embedding into } t \text{ at } u, \vec{w} \\ 0 & \text{otherwise,} \end{cases}$$

where "τ is an embedding into t at $u, \vec{w}$" stands for the $\Delta_0(\mathrm{L_{crsf}})$-formula

$$\tau(\cdot, g(u, f"(u, \vec{w}), \vec{w}), u, \vec{w}) \; : \; g(u, f"(u, \vec{w}), \vec{w}) \preccurlyeq t(u, \vec{w}).$$

It suffices now to show that $\forall x\, \varphi^{\varepsilon,t}(x, f(x,\vec{a}), \vec{a})$. We will use $\Delta_0(\mathrm{L_{crsf}})$-Induction, which is available by Lemma 3.18. Suppose $\forall u{\in}x\, \varphi^{\varepsilon,t}(u, f(u,\vec{a}), \vec{a})$. Let $w = f\text{"}(x,\vec{a})$ and let $v = g(x,w,\vec{a})$. By (6.2) we have $\varphi^{\varepsilon,t}(x,v,\vec{a})$, and in particular $\varepsilon(\cdot,\cdot,v,x,\vec{a}) : v \preccurlyeq t(x,\vec{a})$. Hence also $\tau(\cdot,v,x,\vec{a}) : v \preccurlyeq t(x,\vec{a})$, that is, τ is an embedding into t at $x, \vec{a}$. From the defining axiom for f we conclude that $f(x,\vec{a}) = v$, and thus $\varphi^{\varepsilon,t}(x, f(x,\vec{a}), \vec{a})$. This completes the proof. □

From this we get witnessing for $\mathrm{KP}_1^{\mathrm{u}}(\mathrm{L_{crsf}})$, and a fortiori for $\mathrm{KP}_1^{\mathrm{u}}$:

Theorem 6.10: *Suppose* $\mathrm{KP}_1^{\mathrm{u}}(\mathrm{L_{crsf}}) \vdash \exists y\, \varphi(y,\vec{x})$ *where* φ *is* $\Delta_0(\mathrm{L_{crsf}})$. *Then there is an* $\mathrm{L_{crsf}}$*-function symbol* f *such that* $\mathrm{T_{crsf}} \vdash \exists y \in f(\vec{x})\, \varphi(y,\vec{x})$.

Proof: By Theorem 6.9, $\mathrm{T_{crsf}} \vdash \exists y\, \varphi(y,\vec{x})$. Then apply Theorem 6.3. □

Corollary 6.11: *If* $\mathrm{KP}_1^{\mathrm{u}}(\mathrm{L_{crsf}}) \vdash \exists! y\, \varphi(y,\vec{x})$ *where* φ *is* $\Sigma_1(\mathrm{L_{crsf}})$, *then there is an* $\mathrm{L_{crsf}}$*-function symbol* g *such that* $\mathrm{T_{crsf}} \vdash \varphi(g(\vec{x}),\vec{x})$.

Proof: Suppose $\mathrm{KP}_1^{\mathrm{u}}(\mathrm{L_{crsf}}) \vdash \exists! y\, \exists v\, \theta(y,v,\vec{x})$ where θ is $\Delta_0(\mathrm{L_{crsf}})$. Using Theorem 6.10 it is not hard to show that $\mathrm{T_{crsf}} \vdash \exists y, v \in f(\vec{x})\, \theta(y,v,\vec{x})$ for some $\mathrm{L_{crsf}}$-function symbol f, and from Theorem 6.9 we get $\mathrm{T_{crsf}} \vdash \exists^{\leqslant 1} y\, \theta(y,v,\vec{x})$. Thus we can define the witnessing function as $g(\vec{x}) := \bigcup\{y \in f(\vec{x}) : \exists v \in f(\vec{x})\, \theta(y,v,\vec{x})\}$. □

Together with the Definability Theorem 5.2, the above implies Corollary 6.1.

6.3. *Witnessing with global choice*

We add to our basic language $\mathrm{L_0}$ and theory $\mathrm{T_0}$ a symbol C for a *global choice* function, with defining axiom

$$(\mathrm{GC}): \quad C(0) = 0 \wedge (x \neq 0 \rightarrow C(x) \in x).$$

We denote the augmented language and theory by L_0^C and T_0^C. We write $\mathrm{L}_{\mathrm{crsf}}^C$ and $\mathrm{T}_{\mathrm{crsf}}^C$ for $\mathrm{L_{crsf}}$ and $\mathrm{T_{crsf}}$ defined using L_0^C and T_0^C in place of $\mathrm{L_0}$ and $\mathrm{T_0}$. The symbols in $\mathrm{L}_{\mathrm{crsf}}^C$ correspond to the functions in CRSF^C, defined like CRSF but with the global choice function $C(x)$ as an additional initial function (cf. Remark 2.10).

We write $\mathrm{KPC}_1^{\preccurlyeq}$ for the corresponding version of $\mathrm{KP}_1^{\preccurlyeq}$, that is, the theory consisting of T_0^C and $\Delta_0(\mathrm{L}_0^C)$-Collection and $\Sigma_1^{\preccurlyeq}(\mathrm{L}_0^C)$-Induction schemes. Similarly $\mathrm{KPC}_1^{\mathrm{u}}$ consists of T_0^C and the $\Delta_0(\mathrm{L}_0^C)$-Collection and uniformly bounded unique $\Sigma_1^{\preccurlyeq}(\mathrm{L}_0^C)$-Induction schemes (where we allow embeddings to be $\Delta_0(\mathrm{L}_0^C)$). Earlier results about the theories without choice carry over to the theories with choice as expected; there is one extra case in Lemma 4.10, taken care of by noting that the identity embedding embeds $C(x) \preccurlyeq x$. In particular, $\mathrm{KPC}_1^{\preccurlyeq}(\mathrm{L}_{\mathrm{crsf}}^C)$ is a $\Sigma_1^{\preccurlyeq}(\mathrm{L}_0^C)$-expansion of $\mathrm{KPC}_1^{\preccurlyeq}$ to the language $\mathrm{L}_{\mathrm{crsf}}^C$ which is equivalent to the theory consisting of $\mathrm{T}_{\mathrm{crsf}}^C$ and the schemes of $\mathrm{KPC}_1^{\preccurlyeq}$ over the language $\mathrm{L}_{\mathrm{crsf}}^C$ (cf. Lemma 5.21). Thus, Corollary 6.2 follows from Theorem 6.13 below, a witnessing theorem for $\mathrm{KPC}_1^{\preccurlyeq}(\mathrm{L}_{\mathrm{crsf}}^C)$. We prove it by showing conservativity over $\mathrm{T}_{\mathrm{crsf}}^C$:

Theorem 6.12: *The theory* $\mathrm{KPC}_1^{\preccurlyeq}(\mathrm{L}_{\mathrm{crsf}}^C)$ *is* $\Pi_2(\mathrm{L}_{\mathrm{crsf}}^C)$*-conservative over* $\mathrm{T}_{\mathrm{crsf}}^C$.

Proof: Let M be an arbitrary $\Delta_0(\mathrm{L}_{\mathrm{crsf}}^C)$-Herbrand saturated model of $\mathrm{T}_{\mathrm{crsf}}^C$. As before, by Lemma 6.7 it is enough to show that M is a model of $\mathrm{KPC}_1^{\preccurlyeq}(\mathrm{L}_{\mathrm{crsf}}^C)$. By exactly the same argument as in the proof of Theorem 6.9, we get that M is a model of $\Delta_0(\mathrm{L}_{\mathrm{crsf}}^C)$-Collection.

It remains to show that $\Sigma_1^{\preccurlyeq}(\mathrm{L}_{\mathrm{crsf}}^C)$-Induction holds in M. By Lemma 4.13, it is enough to show that uniformly bounded $\Sigma_1^{\preccurlyeq}(\mathrm{L}_{\mathrm{crsf}}^C)$-Induction holds. That is, exactly the induction shown for the formula $\exists v\, \varphi^{\varepsilon,t}(u,v)$ in the proof of Theorem 6.9, except without the uniqueness assumption that $M \models \forall u\, \exists^{\leqslant 1} v\, \varphi(u,v)$. Working through that proof, we see that uniqueness is used only in one place, to construct an $\mathrm{L}_{\mathrm{crsf}}$-function symbol g satisfying

$$\forall x, w\, \left(\forall u{\in}x\, \exists v \in w\, \varphi^{\varepsilon,t}(u,v) \to \varphi^{\varepsilon,t}(x, g(x,w,\vec{a}))\right)$$

from an $\mathrm{L}_{\mathrm{crsf}}$-function symbol h satisfying

$$\forall x, w\, \left(\forall u{\in}x\, \exists v \in w\, \varphi^{\varepsilon,t}(u,v) \to \exists v' \in h(x,w,\vec{a})\, \varphi^{\varepsilon,t}(x,v')\right)).$$

In $\mathrm{L}_{\mathrm{crsf}}^C$ this can be done without the assumption, by setting

$$g(x,w,\vec{a}) = C(\{v' \in h(x,w,\vec{a}) : \varphi^{\varepsilon,t}(x,v')\}).$$

The rest of the proof goes through as before. □

Theorem 6.13: *Suppose* $\mathrm{KPC}_1^{\preccurlyeq}(\mathrm{L}_{\mathrm{crsf}}^C) \vdash \exists y\, \varphi(y,\vec{x})$ *where* φ *is* $\Delta_0(\mathrm{L}_{\mathrm{crsf}}^C)$. *Then there is an* $\mathrm{L}_{\mathrm{crsf}}^C$*-function symbol* f *such that* $\mathrm{T}_{\mathrm{crsf}}^C \vdash \varphi(f(\vec{x}),\vec{x})$.

Proof: By Theorem 6.12, $\mathrm{T}_{\mathrm{crsf}}^C \vdash \exists y\, \varphi(y,\vec{x})$. Using Theorem 6.3 for $\mathrm{T}_{\mathrm{crsf}}^C$ there is g in $\mathrm{L}_{\mathrm{crsf}}^C$ such that $\mathrm{T}_{\mathrm{crsf}}^C \vdash \exists y{\in}g(\vec{x})\varphi(y,\vec{x})$. Using Lemma 3.18 for

$\mathrm{T}^C_{\mathrm{crsf}}$ we find h such that $\mathrm{T}^C_{\mathrm{crsf}} \vdash h(\vec{x}) = \{y \in g(\vec{x}) : \varphi(y, \vec{x})\}$. Then choose f such that $\mathrm{T}^C_{\mathrm{crsf}} \vdash f(\vec{x}) = C(h(\vec{x}))$. □

For the theory without choice, we get a weak result in the style of Parikh's theorem [21].

Corollary 6.14: *Suppose* $\mathrm{KP}_1^{\preccurlyeq}(\mathrm{L}_{\mathrm{crsf}}) \vdash \exists y\, \varphi(y, \vec{x})$ *where* φ *is* $\Delta_0(\mathrm{L}_{\mathrm{crsf}})$. *Then in the universe of sets we can bound the complexity of the witness* y *in the following sense: there is a #-term* t *such that* $\forall \vec{x}\, \exists y {\preccurlyeq} t(\vec{x})\, \varphi(y, \vec{x})$ *holds.*

Proof: It is easy to show that for every $\mathrm{L}^C_{\mathrm{crsf}}$-function symbol $f(\vec{x})$ there is an $\mathrm{L}^C_{\mathrm{crsf}}$-function symbol $\tau(z, \vec{x})$ and a #-term $t(\vec{x})$ such that $\tau(\cdot, \vec{x}) : f(\vec{x}) \preccurlyeq t(\vec{x})$, provably in $\mathrm{T}^C_{\mathrm{crsf}}$. For the initial symbols from L^C_0 this is by Lemma 4.10 (extended to cover C). For function symbols obtained by composition we use monotonicity, for replacement we use Lemma 4.11, and for syntactic Cobham recursion we are explicitly given such a bound.

Now suppose the assumption of the corollary holds. Then also $\mathrm{KPC}_1^{\preccurlyeq}(\mathrm{L}^C_{\mathrm{crsf}}) \vdash \exists y\, \varphi(y, \vec{x})$, hence $\mathrm{T}^C_{\mathrm{crsf}} \vdash \varphi(f(\vec{x}), \vec{x})$ for some $\mathrm{L}^C_{\mathrm{crsf}}$-function symbol f by Theorem 6.13. It follows that $\mathrm{T}^C_{\mathrm{crsf}} \vdash \exists y {\preccurlyeq} t(\vec{x})\, \varphi(y, \vec{x})$, by the previous paragraph and using $\Delta_0(\mathrm{L}_{\mathrm{crsf}})$-Separation to get a nonuniform embedding. In ZFC, global choice can be forced without adding new sets (see for example [17]) so we can expand the universe V of sets to a model (V, C) of ZF + (GC) and in particular of $\mathrm{T}^C_{\mathrm{crsf}}$. Then $\forall \vec{x}\, \exists y {\preccurlyeq} t(\vec{x})\, \varphi(y, \vec{x})$ holds in (V, C), and thus also in V, since it does not mention the symbol C. □

6.4. *Uniform Cobham recursion*

We can use our definability and witnessing theorems to partially answer a question that arose from [7]. Namely, the embedding giving the bound on a Cobham recursion is given by a CRSF function. If we only allow simpler embeddings, given by $\Delta_0(\mathrm{L}_0)$-formulas, does the class CRSF change? We show that it does not. This is a partial answer because we only consider what happens if we make this change in our definition of CRSF from Proposition 2.9, which is slightly different from the original definition in [7].

Definition 6.15: In the universe of sets, the $\mathrm{CRSF}_{\mathrm{u}}$ functions are those obtained from the projections, zero, pair, union, conditional, transitive closure, cartesian product, set composition and set smash functions by composition, replacement and "weakly uniform syntactic Cobham recursion". This is the following recursion scheme: suppose $g(x, z, \vec{w})$ is a $\mathrm{CRSF_u}$ function, $\varepsilon(z, z', y, x, \vec{w})$ is a $\Delta_0(\mathrm{L_0})$-formula and $t(x, \vec{w})$ is a #-term. Then $\mathrm{CRSF_u}$ contains the function symbol $f = f_{g,\varepsilon,t}$ defined by

$$f(x, \vec{w}) = \begin{cases} g(x, f\text{"}(x, \vec{w}), \vec{w}) & \text{if } \varepsilon \text{ is an embedding into } t \text{ at } x, \vec{w} \\ 0 & \text{otherwise} \end{cases}$$

where the condition "ε is an embedding into t at $x, \vec{w}$" stands for

$$\varepsilon(\cdot, \cdot, g(x, f\text{"}(x, \vec{w}), \vec{w}), x, \vec{w}) : g(x, f\text{"}(x, \vec{w}), \vec{w}) \preccurlyeq t(x, \vec{w}).$$

The language $\mathrm{L_{crsfu}}$ and theory $\mathrm{T_{crsfu}}$ are defined by changing the syntactic Cobham recursion case in the definitions of $\mathrm{L_{crsf}}$ and $\mathrm{T_{crsf}}$ to match the description above.

Theorem 6.16: *The theory* $\mathrm{KP}_1^{\mathrm{u}}$ *is* $\Pi_2(\mathrm{L_0})$*-conservative over* $\mathrm{T_{crsfu}}$.

Proof: It is straightforward to show that $\mathrm{T_{crsfu}}$ proves $\mathrm{T_0}$, just as $\mathrm{T_{crsf}}$ does. Similarly the results about Herbrand saturation go through for $\mathrm{T_{crsfu}}$. By Lemma 6.7 it is enough to show that any $\Delta_0(\mathrm{L_{crsfu}})$-Herbrand saturated model of $\mathrm{T_{crsfu}}$ is a model of $\Delta_0(\mathrm{L_0})$-Collection and uniformly bounded unique $\Sigma_1^{\preccurlyeq}(\mathrm{L_0})$-Induction. For this we can simply repeat the proof of Theorem 6.9 with $\mathrm{T_{crsfu}}$ in place of $\mathrm{T_{crsf}}$, observing that the proof becomes more direct, since in the application of syntactic Cobham recursion we can use the embedding ε directly without needing to construct the function symbol τ. □

Corollary 6.17: *In the universe of sets,* $\mathrm{CRSF_u} = \mathrm{CRSF}$.

Proof: It is clear that every $\mathrm{CRSF_u}$ function is CRSF. For the other direction, suppose $f(\vec{x})$ is CRSF. Then there is a good definition of f in the sense of Definition 5.6, and in particular there is a $\Delta_0(\mathrm{L_0})$-formula $\varphi(v, \vec{x})$ and an L_0^+-term e such that $\mathrm{KP}_1^{\mathrm{u}}$ proves $\exists! v\, \varphi(v, \vec{x})$ and such that $\varphi(v, \vec{x}) \to f(\vec{x}) = e(v)$ holds in the universe for all sets $v, \vec{x}$. There is a similar version of Herbrand's theorem for $\mathrm{T_{crsfu}}$ as there is for $\mathrm{T_{crsf}}$. Combining this with Theorem 6.16 we get that there is an $\mathrm{L_{crsfu}}$ function symbol g such that $\mathrm{T_{crsfu}}$ proves $\varphi(g(\vec{x}), \vec{x})$. Hence in the universe $f(\vec{x}) = e(g(\vec{x}))$, which is a $\mathrm{CRSF_u}$ function. □

7. Partial Conservativity of Global Choice

Recall from Section 6.3 the versions of our theories with global choice (GC).

Proposition 7.1: T_0^C *is not* $\Pi_2(\mathrm{L}_0)$*-conservative over* ZF.

Proof: The theory T_0^C proves (AC) in the form: for every set x of disjoint, nonempty sets, there is a set z containing exactly one element from every member of x. Indeed, $z := \{v \in \mathrm{tc}(x) : \exists u \in x\ C(u) = v\}$ can be obtained by $\Delta_0(\mathrm{L}_0^C)$-Separation. □

In particular, the extension $\mathrm{KPC}_1^{\preccurlyeq}(\mathrm{L}_{\mathrm{crsf}}^C)$ of $\mathrm{KP}_1^{\preccurlyeq}(\mathrm{L}_{\mathrm{crsf}})$ is not $\Pi_2(\mathrm{L}_0)$-conservative. Informally, we ask how much stronger $\mathrm{KPC}_1^{\preccurlyeq}(\mathrm{L}_{\mathrm{crsf}}^C)$ is compared to $\mathrm{KP}_1^{\preccurlyeq}(\mathrm{L}_{\mathrm{crsf}})$. More formally, we aim to encapsulate the difference in some local choice principles, namely a strong form of (AC) plus a form of dependent choice.[e]

7.1. *Dependent Choice*

The class of ordinals is denoted by $Ord(x)$ in L_0^+ with defining axiom $\forall y{\in}x{\cup}\{x\}\,(\mathrm{tc}(y) = y)$. It is routine to verify in T_0^+ some elementary properties of ordinals, e.g., elements of ordinals are ordinals and, given two distinct ordinals, one is an element of the other. We let $\alpha, \beta, \ldots$ range over ordinals. By this we mean that $\forall\alpha \ldots$ and $\exists\alpha \ldots$ stand for $\forall\alpha\,(Ord(\alpha) \to \ldots)$ and $\exists\alpha\,(Ord(\alpha) \wedge \ldots)$ respectively.

The scheme $\Delta_0(\mathrm{L}_{\mathrm{crsf}})$-*Dependent Choice* gives for every $\Delta_0(\mathrm{L}_{\mathrm{crsf}})$-formula $\varphi(x, y, \vec{x})$

$$\forall x\,\exists y\,\varphi(x,y,\vec{x}) \to \forall\alpha\,\exists z\,(Fct(z) \wedge dom(z) = \alpha \wedge \forall\beta{\in}\alpha\,\varphi(z{\restriction}\beta, z\text{'}\beta, \vec{x})). \tag{7.1}$$

We assume $\mathrm{L}_0^+ \subseteq \mathrm{L}_{\mathrm{crsf}}$ (cf. Lemma 4.5), $Fct(y)$ is a unary relation symbol in L_0^+ expressing that y is a function, and $dom(x)$, $im(x)$, $x{\restriction}y$ are function symbols in $\mathrm{L}_{\mathrm{crsf}}$ such that $\mathrm{KP}_1^{\preccurlyeq}(\mathrm{L}_{\mathrm{crsf}})$ proves $dom(x) = \pi_1\text{"}(x)$, $im(x) = \pi_2\text{"}(x)$ and $x{\restriction}y = \{z \in x : \pi_1(z) \in y\}$.

We further consider the following strong version of (AC) that we refer to as the *well-ordering principle* (WO):

$$\forall x\exists\alpha\exists y(\text{“}y \text{ is a bijection from } \alpha \text{ onto } x\text{”} \wedge \forall\beta,\gamma{\in}\alpha\,(y\text{'}\beta \in y\text{'}\gamma \to \beta \in \gamma)).$$

---

[e] This aims at technical simplicity of the argument rather than the strongest possible result.

The goal of this section is to prove:

Theorem 7.2: *The theory* $\mathrm{KPC}_1^{\preccurlyeq}(\mathrm{L}_{\mathrm{crsf}}^C)$ *is conservative over* $\mathrm{KP}_1^{\preccurlyeq}(\mathrm{L}_{\mathrm{crsf}})$ *plus* $\Delta_0(\mathrm{L}_{\mathrm{crsf}})$*-Dependent Choice plus (WO).*

Note this just says that every $\mathrm{L}_{\mathrm{crsf}}$-formula proved by the former theory is also proved by the latter. But the former theory is not an extension of the latter:

Proposition 7.3: $\mathrm{KPC}_1^{\preccurlyeq}(\mathrm{L}_{\mathrm{crsf}}^C)$ *does not prove (WO).*

Proof: By Theorem 6.13, if $\mathrm{KPC}_1^{\preccurlyeq}(\mathrm{L}_{\mathrm{crsf}}^C)$ proves the existence of an ordinal α bijective to a given x, then it proves $f(x)$ is such an α for some function symbol $f(x)$ in $\mathrm{L}_{\mathrm{crsf}}^C$. Fix a universe of sets with a global choice function C, and view it as a structure interpreting $\mathrm{L}_{\mathrm{crsf}}^C$. There, $f(x)$ denotes a function in CRSF^C. By Theorem 2.5 (for CRSF^C instead CRSF, recall Remark 2.10), there is a #-term $t(x)$ such that $\alpha = f(x) \preccurlyeq t(x)$ holds for all x. Then there is a polynomial p such that the von Neumann rank $\mathrm{rk}(f(x)) = \alpha \geq |x|$ is at most $p(\mathrm{rk}(x))$ (cf. [7, Lemma 2, 4, Proposition 10]). This is false for many x. □

7.2. *The forcing*

Let M be a countable model of $\mathrm{KP}_1^{\preccurlyeq}(\mathrm{L}_{\mathrm{crsf}})$ and $\Delta_0(\mathrm{L}_{\mathrm{crsf}})$-Dependent Choice and (WO). We intend to produce a generic extension of M modelling (GC). Note we do *not* assume that M is standard, in particular, M possibly does not interpret $\in$ by $\in$. While the forcing frame is the class forcing commonly used to force global choice, we use a technically simplified forcing relation avoiding the use of names. This is similar to [17]. The argument that the forcing preserves $\mathrm{KP}_1^{\preccurlyeq}(\mathrm{L}_{\mathrm{crsf}})$ needs some care since this theory and hence M is very weak.

The forcing frame $(\mathbb{P}, \leqslant^{\mathbb{P}})$ is defined as follows: $\mathbb{P} \subseteq M$ contains $p \in M$ if and only if p is a choice function in the sense of M, that is, M satisfies

$$(Fct(p) \wedge \langle 0, 0\rangle \in p \wedge \forall x, y\, (\langle x, y\rangle \in p \wedge x \neq 0 \rightarrow y \in x)).$$

Further, $p \leqslant^{\mathbb{P}} q$ means $M \models q \subseteq p$. Then $(\mathbb{P}, \leqslant^{\mathbb{P}})$ is a partial order. In the following we let $p, q, r, \ldots$ range over *conditions*, i.e., elements of $\mathbb{P}$. A subset X of $\mathbb{P}$ is *dense below* p if for all $q \leqslant^{\mathbb{P}} p$ there is $r \leqslant^{\mathbb{P}} q$ such that $r \in X$. Being *dense* means being dense below $1^{\mathbb{P}} := \{\langle 0, 0\rangle\}$ (calculated in M). A subset X of $\mathbb{P}$ is a *filter* if $p \cup^M q \in X$ whenever $p, q \in X$, and $q \in X$ whenever $p \leqslant^{\mathbb{P}} q$ and $p \in X$. Being *generic* means being a filter that intersects all dense subsets of $\mathbb{P}$ that are definable (with parameters) in M. The *forcing language* is $\mathrm{L}_{\mathrm{crsf}} \cup \{R\}$ for a new binary relation symbol R.

The *forcing relation* $\Vdash$ relates conditions p to sentences of the forcing language with parameters from M. It is defined as follows. For an atomic sentence φ that does not mention R we let $p \Vdash \varphi$ if and only if $M \models \varphi$. For an atomic sentence of the form Rts with closed terms t, s we let

$$p \Vdash Rts \iff M \models (t = \{s\} \vee \langle t, s\rangle \in p).$$

We extend this definition via the recurrence:

$$p \Vdash (\varphi \wedge \psi) \iff p \Vdash \varphi \text{ and } p \Vdash \psi,$$
$$p \Vdash \neg\varphi \iff \text{for all } q \leqslant^{\mathbb{P}} p : q \nVdash \varphi,$$
$$p \Vdash \forall x\, \varphi(x) \iff \text{for all } a \in M : p \Vdash \varphi(a).$$

This defines $p \Vdash \varphi$ for all sentences φ of the forcing language with parameters from M which are written using the logical symbols $\wedge, \neg, \forall$. We freely use the symbols $\vee, \rightarrow, \exists$ understanding these as classical abbreviations. Namely, $(\varphi \vee \psi)$, $(\varphi \rightarrow \psi)$, $\exists x\, \chi(x)$ stand for $\neg(\neg\varphi \wedge \neg\psi)$, $\neg(\varphi \wedge \neg\psi)$, $\neg\forall x\, (\neg\chi(x))$ respectively. Lemma 7.4 (f) below shows that $p \Vdash \varphi$ does not depend of the choice of these abbreviations.

Lemma 7.4: *Let φ be a sentence of the forcing language with parameters from M.*

(a) (Conservativity) *If R does not occur in φ, then $p \Vdash \varphi$ if and only if $M \models \varphi$.*
(b) (Extension) *If $p \leqslant^{\mathbb{P}} q$ and $q \Vdash \varphi$, then $p \Vdash \varphi$.*
(c) (Stability) *$p \Vdash \varphi$ if and only if $p \Vdash \neg\neg\varphi$, that is, if and only if $\{q \mid q \Vdash \varphi\}$ is dense below p.*
(d) (Truth Lemma) *For every generic G there is $R_G \subseteq M^2$ such that for every φ we have that $(M, R_G) \models \varphi$ if and only if some $p \in G$ forces φ.*
(e) (Forcing Completeness) *$p \Vdash \varphi$ if and only if $(M, R_G) \models \varphi$ for every generic filter G containing p.*
(f) *$\{\varphi \mid p \Vdash \varphi\}$ is closed under logical consequence.*

Proof: (a) and (b) obviously hold for atomic φ; for general φ the claim follows by a straightforward induction (see, e.g., [3, Lemma 2.6]). Similarly, it suffices to show (c) for atomic φ. Assume $\varphi = Rts$ for closed terms t, s. The second equivalence is trivial. The forward direction follows from (b): if $p \Vdash \varphi$, then $\{q \mid q \Vdash \varphi\} \supseteq \{q \mid q \leqslant^{\mathbb{P}} p\}$ is dense below p. Conversely, it is enough to find given p with $p \nVdash Rts$ some $q \leqslant^{\mathbb{P}} p$ forcing $\neg Rts$. We have

$M \models t \neq \{s\}$ and $M \models \langle t, s\rangle \notin p$. If $M \models t = 0$, then $M \models s \neq 0$ and no condition forces Rts, so $q := p \Vdash \neg Rts$. If $M \models t \neq 0$, then there is $a \in M$ such that $M \models (s \neq a \wedge a \in t)$ and $q := p \cup \{\langle t, a\rangle\}$ calculated in M is a condition. Then no $r \leqslant^{\mathbb{P}} q$ forces Rts.

The remaining claims can be proved by standard means. We give precise references from [3]. A generic G is generic in the sense of [3, Definition 2.9], and $M[G]$ is defined for every such G (cf. [3, Definition 2.16]). [3, Proposition 2.26] states that up to isomorphism each such model $M[G]$ has the form (M, R_G) as in (d). Then (d), (e), (f) are [3, Theorem 2.19, Corollary 2.20 (2), Corollary 2.20 (3)]. □

It is easy to see that for each $\varphi(\vec{x})$ of the forcing language the set $\{(p, \vec{a}) : p \Vdash \varphi(\vec{a})\}$ is definable in M. There is, however, no good control of the logical complexity of the defining formula. Therefore we use the following auxiliary *strong forcing* relation $\Vvdash$ between conditions and sentences of the forcing language with parameters from M. It is defined via the same recurrence as $\Vdash$ except for the negation clause. Namely, $p \Vvdash \neg\varphi$ is defined as $p \Vdash \neg\varphi$ for atomic φ and otherwise via the recursion:

$$p \Vvdash \neg(\psi \wedge \chi) \iff p \Vvdash \neg\psi \text{ or } p \Vvdash \neg\chi,$$
$$p \Vvdash \neg\neg\psi \iff p \Vvdash \psi,$$
$$p \Vvdash \neg\forall x\, \psi(x) \iff \text{there is } a \in M : p \Vvdash \neg\psi(a).$$

Remark 7.5: One can check that $p \Vvdash \exists x\, \varphi(x)$ if and only if there is $a \in M$ such that $p \Vvdash \varphi(a)$, and $p \Vvdash (\varphi \vee \psi)$ if and only if $p \Vvdash \varphi$ or $p \Vvdash \psi$. Here we understand $\exists, \vee$ by the particular abbreviations mentioned earlier. In this sense $\Vvdash$ commutes with quantifiers and connectives $\wedge, \vee$. The price to pay for these nice properties is that $\Vvdash$ does not behave like a notion of forcing. For example, let $a, b \in M, a \neq b$, and calculate $c := \{a, b\}$ in M; then $1^{\mathbb{P}} \not\Vvdash Rca$ and $1^{\mathbb{P}} \not\Vvdash \neg Rca$, so $1^{\mathbb{P}} \not\Vvdash (Rca \vee \neg Rca)$, and hence Lemma 7.4 (f) fails for $\Vvdash$.

A formula is in *negation normal form* (NNF) if negations appear only in front of atomic subformulas.

Lemma 7.6: *Let φ be a sentence of the forcing language with parameters from M.*

(a) If $p \leqslant^{\mathbb{P}} q$ and $q \Vvdash \varphi$, then $p \Vvdash \varphi$.
(b) If $p \Vvdash \varphi$, then $p \Vdash \varphi$.

(c) *Let $L \subseteq \mathrm{L_{crsf}}$ and $\psi(\vec{x})$ be a $\Delta_0(\mathrm{L}_0^+ \cup L \cup \{R\})$-formula with parameters from M. Then there exists a $\Delta_0(\mathrm{L_O} \cup L)$-formula $\widetilde{\psi}(u, \vec{x})$ with parameters from M such that $\widetilde{\psi}(u, \vec{x})$ defines $\{(p, \vec{a}) : p \Vdash \psi(\vec{a})\}$ in M.*

(d) *If φ is a $\Sigma_1(\mathrm{L_{crsf}} \cup \{R\})$-sentence in NNF with parameters from M and $p \Vdash \varphi$, then there is $q \leqslant^{\mathbb{P}} p$ such that $q \Vvdash \varphi$.*

Proof: (a) and (b) are straightforward. (c) is proved by induction on ψ. We only verify the case when $\psi(\vec{x})$ equals $\neg Rts$ for terms $t = t(\vec{x}), s = s(\vec{x})$. Then define $\widetilde{\psi}(u, \vec{x})$ as

$$u \in \mathbb{P} \wedge \big((t = 0 \wedge s \neq 0) \vee (t \neq 0 \wedge s \notin t) \vee \exists x \in t\, (x \neq s \wedge \langle t, x \rangle \in u)\big).$$

Here, $u \in \mathbb{P}$ abbreviates a suitable $\Delta_0(\mathrm{L_O})$-formula defining $\mathbb{P} \subseteq M$ in M. We have to show that for all $p \in \mathbb{P}$ and $\vec{a}$ from M:

$$M \models \widetilde{\psi}(p, \vec{a}) \Longleftrightarrow \text{for all } q \leqslant^{\mathbb{P}} p : q \nVdash Rt(\vec{a})s(\vec{a}).$$

The direction from left to right is easy to see. Conversely, assume no condition $q \leqslant^{\mathbb{P}} p$ forces $Rt(\vec{a})s(\vec{a})$ and note $M \models p \in \mathbb{P}$. Arguing in M, then $t(\vec{a}) \neq \{s(\vec{a})\}$ and $\langle t(\vec{a}), s(\vec{a})\rangle \notin p$, in particular $t(\vec{a}), s(\vec{a})$ are not both 0. If $t(\vec{a}) = 0$, then $s(\vec{a}) \neq 0$ and $\widetilde{\psi}(p, \vec{a})$ is true. So suppose $t(\vec{a}) \neq 0$. Then $q := p \cup \{\langle t(\vec{a}), s(\vec{a})\rangle\} \notin \mathbb{P}$. Hence $s(\vec{a}) \notin t(\vec{a})$ or there is $a \in t(\vec{a})$ with $a \neq s(\vec{a})$ and $\langle t(\vec{a}), a\rangle \in p$. Both cases imply $\widetilde{\psi}(p, \vec{a})$.

(d). Let $\varphi(\vec{x})$ be a formula of the forcing language with parameters from M. Call $\varphi(\vec{x})$ *good* if for all $\vec{a}$ from M and $p \in \mathbb{P}$: if $p \Vdash \varphi(\vec{a})$, then there is $q \leqslant^{\mathbb{P}} p$ with $q \Vvdash \varphi(\vec{a})$.

Atomic and negated atomic formulas are good, as we can take $q := p$. Good formulas are closed under conjunctions and disjunctions, and $\exists y\, \psi(y, \vec{x})$ is good whenever $\psi(y, \vec{x})$ is good: if $p \Vdash \exists y\, \psi(y, \vec{a})$, then $\bigcup_{b \in M}\{q \mid q \Vdash \psi(b, \vec{a})\}$ is dense below p, so there are $q \leqslant^{\mathbb{P}} p$ and $b \in M$ such that $q \Vdash \psi(b, \vec{a})$; as $\psi(y, \vec{x})$ is good, there is $r \leqslant^{\mathbb{P}} q$ such that $r \Vvdash \psi(b, \vec{a})$ and hence $r \Vvdash \exists y\, \psi(y, \vec{a})$.

Finally, we show that for a good $\Delta_0(\mathrm{L_{crsf}} \cup \{R\})$-formula $\psi(y, \vec{x})$, also $\forall y{\in}t(\vec{x})\, \psi(y, \vec{x})$ is good, where t is a term. If $p \Vdash \forall y{\in}t(\vec{a})\, \psi(y, \vec{a})$, then by Conservativity $p \Vdash \psi(b, \vec{a})$ for all b with $M \models b \in t(\vec{a})$. As $\psi(y, \vec{x})$ is good, we find for every $q \leqslant^{\mathbb{P}} p$ and every such b some $q_b \leqslant^{\mathbb{P}} q$ such that $q_b \Vvdash \psi(b, \vec{a})$. By (WO) we find $s \in M$ which is, in the sense of M, a bijection from an ordinal α onto $t(\vec{a})$. It suffices to find $\pi \in M$ such that π is, in the sense of M, a function with domain α and such that for all $\gamma \in^M \beta \in^M \alpha$:

$$\pi\text{'}\beta \leqslant^{\mathbb{P}} \pi\text{'}\gamma \leqslant^{\mathbb{P}} p \text{ and } \pi\text{'}\beta \Vvdash \psi(s\text{'}\beta, \vec{a}). \tag{7.2}$$

More precisely, the first three ' should read $'^M$. This suffices indeed: by (a), then $q := \bigcup im(\pi)$, calculated in M, is a condition extending p such that $q \Vdash \psi(s'\beta, \vec{a})$ for all $\beta \in^M \alpha$. Thus $q \Vdash \psi(b, \vec{a})$ for all b with $M \models b \in t(\vec{a})$, and hence $q \Vdash \forall y{\in}t(\vec{a})\, \psi(y, \vec{a})$.

To find such π we apply $\Delta_0(\mathrm{L_{crsf}})$-Dependent Choice in M with the following $\Delta_0(\mathrm{L_{crsf}})$-formula $\varphi(x, y)$ with parameters from M:

$$(dom(x) \in \alpha \wedge p \cup \bigcup im(x) \in \mathbb{P} \\ \rightarrow p \cup \bigcup im(x) \subseteq y \wedge y \in \mathbb{P} \wedge \widetilde{\psi}(y, s'dom(x), \vec{a})),$$

where $\widetilde{\psi}$ is as in (c). Since ψ is $\Delta_0(\mathrm{L_{crsf}} \cup \{R\})$, we have $\widetilde{\psi}$ (by (c)) and hence φ in $\Delta_0(\mathrm{L_{crsf}})$. We show that M models $\forall x\, \exists y\, \varphi(x, y)$. Argue in M: given $c \in M$ with $\beta := dom(c) \in \alpha$ and $q := p \cup \bigcup im(c) \in \mathbb{P}$ a witness for y is given by q_b for $b := s'\beta$.

For α as above, choose π witnessing z in (7.1). We claim π satisfies (7.2). It suffices to show $M \models \forall \beta{\in}\alpha\, (p \cup \bigcup im(\pi{\restriction}\beta) \in \mathbb{P})$, or equivalently, $M \models \forall \gamma, \gamma'{\in}\beta\, (p \cup \pi'\gamma \cup \pi'\gamma' \in \mathbb{P})$ for all $\beta \in^M \alpha$. This follows by $\Delta_0(\mathrm{L_0})$-Induction on β and elementary properties of ordinals. □

7.3. *Proof of Theorem 7.2*

It suffices to show that every countable model M of $\mathrm{KP}_1^{\preccurlyeq}(\mathrm{L_{crsf}})$ plus $\Delta_0(\mathrm{L_{crsf}})$-Dependent Choice plus (WO) has an expansion to a model of $\mathrm{KPC}_1^{\preccurlyeq}(\mathrm{L}^C_{\mathrm{crsf}})$. Recall, $\mathrm{KPC}_1^{\preccurlyeq}(\mathrm{L}^C_{\mathrm{crsf}})$ is a $\Sigma_1^{\preccurlyeq}(\mathrm{L}_0^C)$-expansion of $\mathrm{KPC}_1^{\preccurlyeq}$ (cf. Lemma 5.21). It thus suffices to find an expansion of M to a model of $\mathrm{KPC}_1^{\preccurlyeq}$. Using the notation of the Truth Lemma 7.4(d), for every generic G we have that R_G is the graph of a function C satisfying the axiom of global choice (GC). A $\Delta_0(\mathrm{L}_0^C)$-formula in the corresponding expansion is equivalent to a $\Delta_0(\mathrm{L_0} \cup \{R\})$-formula. It thus suffices to show that (M, R_G) satisfies $\Sigma_1^{\preccurlyeq}(\mathrm{L_0} \cup \{R\})$-Induction, $\Delta_0(\mathrm{L_0} \cup \{R\})$-Separation and $\Delta_0(\mathrm{L_0} \cup \{R\})$-Collection.

We start with Induction. So, given a $\Delta_0(\mathrm{L_0} \cup \{R\})$-formula $\psi(y, z)$ with parameters from M, a #-term $t(y)$ with parameters from M and $b \in M$ we have to show that

$$(M, R_G) \models \forall x\, (\forall y{\in}x\, \exists z{\preccurlyeq}t(y)\, \psi(y, z) \rightarrow \exists z{\preccurlyeq}t(x)\, \psi(x, z)) \\ \rightarrow \exists z{\preccurlyeq}t(b)\, \psi(b, z).$$

Recall *IsPair*(x) from Examples 4.6. We define

$$\psi'(y, z) := (\mathit{IsPair}(z) \wedge \pi_1(z) : \pi_2(z) \preccurlyeq t(y) \wedge \psi(y, \pi_2(z))).$$

We can assume that ψ' is in NNF. Recall $\mathrm{L}_0^+ \subseteq \mathrm{L}_{\mathrm{crsf}}$, so M interprets L_0^+. We assume

$$(M, R_G) \models \forall x\,(\forall y{\in}x\,\exists z\,\psi'(y,z) \to \exists z\,\psi'(x,z)) \tag{7.3}$$

and aim to show $(M, R_G) \models \exists z\,\psi'(b,z)$. By the Truth Lemma there exists $p \in G$ such that p forces (7.3). It suffices to show that p forces $\exists z\,\psi'(b,z)$. By Stability it suffices to find, given $p' \leqslant^{\mathbb{P}} p$, some $q \leqslant^{\mathbb{P}} p'$ forcing $\exists z\,\psi'(b,z)$.

By (WO) we find $s \in M$ such that, in the sense of M, s is a bijection from some ordinal α onto $\mathrm{tc}^+(b)$ that respects $\in$, i.e., $M \models (s\text{'}\gamma \in s\text{'}\beta \in \mathrm{tc}^+(b) \to \gamma \in \beta)$. So by Lemma 7.6(b), it suffices to find for every $\beta \in^M \alpha$ a pair $\langle q_\beta, a_\beta\rangle$ (in the sense of M) such that $q_\beta \leqslant^{\mathbb{P}} p'$ and

$$q_\beta \Vdash \psi'(s\text{'}\beta, a_\beta). \tag{7.4}$$

We intend to apply $\Delta_0(\mathrm{L}_{\mathrm{crsf}})$-Dependent Choice with the following formula $\varphi(x,y)$:

$$\begin{aligned}
\varphi(x,y) &:= (\varphi_0(x) \to \varphi_1(x,y)),\\
\varphi_0(x) &:= \mathit{Fct}(x) \wedge \mathit{dom}(x) \in \alpha\\
&\quad\wedge \forall \gamma,\gamma'{\in}\mathit{dom}(x)\,(p' \cup \pi_1(x\text{'}\gamma) \cup \pi_1(x\text{'}\gamma') \in \mathbb{P})\\
&\quad\wedge \forall \gamma{\in}\mathit{dom}(x)\,\big(\mathit{IsPair}(x\text{'}\gamma) \wedge \widetilde{\psi}'(\pi_1(x\text{'}\gamma), s\text{'}\gamma, \pi_2(x\text{'}\gamma))\big),\\
\varphi_1(x,y) &:= \mathit{IsPair}(y) \wedge \forall \gamma{\in}\mathit{dom}(x)\,(p' \cup \pi_1(x\text{'}\gamma) \subseteq \pi_1(y))\\
&\quad\wedge \widetilde{\psi}'(\pi_1(y), s\text{'}\mathit{dom}(x), \pi_2(y)),
\end{aligned}$$

where $\widetilde{\psi}'$ is defined as in Lemma 7.6(c).

We have $\varphi \in \Delta_0(\mathrm{L}_{\mathrm{crsf}})$ by Lemma 7.6(c). We show $M \models \forall x\,\exists y\,\varphi(x,y)$. Let $c \in M$ and assume $M \models \varphi_0(c)$. We have to show $M \models \exists y\,\varphi_1(c,y)$. Compute

$$\begin{aligned}
\beta &:= \mathit{dom}(c)\\
q &:= p' \cup \textstyle\bigcup_{\gamma\in\beta} \pi_1(c\text{'}\gamma)
\end{aligned}$$

in M (this can be done: for $f(x)$ such that $\mathrm{KP}_1^{\preccurlyeq}(\mathrm{L}_{\mathrm{crsf}})$ proves $f(x) = \pi_1(\pi_2(x))$, we have $q = p' \cup \bigcup f\text{"}(c)$ in M). Then $q \in \mathbb{P}$ extends $(\pi_1(c\text{'}\gamma))^M$ for all $\gamma \in^M \beta$. By Lemma 7.6(a), $q \Vdash \psi'(s\text{'}\gamma, \pi_2(c\text{'}\gamma))$ for all $\gamma \in^M \beta$, and hence $q \Vdash \exists z\,\psi'(s\text{'}\gamma, z)$ for all $\gamma \in^M \beta$. This implies $q \Vdash \exists z\,\psi'(d,z)$ for all $d \in M$ with $M \models d \in s\text{'}\beta$. By Lemma 7.6 (b), we see that q forces $\forall y{\in}s\text{'}\beta\,\exists z\,\psi'(y,z)$. But $q \leqslant^{\mathbb{P}} p' \leqslant^{\mathbb{P}} p$, so by Extension q forces (7.3). Plugging $s\text{'}\beta$ for x in (7.3) and recalling Lemma 7.4 (f) we see that $q \Vdash \exists z\,\psi'(s\text{'}\beta, z)$. This is a $\Sigma_1(\mathrm{L}_{\mathrm{crsf}} \cup \{R\})$-sentence in NNF with parameters from M, so Lemma 7.6 (d) gives $q_\beta \leqslant^{\mathbb{P}} q$ and $a_\beta \in M$ such that $q_\beta \Vdash \psi'(s'\beta, a_\beta)$. Then $M \models \varphi_1(c, \langle q_\beta, a_\beta \rangle)$ and thus $M \models \exists y\, \varphi_1(c, y)$.

By Dependent Choice there is $\pi \in M$, in the sense of M a function with domain α, such that $M \models \varphi(\pi \restriction \beta, \pi'\beta)$ for all $\beta \in^M \alpha$.

To show (7.4) it suffices to show $M \models \varphi_0(\pi \restriction \beta)$ for all $\beta \in^M \alpha$, or equivalently

$$\begin{aligned} &\forall \gamma, \gamma' {\in} \beta\, (p' \cup \pi_1(\pi'\gamma) \cup \pi_1(\pi'\gamma') \in \mathbb{P}) \\ &\wedge\ \forall \gamma {\in} \beta\, (\mathit{IsPair}(\pi'\gamma) \wedge \widetilde{\psi}'(\pi_1(\pi'\gamma), s'\gamma, \pi_2(\pi'\gamma))) \end{aligned}$$

holds in M for all $\beta \in^M \alpha$. By Lemma 7.6(c), this can be written $\chi(\beta)$ for a $\Delta_0(\mathrm{L}_0^+)$-formula $\chi(x)$ with parameters from M. Since $\Delta_0(\mathrm{L}_0^+)$-Induction holds in M, it suffices to verify $M \models \chi(\beta)$ assuming $M \models \forall \gamma {\in} \beta\, \chi(\gamma)$. This is easy. Thus (M, R_G) satisfies $\Delta_0(\mathrm{L}_0 \cup \{R\})$-Induction.

We show that (M, R_G) satisfies $\Delta_0(\mathrm{L}_0 \cup \{R\})$-Collection. Let $\psi(y, z)$ be a $\Delta_0(\mathrm{L}_0 \cup \{R\})$-formula with parameters from M and $a \in M$ such that (M, R_G) satisfies

$$\varphi := \forall y {\in} a\, \exists z\, \psi(y, z).$$

By the Truth Lemma, φ is forced by some $p \in G$. Arguing as for (7.4) we can find $q \leqslant^{\mathbb{P}} p$ such that for all $b \in^M a$ there is $c \in M$ such that $q \Vdash \psi(b, c)$ (observe that the proof of (7.4) gave a descending chain of q_β's) – equivalently: $q \Vdash \varphi$. Note $q \Vdash \varphi$ if and only if

$$M \models \forall y {\in} a\, \exists z\, \widetilde{\psi}(q, y, z),$$

where $\widetilde{\psi}$ is $\Delta_0(\mathrm{L}_0)$, chosen according Lemma 7.6(c). Then $\{q : q \Vdash \varphi\}$ is M-definable and dense below p, so we find such q in G. Applying $\Delta_0(\mathrm{L}_0)$-Collection in M we get

$$M \models \exists V\, \forall y {\in} a\, \exists z \in V\, \widetilde{\psi}(q, y, z).$$

But for all $b, c \in M$ we have

$$(M, R_G) \models (\widetilde{\psi}(q, b, c) \rightarrow \psi(b, c)),$$

by Lemma 7.6(b) and the Truth Lemma. Thus (M, R_G) satisfies $\Delta_0(\mathrm{L}_0 \cup \{R\})$-Collection.

We show that (M, R_G) satisfies $\Delta_0(\mathrm{L}_0 \cup \{R\})$-Separation. Let $a \in M$ and $\varphi(x)$ be a $\Delta_0(\mathrm{L}_0 \cup \{R\})$-formula with parameters from M. We can assume $\varphi(x)$ is in NNF. Let $\overline{\varphi}(x)$ be logically equivalent to $\neg\varphi(x)$ and in NNF. By the Truth Lemma it suffices to show $1^{\mathbb{P}} \Vdash \exists z\, (z = \{x \in a : \varphi(x)\})$.

By Stability it suffices to show that for every $p \in \mathbb{P}$ there is $q \leqslant^{\mathbb{P}} p$ such that $q \Vdash \exists z\, (z = \{x \in a : \varphi(x)\})$.

Let $p \in \mathbb{P}$ be given. We claim that it suffices to find $q \leqslant^{\mathbb{P}} p$ that *strongly decides* $\varphi(b)$ for every $b \in^M a$ in the sense that $q \Vvdash \varphi(b)$ or $q \Vvdash \overline{\varphi}(b)$. Indeed, such a q forces $\exists z\, (z = \{x \in a : \varphi(x)\})$. By Forcing Completeness we have to show $(M, R_{G'}) \models \exists z\, (z = \{x \in a : \varphi(x)\})$ for every generic G' containing q. But z is witnessed by $\{x \in a : \tilde{\varphi}(q, x)\}$, a set obtainable in M by $\Delta_0(\mathrm{L}_0)$-Separation (Lemma 7.6(c)). To see this, we verify for every $b \in^M a$:

$$(M, R_{G'}) \models (\varphi(b) \leftrightarrow \tilde{\varphi}(q, b)).$$

The direction from right to left follows from Lemma 7.6(b) and the Truth Lemma. Conversely, assuming $(M, R_{G'}) \models \varphi(b)$ the Truth Lemma gives $r \in G'$ forcing $\varphi(b)$; then $r \cup q \in G'$ since G' is a filter, so $r \cup q$ forces $\varphi(b)$ by Extension, so cannot force $\overline{\varphi}(b)$ by Lemma 7.4(f), so $q \nVdash \overline{\varphi}(b)$ by Extension, so $q \not\Vvdash \overline{\varphi}(b)$ by Lemma 7.6(b), so $q \Vvdash \varphi(b)$ and $(M, R_{G'}) \models \tilde{\varphi}(q, b)$ since q strongly decides $\varphi(b)$.

Thus, given a condition p, we are looking for $q \leqslant^{\mathbb{P}} p$ that strongly decides $\varphi(b)$ for every $b \in^M a$. By (WO) choose $s \in M$ such that, in the sense of M, s is a bijection from α onto a. A condition q as desired is obtained in M as the union of a descending sequence $(q_\beta)_{\beta \in \alpha}$ with $q_0 \leqslant^{\mathbb{P}} p$ such that each q_β strongly decides $\varphi(q_\beta, s\text{'}\beta)$. To get such a sequence in M we apply $\Delta_0(\mathrm{L}_{\mathrm{crsf}})$-Dependent Choice on the following formula $\psi(x, y)$:

$$\begin{aligned}(&Fct(x) \wedge dom(x) \in \alpha \wedge \forall \gamma, \gamma' {\in} dom(x)\, x\text{'}\gamma \cup x\text{'}\gamma' \in \mathbb{P} \\ &\quad \to \forall \gamma {\in} dom(x)\, (p \cup x\text{'}\gamma \subseteq y) \wedge (\tilde{\varphi}(y, s\text{'}dom(x)) \vee \tilde{\overline{\varphi}}(y, s\text{'}dom(x)))).\end{aligned} \tag{7.5}$$

A function π with domain α (in the sense of M) such that $M \models \psi(\pi \upharpoonright \beta, \pi\text{'}\beta)$ for all $\beta \in^M \alpha$, is a sequence as desired. We are left to show

$$M \models \forall x\, \exists y\, \psi(x, y).$$

Let $c \in M$ satisfy the antecedent of (7.5), and compute $\beta := dom(c)$ and $q^0 := \bigcup im(c)$ in M. Then $q^0 \in \mathbb{P}$. There exists $q^1 \leqslant^{\mathbb{P}} q^0$ such that $q^1 \Vdash \varphi(s\text{'}\beta)$ or $q^1 \Vdash \overline{\varphi}(s\text{'}\beta)$. Indeed, by Stability, if $q_0 \nVdash \varphi(s\text{'}\beta)$, then there is $q^1 \leqslant^{\mathbb{P}} q^0$ such that $r \nVdash \varphi(s\text{'}\beta)$ for all $r \leqslant^{\mathbb{P}} q^1$, i.e., $q^1 \Vdash \neg\varphi(s\text{'}\beta)$ and hence $q^1 \Vdash \overline{\varphi}(s\text{'}\beta)$ by Lemma 7.4(f). Lemma 7.6(d) gives $q^2 \leqslant^{\mathbb{P}} q^1$ such that $q^2 \Vvdash \varphi(s\text{'}\beta)$ or $q^2 \Vvdash \overline{\varphi}(s\text{'}\beta)$ respectively. Then $M \models \psi(c, q^2)$.

Acknowledgements

We thank the anonymous referee for a careful reading and detailed comments. Part of this work was done when the first four authors attended the workshop *Sets and Computations* at the Institute for Mathematical Sciences, National University of Singapore.

References

1. T. Arai, *Axiomatizing some small classes of set functions*. Preprint, `arXiv:1503.07932`, March 2015.
2. ——, *Predicatively computable functions on sets*, Archive for Mathematical Logic, 54 (2015), pp. 471–485.
3. A. Atserias and M. Müller, *Partially definable forcing and bounded arithmetic*, Archive for Mathematical Logic, 54 (2015), pp. 1–33.
4. J. Avigad, *Saturated models of universal theories*, Annals of Pure and Applied Logic, 118 (2002), pp. 219–234.
5. J. Barwise, *Admissible Sets and Structures: An Approach to Definability Theory*, Perspectives in Mathematical Logic, Number 7, Springer Verlag, Berlin, 1975.
6. A. Beckmann, S. R. Buss, and S.-D. Friedman, *Safe recursive set functions*, Journal of Symbolic Logic, 80 (2015), pp. 730–762.
7. A. Beckmann, S. R. Buss, S.-D. Friedman, M. Müller, and N. Thapen, *Cobham recursive set functions*, Annals of Pure and Applied Logic, 167 (2016), pp. 335–369.
8. ——, *Subset-bounded recursion and a circuit model for feasible set functions*. Manuscript in preparation, 2016.
9. S. Bellantoni and S. Cook, *A new recursion-theoretic characterization of the polytime functions*, Computational Complexity, 2 (1992), pp. 97–110.
10. A. Blass, Y. Gurevich, and S. Shelah, *Choiceless polynomial time*, Annals of Pure and Applied Logic, 100 (1999), pp. 141–187.
11. S. R. Buss, *Bounded Arithmetic*, Bibliopolis, 1986. Revision of 1985 Princeton University Ph.D. thesis.
12. ——, *First-order proof theory of arithmetic*, in Handbook of Proof Theory, S. R. Buss, ed., North-Holland, 1998, pp. 79–147.
13. A. Cobham, *The intrinsic computational difficulty of functions*, in Logic, Methodology and Philosophy of Science, Proceedings of the Second International Congress, held in Jerusalem, 1964, Y. Bar-Hillel, ed., Amsterdam, 1965, North-Holland, pp. 24–30.
14. S. A. Cook, *Feasibly constructive proofs and the propositional calculus*, in Proceedings of the Seventh Annual ACM Symposium on Theory of Computing, 1975, pp. 83–97.
15. S. A. Cook and P. Nguyen, *Foundations of Proof Complexity: Bounded Arithmetic and Propositional Translations*, ASL and Cambridge University Press, 2010. 496 pages.

16. H.-D. EBBINGHAUS AND J. FLUM, *Finite Model Theory*, Springer Verlag, Berlin, second ed., 1995.
17. U. FELGNER, *Comparison of the axioms of local and universal choice*, Fundamenta Mathematicae, 71 (1971), pp. 43–62.
18. J. D. HAMKINS AND A. LEWIS, *Infinite time Turing machines*, Journal of Symbolic Logic, 65 (2000), pp. 567–604.
19. T. JECH, *Set Theory*, Springer, Berlin, third ed., 2006.
20. R. B. JENSEN AND C. KARP, *Primitive recursive set functions*, in Axiomatic Set Theory, Proceedings of Symposia in Pure Mathematics, Volume XIII, Part I, American Mathematical Society, Providence, R.I., 1971, pp. 143–176.
21. R. J. PARIKH, *Existence and feasibility in arithmetic*, Journal of Symbolic Logic, 36 (1971), pp. 494–508.
22. M. RATHJEN, *A proof-theoretic characterization of the primitive recursive set functions*, Journal of Symbolic Logic, 57 (1992), pp. 954–969.
23. V. Y. SAZONOV, *On bounded set theory*, in Logic and Scientific Methods, M. L. Dalla Chiara, M. L. Doets, K. Mundici, and J. van Benthem, eds., Synthese Library Volume 259, Kluwer Academic, 1997, pp. 85–103.
24. D. ZAMBELLA, *Notes on polynomially bounded arithmetic*, Journal of Symbolic Logic, 61 (1996), pp. 942–966.

HIGHER RANDOMNESS AND LIM-SUP FORCING WITHIN AND BEYOND HYPERARITHMETIC

Takayuki Kihara

Department of Mathematics
University of California, Berkeley
970 Evans Hall, CA 94720-3840, USA
kihara.takayuki.logic@gmail.com

We develop arboreal forcing in the context of hyperarithmetical randomness theory. In any transitive model of the Kripke-Platek set theory obtained as the companion of a Spector pointclass, we show that certain kinds of Σ_1-definable lim-sup tree creature forcings have the Σ_1-continuous reading of names, the Σ_1-fusion property, and so on. In this way, we show that the shape of Cichoń's diagram in hyperarithmetic theory is different from those in computability theory and set theory.

1. Summary

1.1. *Introduction*

In recent years, the theory of *hypercomputation* (it has been widely known as *generalized recursion theory* in the 20th century) have again received a lot of attention in computability theory and theoretical computer science. Of course, the theory of hypercomputation (in particular, recursion theory beyond hyperarithmetic, with many kinds of higher-type and ordinal computation models) was one of the central topics in computability theory in the mid to late 20th century, and an enormous number of mathematically deep researches on hypercomputation have been accomplished in parallel with developments of descriptive set theory, infinitary logic, and fine structure theory of Gödel's constructible universe (see also some textbooks [3, 16, 19, 38, 48] for developments of generalized recursion theory in the second half of the 20th century).

The randomness notion in the hyperarithmetical level was first introduced by Martin-Löf and developed by several researchers (see [21, 47, 52]) in the early days. After a blank period, very recently, algorithmic randomness researchers restarted to develop higher randomness theory based on modern notions and ideas from advanced study on algorithmic randomness ([4, 10, 11, 12, 31, 36, 37]; see also Nies [40] (Chapter 9)). Technically, typical separation results in modern higher randomness theory are obtained by using standard arboreal forcing arguments such as Sacks forcing and clumpy Sacks forcing (infinite equal forcing). Indeed, some parts of modern algorithmic randomness theory have a strong connection to forcing theory, in particular, to the theory of cardinal characteristics (see Rupprecht [45, 46], Brendle et al. [6]). In modern set theory, these kinds of arboreal forcing, idealized forcing, and creature forcing have been deeply investigated in a sophisticatedly systematic manner (see Bartozyński-Judah [2], Rosłanowski-Shelah [44] and Zapletal [57] for instance). Our aim is to reflect the modern development of set theory in higher computability theory. In particular, we develop the general framework of *lim-sup tree forcing* in admissible sets generated by Spector pointclasses. We also propose a list of open problems.

2. Preliminaries

2.1. *Represented spaces*

In this paper, most objects are coded by an element of ω^ω; e.g., a Borel set is coded by a Borel code. A *represented space* (see also [54]) is a set $\mathcal{X}$ equipped with a partial surjection $\rho_\mathcal{X} :\subseteq \omega^\omega \to \mathcal{X}$.[a] For instance, the pair $(\mathcal{B}, \mathrm{BC})$ of Borel sets $\mathcal{B}$ and Borel coding $\mathrm{BC} :\subseteq \omega^\omega \to \mathcal{B}$ forms a represented space.

Example 2.1: The following are examples of represented spaces (see also [29, 28]):

(1) (Hereditarily Countable Sets) The collection $\mathbf{H}_{\aleph_1}$ of all hereditarily countable sets is represented by $\rho_{\mathbf{HC}} :\subseteq \omega^\omega \to \mathbf{H}_{\aleph_1}$ such that $\rho_{\mathbf{HC}}(p) = X$ if and only if

$$(\omega, \{(n,m) \in \omega^2 : p(\langle n,m\rangle) = 1\}) \text{ is isomorphic to } (X, \in).$$

[a] To develop the framework of higher randomness theory, the author [28] used $\mathcal{O}\omega$-representations rather than ω^ω-representations since the paper [28] covers partial objects such as the Martin-Löf representation of Lebesgue null sets. If we wish to cover such a partial object in higher computability theory, we cannot replace $\mathcal{O}\omega$-representability with ω^ω-representability; however, in this article, we only work with total objects such as the Schnorr representation of Lebesgue null sets.

Here, $\langle \cdot, \cdot \rangle$ is a fixed effective bijection between ω^2 and ω.

(2) (Borel Coding) One can consider an ω^ω-representation of the class of all Borel sets (or all $\mathbf{\Sigma}^0_\alpha$ sets for a fixed rank $\alpha < \omega_1$) in a (recursively presented) Polish space $\mathcal{X}$. Formally, we introduce codings (representations) $\sigma^0_n : \omega^\omega \to \mathbf{\Sigma}^0_n(\mathcal{X})$ and $\pi^0_n : \omega^\omega \to \mathbf{\Pi}^0_n(\mathcal{X})$ in the following inductive way:

$$\sigma^0_1(p) = \bigcup_{m\in\omega} B_{p(m)}, \quad \pi^0_n(p) = \mathcal{X} \setminus \sigma^0_n(p), \quad \sigma^0_n(p) = \bigcup_{k\in\omega} \pi^0_n(p^{[k]}).$$

Here, $(B_n)_{n\in\omega}$ is a countable basis of $\mathcal{X}$, and the kth section $p^{[k]}$ is defined by $p^{[k]}(m) = p(\langle k, m\rangle)$.

(3) Let $\mathcal{N}$ be the set of all null G_δ sets, and $\mathcal{M}$ be the set of all meager F_σ sets. Then, $\mathcal{N}$ and $\mathcal{M}$ are represented by the following representations:

$$\rho_{\mathcal{N}}(p) = \pi^0_2(p) = \bigcap_{n\in\omega} \sigma^0_1(p^{[n]}),$$

$$\mathrm{dom}(\rho_{\mathcal{N}}) = \{p : \lambda(\sigma^0_1(p^{[n]})) = 2^{-n} \text{ for all } n \in \omega\}.$$

$$\rho_{\mathcal{M}}(p) = \sigma^0_2(p) = \bigcup_{n\in\omega} \pi^0_1(p^{[n]}),$$

$$\mathrm{dom}(\rho_{\mathcal{M}}) = \{p : \pi^0_1(p^{[n]}) \text{ is nowhere dense for all } n \in \omega\}.$$

Here, λ is the Lebesgue measure on 2^ω.

Hereafter, for a pointclass Γ in a space $\mathcal{X}$ (that is, Γ is a collection of subsets of $\mathcal{X}$), we write $\check{\Gamma}$ for the dual of a pointclass Γ (that is, $\check{\Gamma} = \{A \subseteq \mathcal{X} : \mathcal{X} \setminus A \in \Gamma\}$), and Δ for the ambiguous pointclass $\Gamma \cap \check{\Gamma}$ as usual (see [39]). For a pointclass Γ in ω, we say that a point $x \in \omega^\omega$ is in Δ if $\{\langle m, n\rangle : x(m) = n\} \in \Delta$.

Definition 2.2: (See also [28]) Let $(\mathcal{X}, \rho_{\mathcal{X}})$ be a represented space, and Γ be a pointclass.

(1) We say that a point x in a represented space $(\mathcal{X}, \rho_{\mathcal{X}})$ is *Δ-computable* if x has a Δ-name, that is, there is $p \in \omega^\omega \cap \Delta$ such that $x = \rho_{\mathcal{X}}(p)$.

(2) A *Δ-computation* is a Γ-subset Φ of $\omega^\omega \times \omega^2$. We think of such Φ as a partial Γ-measurable function $\Phi :\subseteq \omega^\omega \to \omega^\omega$ ([28, 39]) by declaring that $\Phi(x) = p$ if the following holds:

$$p(m) = n \iff (x, m, n) \in \Phi.$$

(3) We say that $x \in \mathcal{X}$ *is Δ-reducible to* $y \in \omega^\omega$, or $y \in \omega^\omega$ *Δ-computes* $x \in \mathcal{X}$ (written as $x \leq_\Delta y$) if x has a $\Delta(y)$-name. In other words,

there is a Δ-computation Φ such that $\Phi(y)$ is a name of x, that is, $\Phi(y) \in \rho_{\mathcal{X}}^{-1}\{x\}$.

Note that Δ-reducibility is equivalent to *relative* Δ-*definability* if $\mathcal{X} = 2^\omega$. Many natural reducibility notions in generalized recursion theory are obtained as effective measurable maps with respect to reasonable Σ-pointclasses (typically, Spector pointclasses). A *Spector pointclass* is a collection Γ of pointsets including Σ^0_1, closed under $\wedge$, $\vee$, $\exists^\omega$, and $\forall^\omega$, has the substitution property, ω-parametrized, and normed (see also Moschovakis [39] and Kechris [23, 24]). The notion of a Spector pointclass covers many kinds of generalized computations such as hyperarithmetical definability, (the 1- and 2-envelops of) finite type computability, and infinite-time-Turing-machine computability. For instance, the class of hyperarithmetical reductions is clearly equivalent to that of partial effective Π^1_1-measurable maps. Of course, there are many known natural computability-theoretic and descriptive set-theoretic pointclasses between Π^1_1 and Δ^1_2 (see [7, 8]), and hence, we have many natural reducibility notions between hyperarithmetical (Δ^1_1) reducibility and Δ^1_2-reducibility.

2.2. *Tukey morphism and Muchnik reduction*

A *Vojtáš triple* (see [5, 13]) is a multi-valued map $A : A_- \rightrightarrows A_+$ between represented spaces, that is, A_- and A_+ are represented via surjections $\rho_A^- : \widehat{A}_- \to A_-$ and $\rho_A^+ : \widehat{A}_+ \to A_+$ where $\widehat{A}_-$ and $\widehat{A}_+$ are subsets of ω^ω. The *dual* of A is a multi-valued map $A^\perp : A_+ \rightrightarrows A_-$ defined by $y \in A^\perp(x)$ if and only if $x \notin A(y)$. For Vojtáš triples A and B, a Δ-*Tukey morphism* from A to B is a pair of Γ-measurable maps $H : \widehat{A}_- \to \widehat{B}_-$ and $K : \widehat{B}_+ \to \widehat{A}_+$ such that

if G is a realizer of B, then $K \circ G \circ H$ is a realizer of A.

Here, a *realizer* of a multi-valued map $A : A_- \rightrightarrows A_+$ is a single valued map $\alpha : \widehat{A}_- \to \widehat{A}_+$ such that for a given name x of an instance of A, $\alpha(x)$ returns a name of an A-solution of the instance, that is, $\rho_A^+ \circ \alpha(x) \in A \circ \rho_A^-(x)$ for all $x \in \widehat{A}_-$. If a Δ-Tukey morphism exists from A to B, we write $A \leq_{\Delta \mathrm{sW}} B$ or simply $A \longrightarrow B$. Note that if $A \longrightarrow B$, then $||A|| \leq ||B||$, where $||C||$ is the associated cardinal of a Vojtáš triple $C : C_- \rightrightarrows C_+$ defined by

$$||C|| := \min\{\mathrm{card}(F) : (\forall x \in C_-)(\exists y \in F)\; y \in C(x)\}.$$

For a Vojtáš triple A, let $[A]_\Delta$ be the set of all instances $x \in A_-$ such that $A(x)$ has no Δ-computable solution, that is,

$$[A]_\Delta = \{x \in A_- : (\forall y \in A_+)\ y \leq_\Delta 0 \ \to\ y \notin A(x)]\}.$$

One can easily check that $[A^\perp]_\Delta$ is the set of all $y \in A_+$ such that y is an A-solution to all Δ-computable instances $x \in A_-$, that is,

$$[A^\perp]_\Delta = \{y \in A_+ : (\forall x \in A_-)\ x \leq_\Delta 0 \ \to\ y \in A(x)]\}.$$

For a represented space $\mathcal{X}$, the Δ-*degree spectrum* (see [30]) of $S \subseteq \mathcal{X}$ is the $\leq_\Delta$-upward closure of S, that is, $\{y \in 2^\omega : (\exists x \in S)\ x \leq_\Delta y\}$. The Δ^0_1-degree spectrum of $[A]_{\Delta^0_1}$ is also called the *Turing norm* of A (see Rupprecht [45] (Definition IV.1)). The Turing norm is also studied by Brendle et al. [6]. We say that $P \subseteq \mathcal{X}$ is Δ-*Muchnik reducible to* $Q \subseteq \mathcal{Y}$ (written as $P \leq_{\Delta\mathrm{w}} Q$) if the Δ-degree spectrum of Q is included in that of P, that is,

$$(\forall y \in Q)(\forall q \in 2^\omega)\ [y \leq_\Delta q \ \to\ (\exists x \in P)\ x \leq_\Delta q].$$

Proposition 2.3: *Let A and B be any Vojtáš triples. If $A \leq_{\Delta\mathrm{sW}} B$ holds, then we have $[B]_\Delta \leq_{\Delta\mathrm{w}} [A]_\Delta$ and $[A^\perp]_\Delta \leq_{\Delta\mathrm{w}} [B^\perp]_\Delta$.*

Proof: Assume that $A \leq_{\Delta\mathrm{sW}} B$ via a Δ-Tukey morphism (H, K). Fix a real $q \in 2^\omega$. We first claim that for any $x \leq_\Delta q$, $x \in [A]_\Delta$ implies $H(x) \in [B]_\Delta$, that is, H is a Δ-Muchnik reduction witnessing $[B]_\Delta \leq_{\Delta\mathrm{w}} [A]_\Delta$. If $H(x) \notin [B]_\Delta$, then $B(H(x))$ has a Δ-computable solution y. Since (H, K) is a Δ-Tukey morphism, $K(y)$ is a solution to $A(x)$. Moreover, $K(y)$ is Δ-computable, since $K(y) \leq_\Delta y \leq_\Delta 0$. However, this contradicts our assumption that $A(x)$ is has no Δ-computable solution. We next claim that for any $y \leq_\Delta q$, $y \in [B^\perp]_\Delta$ implies $K(y) \in [A^\perp]_\Delta$, that is, K is a Δ-Muchnik reduction witnessing $[A^\perp]_\Delta \leq_{\Delta\mathrm{w}} [B^\perp]_\Delta$. To see this, let x be a Δ-computable instance of A. Then, $H(x)$ is an Δ-computable instance of B; therefore, y is a solution to $B(H(x))$ since $y \in [B^\perp]_\Delta$. Therefore, $K(y)$ is a solution to $A(x)$ since (H, K) is a Δ-Tukey morphism. Consequently, $K(y)$ is an A-solution to all Δ-computable instances. □

By $\mathcal{M}$ and $\mathcal{N}$ we denote σ-ideals consisting of meager sets and null sets endowed with the representations $\rho_\mathcal{M}$ and $\rho_\mathcal{N}$, respectively. We identify a Vojtáš triple with a triple (A_-, A_+, A), where we think of A as a relation $A \subseteq A_- \times A_+$ rather than a multi-valued map. Let $\mathcal{J}$ be a represented σ-ideal (e.g., $\mathcal{M}$ and $\mathcal{N}$). Consider the following Vojtáš triples (see also

Rupprecht [45] (Examples III.6 and 7)):

$$\begin{aligned} \mathsf{Add}\mathcal{J} &:= (\mathcal{J}, \mathcal{J}, \not\supseteq), & \mathsf{Cof}\mathcal{J} &:= \mathsf{Add}\mathcal{J}^{\perp} = (\mathcal{J}, \mathcal{J}, \subseteq), \\ \mathsf{Cov}\mathcal{J} &:= (2^{\omega}, \mathcal{J}, \in), & \mathsf{Non}\mathcal{J} &:= \mathsf{Cov}\mathcal{J}^{\perp} = (\mathcal{J}, 2^{\omega}, \not\ni), \\ \mathsf{B} &:= (\omega^{\omega}, \omega^{\omega}, \not\geq^{*}), & \mathsf{D} &:= \mathsf{B}^{\perp} = (\omega^{\omega}, \omega^{\omega}, \leq^{*}), \\ \mathsf{IE} &:= (\omega^{\omega}, \omega^{\omega}, =^{\infty}), & \mathsf{ED} &:= \mathsf{IE}^{\perp} = (\omega^{\omega}, \omega^{\omega}, \neg =^{\infty}), \\ \mathsf{IE}_h &:= ([h], [h], =^{\infty}), & \mathsf{ED}_h &:= \mathsf{IE}_h^{\perp} = ([h], [h], \neg =^{\infty}). \end{aligned}$$

Here, $f \leq^* g$ if and only if $f(n) \leq g(n)$ for almost all $n \in \omega$; $f =^\infty g$ if and only if $f(n) = g(n)$ for infinitely many $n \in \omega$, and $[h]$ is the represented subspace of ω^ω consisting of all h-bounded functions, that is, $[h] = \{g \in \omega^\omega : (\forall n \in \omega)\ g(n) < h(n)\}$ whose representation is induced from the identical representation of ω^ω.

Definition 2.4: (See also Brendle et al. [6]) Let $x \in 2^\omega$.

(1) x is *Δ-$\mathcal{J}$-engulfing* if there is $y \leq_\Delta x$ such that $y \in [\mathsf{Add}\mathcal{J}]_\Delta$, that is, x Δ-computes a $\mathcal{J}$-null set which covers all Δ-computable $\mathcal{J}$-null sets.
(2) x is *not Δ-$\mathcal{J}$-low* if there is $y \leq_\Delta x$ such that $y \in [\mathsf{Cof}\mathcal{J}]_\Delta$, that is, x Δ-computes a $\mathcal{J}$-null set which is not covered by any Δ-computable $\mathcal{J}$-null sets.
(3) x is *above-Δ-$\mathcal{J}$-quasigeneric* if there is $y \leq_\Delta x$ such that $y \in [\mathsf{Cov}\mathcal{J}]_\Delta$, that is, x Δ-computes a real which avoids all Δ-computable $\mathcal{J}$-null sets.
(4) x is *weakly Δ-$\mathcal{J}$-engulfing* if there is $y \leq_\Delta x$ such that $y \in [\mathsf{Non}\mathcal{J}]_\Delta$, that is, x Δ-computes a $\mathcal{J}$-null set which covers all Δ-computable reals.
(5) x is *Δ-dominant* if there is $y \leq_\Delta x$ such that $y \in [\mathsf{B}]_\Delta$, that is, x Δ-computes a function $f \in \omega^\omega$ which dominates all Δ-computable functions $g \in \omega^\omega$.
(6) x is *Δ-unbounded* if there is $y \leq_\Delta x$ such that $y \in [\mathsf{D}]_\Delta$, that is, x Δ-computes a function $f \in \omega^\omega$ which not dominated by any Δ-computable function $g \in \omega^\omega$.
(7) x is *Δ-h-eventually different* if there is $y \leq_\Delta x$ such that $y \in [\mathsf{IE}_h]_\Delta$, that is, x Δ-computes a h-bounded function $f \in \omega^\omega$ which eventually disagrees with any Δ-computable h-bounded function $g \in \omega^\omega$. We also say that x is *Δ-eventually different* if there is $y \leq_\Delta x$ such that $y \in [\mathsf{IE}]_\Delta$, that is, it is Δ-h-eventually different for all $h \in \omega^\omega \cap \Delta$.
(8) x is *Δ-h-infinitely often equal* if there is $y \leq_\Delta x$ such that $y \in [\mathsf{ED}_h]_\Delta$, that is, x Δ-computes a h-bounded function $f \in \omega^\omega$ which agrees infinitely often with any Δ-computable h-bounded function $g \in \omega^\omega$. We

also say that x is Δ-*infinitely often equal* if there is $y \leq_\Delta x$ such that $y \in [\mathsf{ED}]_\Delta$, that is, it is Δ-h-infinitely often equal for all $h \in \omega^\omega \cap \Delta$.

2.3. *Background and summary*

Over the past ten years, the notion of lowness properties was one of the central topics in algorithmic randomness theory [14, 40]. One of the important topics in algorithmic and higher randomness theory is the separation of various lowness notions in the sense of Muchnik reducibility. Known separation results of Cichoń's diagram in the context of Δ^0_1-Muchnik reducibility are summarized in [6]. For instance, consider the following Tukey morphisms:

$$\mathsf{Non}\mathcal{M} \longrightarrow \mathsf{Cof}\mathcal{M} \longrightarrow \mathsf{Cof}\mathcal{N}.$$

A standard forcing argument shows that there are no reversal definable morphisms. The same holds true for Δ^0_1-Muchnik reducibility. That is, the two facts that there is a real which is low for Kurtz randomness but not low for Schnorr randomness and that there is a real which is low for the pair of Schnorr randomness and Kurtz-randomness but not low for Kurtz randomness imply the following inequalities:

$$[\mathsf{Cof}\mathcal{N}]_{\Delta^0_1} <_{\Delta^0_1\mathrm{w}} [\mathsf{Cof}\mathcal{M}]_{\Delta^0_1} <_{\Delta^0_1\mathrm{w}} [\mathsf{Non}\mathcal{M}]_{\Delta^0_1}. \tag{2.1}$$

The main objects of higher randomness theory are randomness notions relative to the Spector pointclass Π^1_1 (and its companion model $L_{\omega_1^{\mathrm{CK}}}$). In such a theory, Chong et al. [10] used Sacks forcing to show that there are continuum many Δ^1_1-traceable reals (that is, reals that are low for Δ^1_1-randomness). Moreover, Kjos-Hanssen et al. [31] used forcing with clumpy trees to show that there are continuum many reals that are low for Δ^1_1-Kurtz randomness, but not low for Δ^1_1-randomness. The latter result implies that:

$$[\mathsf{Cof}\mathcal{N}]_{\Delta^1_1} <_{\Delta^1_1\mathrm{w}} [\mathsf{Cof}\mathcal{M}]_{\Delta^1_1} \leq_{\Delta^1_1\mathrm{w}} [\mathsf{Non}\mathcal{M}]_{\Delta^1_1}.$$

Later, we will see that the latter inequality is also proper, that is, we will show that there is a real which is low for the pair of Δ^1_1-randomness and Δ^1_1-Kurtz-randomness, but not low for Δ^1_1-Kurtz randomness (not low for Δ^1_1-generic as well).

We are now interested in the gap between Tukey morphism $A \longrightarrow B$ and Δ-Muchnik reducibility $[B]_\Delta \leq_{\Delta\mathrm{w}} [A]_\Delta$. One of the main technical differences between computability theory and set theory is the use of *time trick* (see [4]). This technique yields several strange phenomena in computability theory. For instance, consider the following Tukey morphisms:

$$\mathsf{Cov}\mathcal{M} \longrightarrow \mathsf{ED} \longrightarrow \mathsf{D},$$

whereas there are *no* reversal definable morphisms. Indeed, Laver forcing (see [2]) over Gödel's constructible universe L provides a model of ZFC satisfying $[\mathsf{D}]_L <_{L\mathrm{w}} [\mathsf{ED}]_L$ (see Proposition 3.6 (1a)), and Zapletal's uncountable dimensional forcing [58] over L provides a model of ZFC satisfying $[\mathsf{ED}]_L <_{L\mathrm{w}} [\mathsf{Cov}\mathcal{M}]_L$. However, it is known that if a real computes a function which is not dominated by any computable function, then it also computes a weakly 1-generic real. This implies that:

$$[\mathsf{Cov}\mathcal{M}]_{\Delta^0_1} \equiv_{\mathrm{w}} [\mathsf{ED}]_{\Delta^0_1} \equiv_{\mathrm{w}} [\mathsf{D}]_{\Delta^0_1}. \tag{2.2}$$

In Section 3.3, we will introduce a new Vojtáš triple $\mathsf{Cov}\mathcal{X}_{0.5}$, which is related to the notion of *coarse computability*. As observed by Andrews et al. [1], the above formula (2.2) for instance implies that

$$[\mathsf{Cov}\mathcal{X}_{0.5}]_{\Delta^0_1} \leq_{\mathrm{w}} [\mathsf{D}]_{\Delta^0_1}, \tag{2.3}$$

whereas we will also see that there is *no* definable Tukey morphism $\mathsf{D} \longrightarrow \mathsf{Cov}\mathcal{X}_{0.5}$ (indeed, $[\mathsf{Cov}\mathcal{X}_{0.5}]_L \not\leq_{L\mathrm{w}} [\mathsf{D}]_L$). One may also observe the similar phenomenon for the following Tukey morphisms (see [2]):

$$\mathsf{Add}\mathcal{N} \longrightarrow \mathsf{Add}\mathcal{M} \longrightarrow \mathsf{B},$$

whereas there are *no* reversal definable morphisms. Indeed, $[\mathsf{B}]_L <_{L\mathrm{w}} [\mathsf{Add}\mathcal{M}]_L$ consistently holds (again in a Laver generic extension $L[G]$ of L; see [2]). However, Rupprecht [46] pointed out that if a real computes a function which dominates all computable functions, it also computes a computably null-engulfing real. This implies that:

$$[\mathsf{Add}\mathcal{N}]_{\Delta^0_1} \equiv_{\mathrm{w}} [\mathsf{Add}\mathcal{M}]_{\Delta^0_1} \equiv_{\mathrm{w}} [\mathsf{B}]_{\Delta^0_1}. \tag{2.4}$$

Thus, it is important to ask about the situation in higher randomness theory. We will show that the Δ^1_1-analog of (2.1) holds true as mentioned before; however, the Δ^1_1-analogues of (2.2) and (2.3) fail. Indeed, we show the main result for more general Spector pointclasses, e.g., $\Gamma \in \{\Sigma^0_{<\omega}, \Pi^1_1, \Game(D_{<\omega}\Sigma^0_1), {}_2\mathrm{env}(\mathrm{E}_n), \Game\Sigma^0_k\}$ where $k \neq 2$. Here $\Game$ is the game quantifier, $D_{<\omega}$ indicates the union of all finite ranks in the difference hierarchy, ${}_2\mathrm{env}$ is the 2-envelop (the collection of all sets of reals which are semi-computable in a given functional), and E_n is the nth normal type-2 functional obtained by iterating Gandy's superjump operator.

Theorem 2.5: *Suppose that* $\Gamma \in \{\Sigma^0_{<\omega}, \Pi^1_1, \Game(D_{<\omega}\Sigma^0_1), {}_2\mathrm{env}(\mathrm{E}_n), \Game\Sigma^0_k\}$

where n and k range over positive integers and $k \neq 2$. Then:

$$[\mathsf{Cof}\mathcal{M}]_\Delta <_{\Delta \mathrm{w}} [\mathsf{Non}\mathcal{M}]_\Delta, \tag{2.5}$$

$$[\mathsf{D}]_\Delta <_{\Delta \mathrm{w}} [\mathsf{ED}]_\Delta, \tag{2.6}$$

$$[\mathsf{Cov}\mathcal{X}_{0.5}]_\Delta \not\leq_{\Delta \mathrm{w}} [\mathsf{D}]_\Delta. \tag{2.7}$$

From this result, we may read that infinitary computability theory is closer to set theory than to finitary computability theory. However, we have the following Δ^1_1-analog of (2.4):

Fact 2.6: (Monin) A function $x \in \omega^\omega$ dominates all Δ^1_1-functions if and only if $y \leq_T x$ for every hyperarithmetical real y. In particular, we have the following equivalences:

$$[\mathsf{Add}\mathcal{N}]_{\Delta^1_1} \equiv_{\Delta^1_1 \mathrm{w}} [\mathsf{Add}\mathcal{M}]_{\Delta^1_1} \equiv_{\Delta^1_1 \mathrm{w}} [\mathsf{B}]_{\Delta^1_1}. \tag{2.8}$$

Furthermore, by the same argument as in Fact 2.6, one can also show that x dominates all arithmetically definable functions if and only if $\emptyset^{(n)} \leq_T x$ for every $n \in \omega$, and therefore, arithmetical dominance is also equivalent to being arithmetically meager engulfing (arithmetically null engulfing, resp.) Note that there is a Δ^1_1-dominant $x <_h \mathcal{O}$ (see Enderton-Putnam [15]) whereas x is arithmetically dominant if and only if $\emptyset^{(\omega)} \leq_a x$ since $\emptyset^{(\omega)}$ is a 2-least upper bound of $\{\emptyset^{(n)}\}_{n\in\omega}$ (see Enderton-Putnam [15]).

The above contrasting two results (Theorem 2.5 and Fact 2.6) can be explained in the context of forcing theory. The failure of (2.2) in the L-degrees can be witnessed by *lim-sup* tree forcing such as rational perfect set forcing $\mathbb{PT}$, whereas we need *lim-inf* tree forcing such as Laver forcing $\mathbb{LT}$ to show the failure of (2.4). The equivalences in (2.8) suggest that lim-inf tree forcing (in particular, Laver forcing $\mathbb{LT}$) does not work at $L_{\omega_1^{\mathrm{CK}}}$ (the companion model of Π^1_1). In this paper, we develop a general theory of lim-sup forcing over the companion model of a Spector pointclass to show Theorem 2.5.

3. Basic Properties

3.1. *Traceability*

To prove Theorem 2.5, we need a technical notion called traceability.

Definition 3.1: (See also [14, 28, 29, 40]) A *slalom* is a sequence $(T_n)_{n\in\omega}$ of finite subsets of ω such that $|T_n| \leq n$. It is a Γ-*slalom* if $\{(n, m) : m \in$

$T_n\} \in \Gamma$. It is a Δ-*slalom* if the sequence of the canonical indices of T_n is in Δ, or equivalently, it is a Γ-slalom and $\lambda n.|T_n| \in \Delta$.

We say that a function $f \in \omega^\omega$ is *traced* (*infinitely often traced*, respectively) by a slalom $(T_n)_{n\in\omega}$ if $f(n) \in T_n$ for all $n \in \omega$ (for infinitely many $n \in \omega$, respectively). We also say that a function $f \in \omega^\omega$ is Δ-*often traced* by a slalom $(T_n)_{n\in\omega}$ if there is a Δ-function $h \in \omega^\omega$ such that for every $n \in \omega$, $f(k) \in T_k$ for some $k \in [h(n), h(n+1))$.

A real x is Γ-*traceable* (Δ-*traceable*, respectively) if every $f \leq_\Delta x$ is traced by a Γ-slalom (Δ-slalom, respectively). One can also define infinitely often (abbreviated as i.o.) Δ-traceability, Δ-often Δ-traceability, etc. in a straightforward manner (see also Kihara-Miyabe [28, 29]).

As pointed out by Kjos-Hanssen et al. [31] (Lemma 4.5), (i.o.) Π^1_1-traceability is equivalent to (i.o.) Δ^1_1-traceability. Generally, if Γ is a Spector pointclass, then, by the Spector criterion, we have the equivalence of (i.o.) Γ-traceability and (i.o.) Δ-traceability. The notion of traceability characterizes (transitive-)additivity in set theory and lowness for randomness in computability theory (see [28]):

(1) x is Δ-traceable if and only if $x \notin [\mathsf{Cof}\mathcal{N}]_\Delta$ (i.e., x is low for Δ-randomness).
(2) x is Δ-often Δ-traceable if and only if $x \notin [\mathsf{Cof}\mathcal{M}]_\Delta$ (i.e., low for Δ-Kurtz randomness).
(3) x is i.o. Δ-traceable if and only if $x \notin [\mathsf{Non}\mathcal{M}_\Delta]$ (i.e., low for the pair of Δ-randomness and Δ-Kurtz randomness).

We use the following traceability notion, which is related to the so-called *Laver property* of Laver forcing ([2] (Definition 6.3.27); see also Section 4.4).

Definition 3.2: Recall that a function $f \in \omega^\omega$ is Δ-bounded if there is $h \in \omega^\omega \cap \Delta$ such that $f(n) < h(n)$ for all $n \in \omega$. A real $x \in \omega^\omega$ is Γ-*Laver traceable* (Δ-*Laver traceable*) if for every Δ-bounded function $f \leq_\Delta x$, there is a Γ-slalom (a Δ-slalom) $(T_n)_{n\in\omega}$ such that $f(n) \in T_n$ for all $n \in \omega$.

Let λ^x_Γ be the supremum of all ordinals $\alpha \leq_\Delta x$. Then $\lambda_\Gamma := \lambda^\emptyset_\Gamma$ is the supremum of all Δ-ordinals. By modifying the argument in Kjos-Hanssen et al. [31] (Lemma 4.5), one can easily show the following:

Lemma 3.3: *Let Γ be a Spector pointclass. Suppose that a real $x \in \omega^\omega$ satisfies $\lambda^x_\Gamma = \lambda_\Gamma$. Then, x is Γ-Laver traceable if and only if x is Δ-Laver traceable.*

Note that most known traceability notions can be easily interpreted as properties concerning Kolmogorov complexity. A *machine* is a partial function $G :\subseteq 2^{<\omega} \to 2^{<\omega}$ with a prefix-free domain (here, we do not require a machine to be computable). A Δ-*machine* is a machine which has a Δ-graph (i.e., the graph is a Δ subset of $2^{<\omega} \times 2^{<\omega}$). The *Kolmogorov complexity* of a binary string $\sigma \in 2^{<\omega}$ with respect to a machine G is defined by $K_G(\sigma) = \min\{|\tau| : G(\tau) = \sigma\}$.

Definition 3.4: (See also [28]) A real x is Δ-*trivial* if for every Δ-machine H, there exists a Δ-machine G such that $K_G(x \upharpoonright n) \leq K_H(n) + O(1)$ holds.

Note that this definition is a modification of *Schnorr triviality*. Note that Franklin-Stephan [17] found a traceability characterizatin of Schnorr triviality. Generally, a real x is Δ-trivial if and only if x is Δ-tt-traceable, that is, for every $h \in \omega^\omega \cap \Delta$, there is a Δ-slalom $(T_n)_{n\in\omega}$ such that $x \upharpoonright h(n) \in T_n$ for all $n \in \omega$. The exactly same traceability notion is also used to characterize null-additivity in set theory. Therefore, Δ-triviality is equivalent to Δ-null-additivity (see [28]).

Proposition 3.5: *If $x \in \omega^\omega$ is Δ-Laver traceable, then every $y \leq_\Delta x$ with $y \in 2^\omega$ is Δ-trivial.*

In our main theorem, we will construct a real $x \in \omega^\omega$ such that (i) x is not Δ-often Δ-traceable; in particular, x is not Δ-traceable; (ii) x is i.o. Δ-traceable; (iii) x is Δ-Laver traceable; hence, every $y \leq_\Delta x$ with $y \in 2^\omega$ is Δ-tt-traceable.

3.2. *Computable Tukey morphism*

It is known that all implications in Cichoń's diagram are witnessed by computable Tukey morphisms (see also [41]). In this section, we will see some examples of computable Tukey morphisms outside Cichoń's diagram. A slalom $(T_n)_{n\in\omega}$ is h-*bounded* if $T_n \subseteq h(n)$ for all $n \in \omega$. Let $\mathcal{T}_h$ be the collection of all h-bounded slaloms, and $\mathsf{Non}\mathcal{T}_h$ be the Vojtáš triple $(\mathcal{T}_h, [h], \not\ni)$.

Proposition 3.6: *We have the following computable Tukey morphisms:*

(1) *Let h be a function such that $h(n) > n$ for all n. Then, there is a computable Tukey morphism from $\mathsf{Non}\mathcal{T}_h$ to IE_h. In particular, we have the following:*

(a) *No Δ-Laver-traceable real Δ-computes an h-infinitely often equal real for any Δ-order h.*
(b) *Every weakly Δ-Laver-trace-engulfing real Δ-computes an h-eventually different real for any Δ-order h.*

(2) *(see also Rupprecht [46] (Proposition 16)) There is a computable Tukey morphism from $\mathsf{Cov}_{\mathcal{N}}$ to $\mathsf{IE}_{\lambda n.2^n}$. In particular, we have the following:*
(a) *If x Δ-computes a $(\lambda n.2^n)$-infinitely often equal real over Δ, then x is weakly null-engulfing over Δ.*
(b) *Every Δ-Schnorr random real Δ-computes a $(\lambda n.2^n)$-eventually different real over Δ.*

Proof: (1) Let T_n be an h-bounded slalom with $|T_n| \leq n$. Then, we can choose some $f(n) < h(n)$ such that $f(n) \notin T_n$. Define $H(T_n : n \in \omega) = f$. If g is infinitely often equal to f, then it is easy to see that $K(g) := g$ is not traced by $(T_n)_{n\in\omega}$.

(2) Suppose that a real y is given. Let J_n be the interval $[\sum_{k=0}^{n-1} k, \sum_{k=0}^{n} k)$ of length n. Consider $H(y) := y \upharpoonright J_n$. If x is a $(\lambda n.2^n)$-infinite often equal real, then $x(n) = y \upharpoonright J_n$ for infinitely many n. Put $W_n = \bigcup_{m\geq n}[x(m)]$. Clearly, $K(x) := (W_n)_{n\in\omega}$ is a uniform sequence of open sets such that $\mu(W_n) = 2^{-n-1}$ for every n. Moreover, we have $y \in \bigcap_n W_n$.

The items (a) and (b) follow from Proposition 2.3. □

3.3. *Infinite equality and asymptotic density*

We consider the asymptotic density version of infinite often equality. For a set $A \subseteq \omega$, the *lower density of A* is defined as follows:

$$\underline{\rho}(A) = \liminf_{n\to\infty} \frac{|\{k < n : k \in A\}|}{n}.$$

By using the asymptotic density, we consider the following "small" set for any $A \subseteq \omega$.

$$\langle A \rangle_r = \{B \subseteq \omega : \underline{\rho}(\{n \in \omega : A(n) = B(n)\}) \geq r\},$$

and then let $\mathcal{X}_r$ be the σ-ideal generated by $\{\langle A \rangle_r : A \subseteq \omega\}$. We then define the *coarse Δ-equality bounds* of $B \subseteq \omega$ as follows:

$$\mathfrak{g}_\Delta(B) = \sup\{r \in [0,1] : (\exists A \in \Delta)\ B \in \langle A \rangle_r\},$$
$$\mathfrak{G}_\Delta(x) = \inf\{\mathfrak{g}_\Delta(B) : B \leq_\Delta x\}.$$

The *upper density of* A, $\overline{\rho}(A)$, is defined in a similar manner. The *upper coarse* Δ*-equality bounds* $\overline{\mathfrak{g}}_\Delta$ and $\overline{\mathfrak{G}}_\Delta$ are also defined by the similar way. The above notions have already been studied in the computability theoretic context (see [1, 20]). In their terminology, $\mathfrak{g}_{\Delta^0_1}$ and $\mathfrak{G}_{\Delta^0_1}$ are denoted by γ and Γ, respectively.

Proposition 3.7:

(1) $\mathfrak{G}_\Delta(x) < r$ *if and only if there is* $y \leq_\Delta x$ *such that* $y \in [\mathsf{Cov}\mathcal{X}_r]_\Delta$.
(2) $\overline{\mathfrak{G}}_\Delta(x) \leq 1 - r$ *if and only if there is* $y \leq_\Delta x$ *such that* $y \in [\mathsf{Non}\mathcal{X}_r]_\Delta$.

Proof: The set $\langle A\rangle_r$ does not cover all computable reals if and only if there is a computable real B such that $\underline{\rho}(A = B) < r$. Consider the complement $C = \omega \setminus B$. Then,

$$\frac{|\{k < n : A(n) = C(n)\}|}{n} = 1 - \frac{|\{k < n : A(n) = B(n)\}|}{n}.$$

Since the infimum limit of the right-hand fraction is less than r, the supremum limit of the left-side value is greater than $1 - r$. In other words, $\overline{\mathfrak{g}}_\Delta(A) > 1 - r$. □

Next, we see the relationship between infinite often equality and lower asymptotic density.

Proposition 3.8: *We have the following computable Tukey morphisms:*

(1) (see also Monin-Nies [37]) There is a computable Tukey morphism from $\mathsf{Non}\mathcal{X}_{1/c}$ *to* $\mathsf{IE}_{\lambda n.2^{c^n}}$. *In particular, we have the following:*
  (a) If x *is* $(\lambda n.2^{c^n})$*-infinitely often equal over* Δ, *then* $\mathfrak{G}_\Delta(x) \leq 1/c$. *Hence, if* x *is* $(\lambda n.2^{n^n})$*-infinitely often equal over* Δ, *then* $\mathfrak{G}_\Delta(x) = 0$.
  (b) If x *is not* $(\lambda n.2^{c^n})$*-eventually different over* Δ, *then* $\overline{\mathfrak{G}}_\Delta(x) \geq 1 - 1/c$. *Hence, if* x *is not* $(\lambda n.2^{n^n})$*-eventually different over* Δ, *then* $\overline{\mathfrak{G}}_\Delta(x) = 1$.

(2) (see also Andrews et al. [1]) There is a computable Tukey morphism from $\mathsf{Cov}\mathcal{X}_{1/2}$ *to* $\mathsf{Cof}\mathcal{N}$. *In particular, we have the following:*
  (a) If A *is* Δ*-traceable, then* $\mathfrak{G}_\Delta(A) \geq 1/2$.
  (b) If D *is null-engulfing over* Δ, *then* $\overline{\mathfrak{G}}_\Delta(D) \leq 1/2$.

Proof: (1) Let B be any real, and J_n be the interval $[c^+_{n-1}, c^+_n)$, where $c^+_n = \sum_{k\leq n} c^k$. Define $H(B) := \lambda n.(\omega \setminus B) \upharpoonright J_n$. Note that $H(B)$ is 2^{c^n}-

bounded. Let g be a function which is infinitely often equal to $H(B)$. Consider $K(g) := A = g(0)^\frown g(1)^\frown \dots$. For $n \in \omega$ such that $g(n) = H(B)(n)$, we have

$$\frac{\{k < c_n^+ : A(k) = B(k)\}}{c_n^+} \le \frac{c_n^+ - c^n}{c_n^+} = \frac{c^n - 1}{c^{n+1} - 1}.$$

Clearly, the rightmost value converges to $1/c$ as n tends to infinity.

(2) We give a sketch of proof. Let A be a real, For $H(A) := \lambda n.(\omega \setminus A) \upharpoonright J_n$, let $(T_n)_{n\in\omega}$ be any slalom such that $(\omega \setminus A) \upharpoonright J_n \in T_n$ for almost all $n \in \omega$. By a probabilistic argument in [1] (proof of Theorem 1.10), we can construct $K(T_n : n \in \omega) := B$ such that $B \upharpoonright J_n$ is as close to T_n as possible. Let us consider $C = \{k \in \omega : B(k) \neq A(k)\}$. By [1] (Lemmas 2.2), $\liminf_n \rho(C|J_n) \geq 1/2$, and indeed, $\underline{\rho}(C) \geq 1/2$. Then, $\overline{\rho}(\omega \setminus C) = 1 - \underline{\rho}(C) \leq 1/2$. Consequently, $\overline{\mathfrak{g}}_\Delta(B) \leq 1/2$. Therefore, B is a solution to the $\mathsf{Cov}\mathcal{X}_{1/2}$-instance generated from A by (the proof of) Proposition 3.7. □

4. Forcing Argument

4.1. *Models and definability*

We assume that any model $\mathbf{M}$ discussed in this paper is the *companion* of a Spector pointclass (see Moschovakis [38] (Theorem 9E.1)). Roughly speaking, the companion $\mathbf{M}$ of a Spector pointclass Γ is the set M_Δ of (Mostowski collapses of) all Δ-coded hereditarily countable sets (w.r.t. our representation of $\mathbf{H}_{\aleph_1}$; Example 2.1) with a certain relation R on M_Δ which for instance satisfies that a subset of ω is Γ if and only if it is Σ_1 on $\mathbf{M} = \langle M_\Delta, \in, R\rangle$. Indeed, a companion $\mathbf{M}$ is always of the form $\langle L_\alpha[R], \in, R\rangle$ where α is the supremum of M_Δ-ordinals (see also [16] (Section 5.4)). Note that α is a countable R-admissible ordinal.

For a companion $\mathbf{M} = \langle L_\alpha[R], \in, R\rangle$ and a given real x, we consider the relativization $\mathbf{M}[x] := \langle L_\alpha[x, R], \in, R\rangle$. Note that $\mathbf{M}[x]$ may not be a companion of $\Gamma(x)$ even if $\mathbf{M}$ is a companion of Γ. For instance, if $\omega_1^x > \omega_1^{\mathrm{CK}}$ then $L_{\omega_1^{\mathrm{CK}}}[x]$ is not a companion of the pointclass $\Pi^1_1(x)$.

Definition 4.1: A partial function $\Phi :\subseteq \omega^\omega \to \omega^\omega$ is $\Sigma_1(\mathbf{M})$-*definable* if there is a $\Sigma_1(\mathbf{M})$-formula $\varphi(\sigma, x, \vec{v})$ with a parameter $\vec{v} \in \mathbf{M}$ such that for any $\sigma \in \omega^{<\omega}$ and $x \in \omega^\omega$

$$\Phi^{-1}[\sigma] = \mathrm{dom}(f) \cap \{x \in \omega^\omega : \mathbf{M}[x] \models \varphi(\sigma, x, \vec{v})\}.$$

A real $y \in \omega^\omega$ is $\mathbf{M}$-*reducible* to $x \in \omega^\omega$ (written as $y \leq_\mathbf{M} x$) if there is a partial $\Sigma_1(\mathbf{M})$-definable function $\Phi :\subseteq \omega^\omega \to \omega^\omega$ such that $x \in \mathrm{dom}(\Phi)$ and $\Phi(x) = y$.

Note that, even if $\mathbf{M}$ is the companion of a Spector pointclass Γ, the reducibillity notion $\leq_{\mathbf{M}}$ does not coincide with $\leq_\Delta$ (e.g., consider $\mathbf{M} = L_{\omega_1^{\mathrm{CK}}}$ and $\Gamma = \Pi^1_1$).

4.2. *Forcing and fusion*

The notions of set-forcing (i.e., a forcing $\mathbb{P}$ with $\mathbb{P} \in \mathbf{M}$) in admissible recursion theory have been extensively studied in, for instance, Mathias [35], Sacks [48], and Zarach [59]. In particular, it is shown that any set-forcing preserves admissibility. However, most forcing notions which we are interested in are class-forcing notions in our companion model except for Cohen forcing. For instance, if $\mathbb{S}$ is the Sacks forcing notion (i.e., the set of all perfect subtrees of $2^{<\omega}$), we obviously have $\mathbb{S} \cap L_{\omega_1^{\mathrm{CK}}} \notin L_{\omega_1^{\mathrm{CK}}}$. To avoid some technical issues concerning class forcing, we will only consider a restricted class of forcing notions.

A *poset* is a triple $(\mathbb{P}, \leq_{\mathbb{P}}, \mathbf{1}_{\mathbb{P}})$ of a partial order $\leq_{\mathbb{P}}$ on a set $\mathbb{P}$ with a $\leq_{\mathbb{P}}$-greatest element $\mathbf{1}_{\mathbb{P}} \in \mathbb{P}$. In this paper, we always assume that a poset is *arboreal*, that is, $\mathbf{1}_{\mathbb{P}} := H$ is a perfect subtree of $\omega^{<\omega}$, each condition $p \in \mathbb{P}$ is a perfect subtree of $\mathbf{P}$, and $\leq_{\mathbb{P}}$ is the inclusion relation $\subseteq$. We also say that a poset is *strongly arboreal* if it is arboreal, and if $p \in \mathbb{P}$ and $\sigma \in p$ implies $p \upharpoonright \sigma \in \mathbb{P}$, where $p \upharpoonright \sigma = \{\tau \in p : \sigma \preceq \tau \text{ or } \tau \preceq \sigma\}$. For instance, Cohen forcing, random forcing, Hechler forcing, Mathias forcing, and Laver forcing are all strongly arboreal forcing notions (see [2, 57] for arboreal and idealized forcing). We define the relativization $\mathbb{P}^{\mathbf{M}} = \mathbb{P} \cap \mathbf{M}$. We say that $\mathbb{P}$ is a Σ_1*-forcing over* $\mathbf{M}$ if $\mathbb{P}^{\mathbf{M}}$, $\{(p, q) \in \mathbb{P}^{\mathbf{M}} : p \leq q\}$, and $\{(p, q) \in \mathbb{P}^{\mathbf{M}} : p \not\leq q\}$ are Σ_1-definable over $\mathbf{M}$ (see also Kechris [22]).

A set $U \subseteq \mathbb{P}$ is $\mathbb{P}$*-open* if $q \leq p \in U$ implies $q \in U$. A set $U \subseteq \mathbb{P}$ is $\mathbf{M}$-$\mathbb{P}$*-dense below* p if for every $q \in \mathbb{P}^{\mathbf{M}}$ with $q \leq p$ there is $r \in \mathbb{P}^{\mathbf{M}}$ such that $r \leq q$ and $r \in U$. A real $x \in [\mathbf{H}]$ is $\mathbb{P}$*-generic over* $\mathbf{M}$ if for every $\Sigma_n(\mathbf{M})$-definable $\mathbf{M}$-$\mathbb{P}$-dense $\mathbb{P}$-open set $D \subseteq \mathbb{P}$, there is $p \in \mathbb{P}^{\mathbf{M}}$ such that $x \in [p]$ and $p \in D$, where $[p]$ denotes the set of all infinite paths through a tree $p \subseteq H$.

We now introduce the forcing language $\mathcal{L}_{\mathbf{M}}$ (see also Sacks [48] (Section 3.4)). Our model $\mathbf{M} = \langle L_\alpha[R], \in, R\rangle$ is the companion (Moschovakis [38] (Section 9E)) of a Spector pointclass, and therefore it is *resolvable*, that is, there is a $\Delta_1(\mathbf{M})$ function $\tau : \alpha \to \mathbf{M}$ such that $\mathbf{M} = \bigcup_\xi \tau(\xi)$. Our forcing language $\mathcal{L}_{\mathbf{M}}$ consists of two relation symbols $\in$ and R, a constant symbol

$\dot{r}_{gen}$ for a generic real, a constant symbol $\check{a}$ for every $a \in \mathbf{M}$, ranked variables $x^\xi, y^\xi, \ldots$ (ranging over $\bigcup_{\zeta<\xi} \tau(\zeta)$) for $\xi < \alpha$ and unranked variables $x, y, \ldots$ In particular, $\mathcal{L}_\mathbf{M}$ contains a name $\check{\omega}$ of ω since any companion $\mathbf{M}$ of a Spector pointclass is an ω-model. An $\mathcal{L}_\mathbf{M}$-formula is *ranked* if all of its variables are ranked. An $\mathcal{L}_\mathbf{M}$-formula is Σ_1 if it is of the form $\exists y \varphi(y, \check{\mathbf{v}})$ where φ has no quantification over unranked variables.

Fix an arboreal forcing notion $\mathbb{P}$. We define the (strong) forcing relation $\Vdash_\mathbb{P}$ by the following way (see also Sacks [48] (Section 4.4.1)):

Suppose that $\varphi(\dot{r}_{gen}, \check{\mathbf{v}})$ is ranked.

(1) $p \Vdash_\mathbb{P} \varphi(\dot{r}_{gen}, \check{\mathbf{v}})$ if $[p] \subseteq \{x \in \omega^\omega : \mathbf{M}[x] \models \varphi(x, \mathbf{v})\}$.

Suppose that $\varphi(\dot{r}_{gen}, \check{\mathbf{v}})$ is unranked.

(2) $p \Vdash_\mathbb{P} \exists y^\xi\, \varphi(\dot{r}_{gen}, y^\xi, \check{\mathbf{v}})$ if $p \Vdash_\mathbb{P} \varphi(\dot{r}_{gen}, z, \check{\mathbf{v}})$ for some $z \in \bigcup_{\zeta<\xi} \tau(\zeta)$.
(3) $p \Vdash_\mathbb{P} \exists y\, \varphi(\dot{r}_{gen}, y, \check{\mathbf{v}})$ if $p \Vdash_\mathbb{P} \exists y^\xi\, \varphi(\dot{r}_{gen}, y^\xi, \check{\mathbf{v}})$ for some $\xi < \alpha$.
(4) $p \Vdash_\mathbb{P} \varphi \wedge \psi$ if $p \Vdash_\mathbb{P} \varphi$ and $p \Vdash_\mathbb{P} \psi$.
(5) $p \Vdash_\mathbb{P} \neg\varphi$ if $q \not\Vdash_\mathbb{P} \varphi$ for every $q \leq p$.

Note that for every $p \in \mathbb{P}$, the formula $\dot{r}_{gen} \in [\check{p}]$ is Δ_0 in $\mathbf{M}$ since $\Pi^1_1(u) \subseteq \Sigma_1(\mathbf{M})$ for all $u \in \mathbf{M}$. We abbreviate $\Vdash_\mathbb{P}$ as $\Vdash$ if $\mathbb{P}$ is clear from the context. As in Shinoda [49] (Lemma 4.3), one can easily show that the strong forcing relation for Σ_1-formulas on $\mathbf{M}$ is $\Sigma_1(\mathbf{M})$-definable (since a companion $\mathbf{M}$ is Σ_1-projectible to ω; see [16, 38]).

Lemma 4.2: *(Definability; see Shinoda* [49] *(Lemma 4.3)) Suppose that M is a companion model, $\mathbb{P}$ is a Σ_1-forcing notion, and $\varphi(\dot{r}_{gen}, \check{\mathbf{v}})$ is a Σ_1 formula with parameters from M. Then, $p \Vdash \varphi(\dot{r}_{gen}, \check{\mathbf{v}})$ is Σ_1-definable over M uniformly in p, φ, and $\mathbf{v}$.*

In set theory, the fusion argument is a key method to show various properties of forcing notions. For an arboreal forcing $\mathbb{P}$, we say that a collection $(q_i)_{i\in\omega} \subseteq \mathbb{P}$ is separated if $q_\eta \cap q_\lambda$ is finite (therefore, there is a finitely generated clopen (in the Baire topology) set $C \subseteq [H]$ such that $[q_\eta] \subseteq C$ and $C \cap [q_\lambda] = \emptyset$) whenever $\eta \neq \lambda$. A $\mathbb{P}$-open set U is Σ_1*-pre-regular* if U is closed under separated Σ_1-union, that is, the union of a separated $\Sigma_1(\mathbf{M})$-definable sequence $(q_i)_{i\in\omega} \subseteq U$ belongs to U.

Lemma 4.3: *Let φ be a $\Sigma_1(\mathbf{M})$-formula. Then $\{q : q \Vdash \varphi\}$ is Σ_1-pre-regular.*

Proof: If φ is ranked, by definition, it is clear that $\{q : q \Vdash \varphi\}$ is closed under arbitrary union. Assume that φ is an unranked formula of the form

$\varphi \equiv \exists y \psi(y)$ such that ψ is ranked. Let q be the union of a $\Sigma_1(\mathbf{M})$-definable sequence $(q_i)_{i \in \omega}$ of forcing conditions such that $q_i \Vdash \varphi$ for each $i \in \omega$. Then, for any $i \in \omega$ there is $\xi(i) < \alpha$ such that $q_i \Vdash \exists y^{\xi(i)} \psi(y^{\xi(i)})$. By Lemma 4.2 and by admissibility of $\mathbf{M}$, the map $i \mapsto \xi(i)$ is $\Sigma_1(\mathbf{M})$, and therefore, there is an ordinal $\xi < \alpha$ such that $\xi(i) < \xi$ for all $i \in \omega$. Then $q \Vdash \exists y^{\xi} \psi(y^{\xi})$ since ψ is ranked and so is $\exists y^{\xi} \psi(y^{\xi})$. Consequently $q \Vdash \varphi$. □

Definition 4.4: Suppose that $\mathbb{P}$ is an arboreal Σ_1-forcing on $H \in \mathbf{M}$. We say that $\mathbb{P}$ satisfies the *Σ_1-fusion property over* $\mathbf{M}$ if for any $\Sigma_1(\mathbf{M})$-definable sequence $(D_n)_{n \in \omega}$ of Σ_1-pre-regular $\mathbf{M}$-$\mathbb{P}$-dense-below-p $\mathbb{P}$-open subsets of $\mathbb{P}$, their intersection $\mathbb{P}^{\mathbf{M}} \cap \bigcap_n D_n$ is nonempty below p.

By Lemmas 4.2 and 4.3, the Σ_1-fusion property enables us to use the following basic fusion argument: If $\varphi(n, x)$ is a $\Sigma_1(\mathbf{M})$-formula, and $q \Vdash \varphi(n, \dot{r}_{gen})$ is dense below p for any $n \in \omega$, then there is a condition $r \leq p$ that forces $\varphi(n, \dot{r}_{gen})$ for all $n \in \omega$ (consider $D_n = \{q : q \Vdash \varphi(n, \dot{r}_{gen})\}$). For a forcing $\mathbb{P}$ with the Σ_1-fusion property, one can easily check the following basic properties by using the standard argument.

Lemma 4.5: *(See Sacks [48] (Chapter IV)) Suppose that $\mathbb{P}$ is a strongly arboreal Σ_1-forcing notion which satisfies the Σ_1-fusion property.*

(1) (Quasi-completeness) For any formula φ and condition $p \in \mathbb{P}$, there is a condition $q \leq p$ such that either $q \Vdash \varphi$ or $q \Vdash \neg\varphi$ holds.
(2) (Forcing = Truth) For a $\mathbb{P}$-generic real z, $\mathbf{M}[z] \models \varphi(z)$ if and only if there exists a condition p such that $z \in [p]$ and $p \Vdash \varphi(\dot{r}_{gen})$.
(3) If z is a $\mathbb{P}$-generic real, then, $\mathbf{M}[z]$ is admissible.

As a consequence of Lemma 4.5 (3), such a forcing $\mathbb{P}$ satisfies the *"Borel reading of names"* in the following sense: For any $\mathbf{M}$-coded partial Π^1_1-measurable function $h :\subseteq [H] \to \omega^\omega$, if $z \in [H]$ is sufficiently $\mathbb{P}$-generic and $z \in \mathrm{dom}(h)$, then there is a partial $\Sigma_1(\mathbf{M})$-definable (in particular, Borel) function $g :\subseteq [H] \to \omega^\omega$ with $z \in \mathrm{dom}(g)$ such that $g(z) = h(z)$.

4.3. *Creature forcing*

In this section, we deal with certain kinds of arboreal forcing notions induced by *norms* (see Rosłanowski-Shelah [44]). To simplify our argument, we only consider a very restricted class of lim-sup tree-creating creature forcings. However, most of our results can be easily generalized to a slightly larger class of lim-sup creature forcings (including several important finitary forgetful creature forcings such as Silver forcing).

Definition 4.6: (See also Rosłanowski-Shelah [44]) Let H be a subtree of $\omega^{<\omega}$. A *norm on* H is an ω-valued Borel function defined on all pairs (σ, A), where $\sigma \in H$ and A is a subset of $\{\sigma^\frown i \in H : i \in \omega\}$. Given a norm **nor** on H, we say that a tree $T \subseteq H$ is **nor**-*perfect* if for any $n \in \omega$ and $\sigma \in T$, there is $\tau \in T$ extending σ such that $\mathbf{nor}(\tau, \mathrm{succ}_T(\tau)) > n$, where $\mathrm{succ}_T(\tau)$ is the set of all immediate successors of τ in T. Then, $\mathbb{Q}^{\mathrm{w}\infty}_{\mathbf{nor}}$ is the *forcing notion with* **nor**-*perfect trees*, that is,

(1) A condition $p \in \mathbb{Q}^{\mathrm{w}\infty}_{\mathbf{nor}}$ is a **nor**-perfect subtree of H.
(2) The order $p \leq q$ on $\mathbb{Q}^{\mathrm{w}\infty}_{\mathbf{nor}}$ is introduced by the inclusion $p \subseteq q$.

We say that $\mathbb{Q}$ is *a lim-sup tree forcing* if it is of the form $\mathbb{Q}^{\mathrm{w}\infty}_{\mathbf{nor}}$ where **nor** satisfies that $\mathbf{nor}(\sigma, A) = 0$ for any $\sigma \in H$ and $|A| \leq 1$ (this implies that every **nor**-perfect tree is perfect), and moreover, that H is **nor**-perfect. We also write $\mathbf{nor}_T(\sigma)$ for $\mathbf{nor}(\sigma, \mathrm{succ}_T(\sigma))$.

Let $\mathbb{Q}^{\mathrm{w}\infty+}_{\mathbf{nor}}$ be the suborder of $\mathbb{Q}^{\mathrm{w}\infty}_{\mathbf{nor}}$ defined as follows:

$$\mathbb{Q}^{\mathrm{w}\infty+}_{\mathbf{nor}} = \{p \in \mathbb{Q}^{\mathrm{w}\infty}_{\mathbf{nor}} : \limsup_{n\to\infty} \mathbf{nor}_p(x \upharpoonright n) = \infty \text{ for all } x \in [p]\}.$$

Clearly, $\mathbb{Q}^{\mathrm{w}\infty+}_{\mathbf{nor}}$ is a dense suborder of $\mathbb{Q}^{\mathrm{w}\infty}_{\mathbf{nor}}$ (see the proof of Lemma 4.12). Therefore, we may assume that our lim-sup tree forcing notion is of the form $\mathbb{Q}^{\mathrm{w}\infty+}_{\mathbf{nor}}$.

Example 4.7: The following are examples of some lim-sup tree forcings.

(1) Let $H = 2^{<\omega}$, and define $\mathbf{nor}_T(\sigma) = |\{\tau \prec \sigma : \mathbf{nor}_T(\sigma) > 0\}| + 1$ if $|\mathrm{succ}_T(\sigma)| \geq 2$ (i.e., σ is branching), and $\mathbf{nor}_T(\sigma) = 0$ otherwise. Then, $\mathbb{Q}^{\mathrm{w}\infty}_{\mathbf{nor}}$ is *Sacks forcing*.
(2) Let $H = \prod_n 2^n$, and define $\mathbf{nor}_T(\sigma) = |\{\tau \prec \sigma : \mathbf{nor}_T(\sigma) > 0\}| + 1$ if $|\mathrm{succ}_T(\sigma)| = 2^{|\sigma|}$ (i.e., σ is full-branching), and $\mathbf{nor}_T(\sigma) = 0$ otherwise. Then, $\mathbb{Q}^{\mathrm{w}\infty}_{\mathbf{nor}}$ is *infinitely equal forcing*.
(3) Let $H = \omega^{<\omega}$, and define $\mathbf{nor}_T(\sigma) = |\{\tau \prec \sigma : \mathbf{nor}_T(\sigma) > 0\}| + 1$ if $|\mathrm{succ}_T(\sigma)| = \omega$ (i.e., σ has infinitely many immediate successors), and $\mathbf{nor}_T(\sigma) = 0$ otherwise. Then, $\mathbb{Q}^{\mathrm{w}\infty}_{\mathbf{nor}}$ is *rational perfect set forcing*.
(4) A limsup tree creature forcing with a nontrivial norm has been used by Goldstern-Shelah [18] (Definition 2.6) and others.

If $\mathbb{Q}$ is a lim-sup tree forcing generated by $H, \mathbf{nor} \in \mathbf{M}$, we say that $\mathbb{Q}$ is a lim-sup tree forcing over $\mathbf{M}$.

Proposition 4.8: *Suppose that* $\mathbb{Q}$ *is a lim-sup tree forcing over* $\mathbf{M}$. *Then,* $\mathbb{Q}^{\mathbf{M}}$ *is a strongly arboreal* Σ_1*-forcing over* $\mathbf{M}$.

For a tree p and $\sigma \in \omega^{<\omega}$, let $p(\sigma)$ be the σth branching node of p if such a node exists. In other words, $p(\emptyset)$ is the stem (the first branching node) of p, and $p(\sigma^\frown i)$ is the first branching node of p extending the ith immediate successor of $p(\sigma)$ if such a node exists. Given σ, let $\tau_{p,n}(\sigma)$ be the lexicographically least $\tau \succeq \sigma$ such that $\mathbf{nor}_p(p(\tau)) \geq n$ if such a node exists; otherwise $\tau_{p,n}(\sigma)$ is undefined. Then, we define

$$B_n(p) := B_{n-1}(p) \cup \{\tau : (\exists \sigma \in n^n)\, \tau \preceq \tau_{p,n}(\sigma)\},$$
$$\overline{B}_n(p) := \{\tau : (\exists \sigma \in B_n(p))\, \tau \preceq p(\sigma)\}$$

where $B_{-1}(p) = \emptyset$ and n^n is the set of all strings σ of length n such that $\sigma(k) < n$ for all $k < n$. Roughly speaking, at stage n, for a given $\sigma \in n^n$, enumerate the first string τ extending σ whose norm is greater than or equal to n, and then define $\overline{B}_n(p)$ as the downward closure of such a collection. Note that $(\overline{B}_n(p))_{n\in\omega}$ is an increasing sequence of finite subtrees of p.

Definition 4.9: Let $\mathbb{Q}$ be a lim-sup tree forcing on H generated by **nor**. We define the order $\leq_n$ on $\mathbb{Q}$ as follows:

$$q \leq_n p \iff q \leq p \;\&\; (\forall \sigma \in \overline{B}_n(p))\ \mathrm{succ}_p(\sigma) = \mathrm{succ}_q(\sigma).$$

Note that $q \leq_0 p$ if $q \leq p$ and q has the same stem with p. It is easy to see that $q \leq_{n+1} p$ implies $q \leq_n p$. It is obvious that $\leq_n$ is $\Sigma_1(\mathbf{M})$ uniformly in $n \in \omega$ if $\mathbb{Q}$ is a lim-sup tree forcing over $\mathbf{M}$ by Proposition 4.8. The following is a key property of the ordering $\leq_n$.

Lemma 4.10: *(Fusion) For every $(p_n)_{n\in\omega} \in \mathbb{Q}^\omega$ with $p_{n+1} \leq_n p_n$, then there exists $p \in \mathbb{Q}$ such that $p \leq_n p_n$ for all n.*

Proof: We define $p = \bigcap_s p_s$, and we will show that $p \in \mathbb{Q}$ and $p \leq_n p_n$ for all $n \in \omega$. To see this, we first note that if $\sigma \in p$, then $\sigma \in p_s$ for all $s \in \omega$; therefore, $\sigma \preceq p_0(\eta)$ for some $\eta \in m^m$ since p_0 is perfect. Fix any $n \geq m$. This implies that for any $s \in \omega$, $\sigma \preceq p_s(\eta_s)$ for some $\eta_s \in n^n$. By the definition of $\leq_n$, it is not hard to see that $\sigma \preceq p(\tau_{p_n,n}(\eta_n)) \in p$ and $\mathbf{nor}_p(p(\tau_{p_n,n}(\eta_n))) \geq n$. Consequently, p is **nor**-perfect. □

Here we discuss common properties of lim-sup tree forcing notions.

Definition 4.11: A Σ_1-forcing $\mathbb{P}$ on H satisfies the *Σ_1-continuous reading of names over* $\mathbf{M}$ (see also Zapletal [57]) if, for every partial $\Sigma_1(M)$-definable function $\Phi :\subseteq [H] \to \omega^\omega$ such that $\mathrm{dom}(\Phi)$ contains a $\mathbb{P}$-generic point z, there is a $\Sigma_1(M)$-continuous function $f :\subseteq [H] \to \omega^\omega$ such that $z \in \mathrm{dom}(f)$ and $f(z) = \Phi(z)$.

Lemma 4.12: *Suppose that $\mathbb{Q}$ is a lim-sup forcing over* $\mathbf{M}$. *Then, we have the following:*

(1) $\mathbb{P}$ satisfies the Σ_1-fusion property over $\mathbf{M}$.
(2) $\mathbb{P}$ satisfies the Σ_1-continuous reading of names over $\mathbf{M}$.

Proof: (1) Let $(D_n)_{n\in\omega}$ be a $\Sigma_1(\mathbf{M})$-sequence of Σ_1-pre-regular $\mathbf{M}$-$\mathbb{P}$-dense $\mathbb{P}$-open set. We will construct a fusion sequence $(p_n)_{n\in\omega}$ such that $p_{n+1} \leq_n p_n$ for all $n \in \omega$. Put $p_0 = p$, and assume that p_n is given. Inductively we assume that $\tau_{p_n,n}(\sigma) = \sigma$ for any $\sigma \in n^n$, that is, $B_n(p_n) = \bigcup_{m\leq n} m^m$. We inductively define a sequence $(p_n^k)_{k\leq n+2}$ with $p_n^0 = p_n$. Given $k \leq n+1$, define $I_n^k := (n+1)^{k+1} \setminus \bigcup_{m\leq n} m^{\leq m}$. By $\mathbb{P}$-density of D_m for any m, for every $\sigma \in I_n^k$, one can find $q_\sigma \leq p_n^k \upharpoonright p_n^k(\sigma)$ such that $q_\sigma \in D_m$ for all $m \leq n$. Moreover, we may assume that the norm of the stem of q_σ is greater than the length of σ by replacing q_σ with $q_\sigma \upharpoonright q_\sigma(\tau_{q_\sigma,n+1}(\emptyset))$. Then, define p_n^{k+1} as the union of q_σ for $\sigma \in I_n^k$ and $p_n^k \upharpoonright p_n^k(\sigma)$ for σ such that $\sigma \not\succeq \tau$ for any $\tau \in I_n^k$. Finally, we define $p_{n+1} = p_n^{n+2}$. It is easy to see that $p_{n+1} \leq_n p_n$, and $\tau_{p_{n+1},n+1}(\sigma) = \sigma$ for any $\sigma \in (n+1)^{n+1}$. Therefore, one can find a fusion $r \leq_n p_n$ for all $n \in \omega$ by Lemma 4.10 by an $\mathbf{M}$-computable way.

For any $n \in \omega$ and $x \in \omega^\omega$, there are $m \geq n$ and $k \leq m+1$ such that $x \upharpoonright k+1 \in I_m^k$. We define $\sigma(x) = x \upharpoonright k+1$ for minimal such k. Then, clearly there is a computable enumeration $(\sigma_i)_{i\in\omega}$ of the range of σ, and by minimality of $\sigma(x)$, $\{\sigma_i : i \in \omega\} \subseteq \omega^{<\omega}$ forms an antichain. If $r(\sigma_i)$ is defined, then q_{σ_i} in our construction belongs to D_n. Moreover, it is easy to see that $r(\sigma_i) \leq q_{\sigma_i}$, and therefore, $r(\sigma_i) \in D_n$ since D_n is $\mathbb{P}$-open. Then r is the Σ_1-separated union of $(r \upharpoonright r(\sigma_i))_{i\in\omega}$ and $r(\sigma_i) \in D_n$ for all $i \in \omega$. Since D_n is Σ_1-pre-regular, we have $r \in D_n$. Consequently, $r \in \bigcap_n D_n$.

(2) Let Φ be a $\Sigma_1(\mathbf{M})$-definable function. Suppose that $p \Vdash \dot{r}_{gen} \in \mathrm{dom}(\check{\Phi})$, that is, $p \Vdash (\forall n)(\exists k)\ \check{\Phi}(\dot{r}_{gen})(n) \downarrow= k$. As in the above argument, for any $\sigma \in I_n^k$, one can find $q_\sigma \leq p_n^k \upharpoonright p_n^k(\sigma)$ and $v_\sigma \in \omega^n$ such that $q_\sigma \Vdash \check{\Phi}(\dot{r}_{gen}) \upharpoonright n = v_\sigma$. We can find such q_σ and v_σ by an $\mathbf{M}$-computable way. Define $h(\sigma) = v_\sigma$, and $p_{n+1} = p_n^{n+2}$. As before, we get a fusion $r \leq_n p_n$ for all $n \in \omega$ by Lemma 4.10. We define $\hat{h}(x) = \bigcup_{\sigma\prec x} h(\sigma)$ for all $x \in [r]$. Clearly, $\hat{h}$ is a continuous function defined on $[r]$. If z is an infinite path through r, then for any n there is $\sigma_n \prec z$ and $\sigma_n \in I_m^k$ for some $m \geq n$ and k. Hence, $r \upharpoonright \sigma_n \Vdash \check{\Phi}(\dot{r}_{gen}) \upharpoonright k = h(\sigma_n)$. Therefore, by Lemma 4.5 (2), if z is a $\mathbb{Q}$-generic point in $[r]$, then $\Phi(z) = \hat{h}(z)$ as desired. □

We say that a real $z \in \omega^\omega$ is *weakly meager engulfing over* $\mathbf{M}$ (see also

Definition 2.4) if z **M**-computes a meager set that covers all **M**-coded reals.

Lemma 4.13: *Suppose that $\mathbb{Q}$ is a lim-sup tree forcing over* **M**. *Then, every $\mathbb{Q}$-generic real z over* **M** *is not weakly meager engulfing over* **M**.

Proof: For any real $x \in 2^\omega$ and order $h \in \omega^\omega$, let $E(h,x)$ denote the meager set defined by

$$E(h,x) = \{y \in 2^\omega : (\forall^\infty k)(\exists n \in [h(k), h(k+1)))\ y(n) \neq x(n)\}.$$

Note that every meager set in 2^ω is covered by a set of the form $E(h,x)$ (see [2] (Theorem 2.2.4)). Assume that there are two M-computations Φ, ψ such that $E(\Phi(z), \psi(z))$ is meager. Then, there is a condition $p \in \mathbb{P}$ such that

$$p \Vdash \check{\Phi}(\dot{r}_{gen}) \in \omega^\omega \ \wedge\ \check{\psi}(\dot{r}_{gen}) \in 2^\omega.$$

By continuous reading of names (Lemma 4.12), there is $q \leq p$ such that for every n and $\sigma \in n^{\leq n}$, $q(\sigma)$ decides the values $\check{\Phi}(\dot{r}_{gen}) \upharpoonright n$ and $\check{\psi}(\dot{r}_{gen}) \upharpoonright n$. In other words, we have monotone functions $h : q \to \omega^{<\omega}$ and $x : q \to 2^{<\omega}$ in M such that for every n and $\sigma \in n^{\leq n}$,

$$q \upharpoonright q(\sigma) \Vdash \check{\Phi}(\dot{r}_{gen}) \upharpoonright n = h(q(\sigma)) \upharpoonright n \ \wedge\ \check{\psi}(\dot{r}_{gen}) \upharpoonright n = x(q(\sigma)) \upharpoonright n.$$

We define $h(y) = \bigcup_n h(y \upharpoonright n)$ and $x(y) = \bigcup_n x(y \upharpoonright n)$ for $y \in 2^\omega$. We say that r is *q-rational* if r is the leftmost path through q which extends $q(\sigma)$ for some $\sigma \in \omega^{<\omega}$. Note that the set $\mathbb{Q}_q$ of all q-rationals can be coded in **M**. Hence, $E_q = \bigcup_{y \in \mathbb{Q}_q} E(h(y), x(y))$ is an **M**-coded meager set. Therefore, there is an **M**-coded real $c \notin E_q$.

Let z be a $\mathbb{Q}$-generic real over **M**. It suffices to show $c \notin E(\Phi(z), \psi(z))$. We construct a decreasing fusion sequence $q = q_0 \geq_0 q_1 \geq_1 \cdots$ such that each $[q_n]$ is a clopen subset of $[q]$; therefore, $\mathbb{Q}_q$ is dense in $[q_n]$ for any $n \in \omega$. Suppose that q_n is already constructed. Let S_n be the set of all successors of a level n node in q_n. For every $\tau \in S_n$, we choose a q-rational v_τ extending τ in $[q_n]$. Then $c \notin E(h(v_\tau), x(v_\tau))$ since $E(h(v_\tau), x(v_\tau)) \subseteq E_q$. In other words,

$$(\exists^\infty k)\ x(v_\tau) \upharpoonright [h(v_\tau)(k), h(v_\tau)(k+1)) = c \upharpoonright [h(v_\tau)(k), h(v_\tau)(k+1)).$$

Thus, one can find a sufficiently long initial segment $\eta_\tau \prec v_\tau$ such that

$$(\exists k > n)\ x(\eta_\tau) \upharpoonright [h(\eta_\tau)(k), h(\eta_\tau)(k+1)) = c \upharpoonright [h(\eta_\tau)(k), h(\eta_\tau)(k+1)).$$

Since $\limsup_{\eta \prec v_\tau} \mathbf{nor}(\eta) = \infty$, there must exist η^*_τ such that $\eta_\tau \preceq \eta^*_\tau \prec v_\tau$ and $\mathbf{nor}(\eta^*_\tau) > n$. Then, we define

$$q_{n+1} = \bigcup_{\tau \in S_n} q_n \upharpoonright \eta^*_\tau.$$

Clearly, q_{n+1} is a clopen subset of q_n. Let r be the fused condition obtained from the sequence $\langle q_n \rangle_{n\in\omega}$. Then,

$$r \Vdash \exists^\infty k\, \check{x}(\dot{r}_{gen}) \upharpoonright [\check{h}(\dot{r}_{gen})(k), \check{h}(\dot{r}_{gen})(k+1)) = \check{c} \upharpoonright [\check{h}(\dot{r}_{gen})(k), \check{h}(\dot{r}_{gen})(k+1)).$$

Hence, this implies $c \notin E(\Phi(z), \psi(z))$. □

Corollary 4.14: *Let $\mathbb{Q}$ be a lim-sup tree forcing over $L_{\omega_1^{\mathrm{CK}}}$. If z is $\mathbb{Q}$-generic over $L_{\omega_1^{\mathrm{CK}}}$, then z is i.o. Δ^1_1-traceable.*

Proof: Since $\mathbb{Q}$ is a lim-sup tree forcing over $L_{\omega_1^{\mathrm{CK}}}$, $\mathbb{Q}$ satisfies the $\Sigma_1(L_{\omega_1^{\mathrm{CK}}})$-fusion property by Lemma 4.12. Therefore, $\omega_1^{\mathrm{CK},z} = \omega_1^{\mathrm{CK}}$ by Lemma 4.5. By Lemma 4.4, z is not weakly meager engulfing over $L_{\omega_1^{\mathrm{CK}}}$. Therefore, z is not Δ^1_1-weakly meager engulfing since $\omega_1^{\mathrm{CK},z} = \omega_1^{\mathrm{CK}}$. Hence, z is i.o. Δ^1_1-traceable as mentioned in Section 3.1. □

4.4. *Examples of lim-sup forcings*

Here, we see properties of some examples of lim-sup forcings. A notion $\mathbb{P}$ of forcing has the Σ_1*-Laver property over* $\mathbf{M}$ if for every $\mathbf{M}$-computation Φ and for every bound $f \in \mathbf{M} \cap \omega^\omega$, if

$$\Vdash_{\mathbb{P}} (\forall n \in \check{\omega})\ \check{\Phi}(\dot{r}_{gen})(n) \downarrow < \check{f}(n),$$

then there exists a $\Sigma_1(\mathbf{M})$-slalom $\{T_n\}_{n\in\omega}$ with $|T_n| \leq n$ such that

$$\Vdash_{\mathbb{P}} (\forall n \in \check{\omega})\ \check{\Phi}(\dot{r}_{gen})(n) \in \check{T}_n.$$

The items (1) and (2) in the following lemma can be proved by using the straightforward modification of the standard argument (see Bartoszyński-Judah [2]).

Lemma 4.15: *Let $\mathbb{PT}$ be the rational perfect set forcing (Example 4.7 (3)).*

(1) Every $\mathbb{PT}$-generic real over $\mathbf{M}$ is $\mathbf{M}$-unbounded.
(2) The forcing $\mathbb{PT}$ satisfies the Σ_1-Laver property; therefore, every $\mathbb{PT}$-generic real over $\mathbf{M}$ is $\Sigma_1(\mathbf{M})$-Laver traceable.
(3) If z is $\mathbb{PT}$-generic real over $\mathbf{M}$, then we have $\mathfrak{G}_{\mathbf{M}}(z) = 1/2$.

Proof: (2) Suppose that $f \in \omega^\omega \cap \mathbf{M}$ is given, and assume that $p \Vdash \check{\Phi}(\dot{r}_{gen})(n) < f(n)$. By continuous reading of names (Lemma 4.12), there is $q \leq p$ such that for every n and $\sigma \in n^n$, $q(\sigma)$ decides the values $\check{\Phi}(\dot{r}_{gen}) \restriction n$. In other words, we have a monotone function $h : q \to \omega^{<\omega}$ in M such that for every n and $\sigma \in n^n$,

$$q \restriction q(\sigma) \Vdash \check{\Phi}(\dot{r}_{gen}) \restriction n = h(q(\sigma)) \restriction n.$$

Moreover, without loss of generality, we may assume that $h(\tau)(n) < f(n)$ for all $\tau \in q$ and $n \in \omega$, and every branching node of q is infinitely branching. We will define a decreasing sequence $(q_n)_{n \in \omega} \subseteq \mathbb{PT}$ such that $q_0 = q$ and $q_{n+1} \leq_n q_n$. Given q_n and $\sigma \in n^{\leq n}$, since $q_n(\sigma)$ is infinitely branching, by the pigeonhole principle, there is $k_\sigma < f(n)$, there are infinitely many j such that $h(q_n(\sigma^\frown j))(n) = k$. Let J_σ be all such j's, and define $S = \{\sigma^\frown j : \sigma \in n^{\leq n},\ j \in J_\sigma \text{ and } \sigma^\frown j \notin n^{\leq n}\}$, and $T_n = \{k_\sigma : \sigma \in n^{\leq n}\}$. Then, define $q_{n+1} = \bigcup_{\sigma^\frown j \in S} q_n(\sigma^\frown j)$. It is not hard to see that $q_n \geq_n q_{n+1} \in \mathbb{PT}$. Let r be a fusion such that $r \leq_n q_n$ for all $n \in \omega$. One can check that $h(z)(n) \in T_n$ for all $z \in [r]$ and $n \in \omega$.

(3) We give a sketch of the proof. We follow the argument of Andrews et al. [1] which shows that computably traceability of x implies $\mathfrak{G}_{\Delta^0_1}(x) \geq 1/2$. Let z be a $\mathbb{PT}$-generic real over $\mathbf{M}$. By the Σ_1-Laver property, for every $A \leq_{\mathbf{M}} z$, there is a $\mathbf{M}$-coded slalom $(T_n)_{n \in \omega}$ such that $A \restriction J_n \in T_n$ for almost all n. Thus, as in Andrews et al. [1], we can find a set $B \subseteq \omega$ in $\mathbf{M}$ such that $\underline{\rho}_{\mathbf{M}}(\{n : A(n) = B(n)\}) \geq 1/2$. Thus, $\mathfrak{G}_{\mathbf{M}}(z) \geq 1/2$. Conversely, $\mathfrak{G}_{\mathbf{M}}(z) \leq 1/2$ since $z \notin \mathbf{M}$ (see [1]). Hence, $\mathfrak{G}_{\mathbf{M}}(z) = 1/2$. □

As in the argument in Corollary 4.14, we can see the following:

Corollary 4.16: *Let z be a $\mathbb{PT}$-generic real over $L_{\omega_1^{\mathrm{CK}}}$. Then, z is Δ^1_1-unbounded, Δ^1_1-Laver traceable (hence, not Δ^1_1-infinitely often equal), and $\mathfrak{G}_{\Delta^1_1}(z) = 1/2$.*

Proof: We only show that z is not Δ^1_1-infinitely often equal. As in Corollary 4.14, $\omega_1^{\mathrm{CK},z} = \omega_1^{\mathrm{CK}}$ by Lemmata 4.12 and 4.5. By Lemma 4.15, z is $\Sigma_1(L_{\omega_1^{\mathrm{CK}}})$-Laver traceable. Therefore, z is Π^1_1-Laver traceable since $\omega_1^{\mathrm{CK},z} = \omega_1^{\mathrm{CK}}$. By Lemma 3.3, z is Δ^1_1-Laver traceable. Consequently, by Proposition 3.6 (1a), z is not Δ^1_1-infinitely often equal. □

We are now ready to prove our main theorem for $\Gamma = \Pi^1_1$.

Proof: [Proof of Theorem 2.5 for $\Gamma = \Pi^1_1$] Let z be a $\mathbb{PT}$-generic real over $L_{\omega_1^{\mathrm{CK}}}$. By Corollary 4.16, $z \in [\mathsf{D}]_\Delta$ (i.e., z is Δ^1_1-bounded), but $z \notin [\mathsf{ED}]_\Delta$

(i.e., z is not Δ-infinitely often equal) and $z \notin [\mathsf{Cov}\mathcal{X}_{0.5}]_\Delta$ (since $\mathfrak{G}_{\Delta^1_1}(z) = 1/2$). Therefore, we have $[\mathsf{ED}]_{\Delta^1_1} \not\leq_{\Delta^1_1\mathrm{w}} [\mathsf{D}]_{\Delta^1_1}$ and $[\mathsf{Cov}\mathcal{X}_{0.5}]_{\Delta^1_1} \not\leq_{\Delta^1_1\mathrm{w}} [\mathsf{D}]_{\Delta^1_1}$. Moreover, since there is a computable Tukey morphism $\mathsf{D} \longrightarrow \mathsf{Cof}\mathcal{M}$, we have $[\mathsf{D}]_{\Delta^1_1} \subseteq [\mathsf{Cof}\mathcal{M}]_{\Delta^1_1}$. Therefore $z \in [\mathsf{Cof}\mathcal{M}]_{\Delta^1_1}$ (i.e., z is not low for Δ^1_1-generic). Finally, by Lemma and by $\omega_1^{\mathrm{CK},z} = \omega_1^{\mathrm{CK}}$, z is not weakly Δ^1_1-meager engulfing, that is, $z \notin [\mathsf{Non}\mathcal{M}]_{\Delta^1_1}$. Consequently, we have $[\mathsf{Non}\mathcal{M}]_{\Delta^1_1} \not\leq_{\Delta^1_1} [\mathsf{Cof}\mathcal{M}]_{\Delta^1_1}$. □

The infinite equal forcing $\mathbb{EE}$ (Example 4.7 (4)) is useful to show the properness of the Muchnik reduction $[\mathsf{Non}\mathcal{N}]_L <_{L\mathrm{w}} [\mathsf{Cof}\mathcal{M}]_L$ in Cichoń's diagram. As in the usual argument (see Bartzyński-Judah [2] (Section 7.4.C), Kjos-Hanssen et al. [31]), one can easily see that Every $\mathbb{EE}$-generic real z over $\mathbf{M}$ is an $(\lambda n.2^n)$-infinitely often equal real over $\mathbf{M}$ (in particular, z is weakly null-engulfing over $\mathbf{M}$ by Proposition 3.6), and every $\mathbb{EE}$-generic is $\mathbf{M}$-often $\mathbf{M}$-traceable. Hence, every $\mathbb{EE}$-generic real over $L_{\omega_1^{\mathrm{CK}}}$ is Δ^1_1-weakly null-engulfing, and Δ^1_1-often Δ^1_1-traceable (i.e., low for Δ^1_1-Kurtz randomness).

4.5. *Arithmetical definability*

One can also ask whether the similar results also hold for the arithmetical degrees. Of course, we cannot use the forcing relation introduced in Section 4 since the universe $L_{\omega+1}$ of arithmetically coded sets is not admissible, and the definition of $\Vdash_{\mathbb{Q}}$ is Π^1_1. To overcome this difficulty, we consider the following local Cohen forcing argument: If $\mathbb{Q}$ is lim-sup tree forcing over $\mathbf{M} = L_\alpha[R]$, then the forcing relation $\Vdash_{\mathbb{Q}}$ for ranked formulas can be characterized by using *local Cohen forcing.* Indeed, by using the usual fusion argument, we can show that the following two conditions are equivalent for any ranked formula φ (see also Lemma 4.27):

$$(\exists q \leq p)\; q \Vdash_{\mathbb{Q}} \varphi(\dot{r}_{gen}, \vec{v}), \tag{4.1}$$

$$(\exists r \leq p)\; [r \in \mathbb{Q} \;\&\; \mathbf{1} \Vdash_{\mathbb{C}(r)} \varphi(\dot{r}_{gen}, \vec{v})]. \tag{4.2}$$

Here $\mathbb{C}(r)$ is the local Cohen forcing inside the tree $r \subseteq H$, that is, each forcing condition is a string in r, and $p \leq q$ if and only if p extends q. The implication from (4.1) to (4.2) is obvious by the definition of the forcing relation $\Vdash_{\mathbb{Q}}$ for ranked formulas. For the converse direction, assume $\mathbf{1} \Vdash_{\mathbb{C}(r)} \varphi(\dot{r}_{gen}, \vec{v})$, and find a sufficiently large rank $\gamma \in \mathbf{M}$ such that $L_\gamma[R]$ contains r, φ and $\vec{v}$. By fusion argument inside r (see also Lemmas 4.27 for the detail), we can find a $\mathbb{Q}$-condition $q \leq r$ such that every element of $[q]$ is

$\mathbb{C}(r)$-generic over $L_{\gamma+1}[R]$. Then, every $x \in [q]$ satisfies $\varphi(x, \vec{v})$ since x is an r-local Cohen real, and $\mathbf{1} \Vdash_{\mathbb{C}(r)} \varphi(\dot{r}_{gen}, \vec{v})$. Hence, $[q] \subseteq \{x \in \omega^\omega : \varphi(x, \mathbf{v})\}$, and $q \Vdash_{\mathbb{Q}} \varphi(\dot{r}_{gen}, \vec{v})$ by the definition.

It is well-known that the Cohen forcing relation $\Vdash_{\mathbb{C}(r)}$ for Σ^0_n formulas is Σ^0_n-definable relative to r. Thus, by using the condition (4.2) instead of the condition (4.1), we have the similar results for arithmetical degrees: Every arithmetically $\mathbb{PT}$-generic real z is arithmetically unbounded, arithmetically i.o. traceable, arithmetically Laver traceable (hence, not arithmetically infinitely often equal), and the arithmetical $\mathfrak{G}$-value of z is $1/2$.

4.6. *Infinite time register machine computability*

The next Spector pointclasses above Π^1_1 are obtained as the Nikodym hierarchy of Selivanovskij's $\mathcal{C}$-sets (that is, the hierarchy obtained by iterating Suslin's $\mathcal{A}$-operation). This Nikodym hierarchy has a logical representation by using the game quantifier $\Game(D_\alpha\Sigma^0_1)$ for the αth level $D_\alpha\Sigma^0_1$ of the (lightface) Hausdorff difference hierarchy starting from (c.e.) open sets (see Burgess [7], and Tanaka [53]), where recall that $\Game\Sigma^0_1 = \Pi^1_1$ and $\Game\Pi^0_1 = \Sigma^1_1$. Of course, the companions of these pointclasses are an initial segment of admissible sets $L_{\omega_1^{\mathrm{CK}}}, L_{\omega_2^{\mathrm{CK}}}, L_{\omega_3^{\mathrm{CK}}}, \dots$. The first ω admissible ordinals and their limit also naturally occur in the context of *infinite time register machines.*

At the beginning of the 1960s, computability theorists (proof theorists, and set theorists) started to develop ordinal-length computations. Earlier developments (the 1960s–1980s) on computability on ordinals have been summarized in several classical textbooks such as Sacks [48] (for α-recursion theory) and Hinman [19] (Chapter 8; for ordinal recursion theory). *Infinite time Turing machines* (ITTMs) and *infinite time register machines* (ITRMs) are special kinds of computation models on ordinals. An ITTM/ITRM is designed as the same as a usual TM/RM, but an ITTM/ITRM-computation is allowed to run for ordinal steps, while the memory storage of a machine is limited to ω as a usual TM/RM (see Koepke [32], Koepke-Miller [33] and Carl-Schlicht [9]).

The theory of ITTMs has been found to be very interesting because it involves a new kind of large countable ordinals (see Welch [56]) which has not been discovered in the 1980s. The ordinals associated with ITRMs are relatively small as described below. We use the symbol $\omega_\alpha^{\mathrm{CK}}$ (and $\omega_\alpha^{\mathrm{CK},z}$) to denote the αth ordinal which is admissible or the limit of admissible ordinals (relative to z). Note that $\omega_\omega^{\mathrm{CK}}$ itself is not admissible, and clearly, it is much smaller than the first recursively inaccessible ordinal $\omega_1^{\mathrm{E}_1}$. (Indeed, $\omega_1^{\mathrm{E}_1}$ is the least ordinal α such that $\alpha = \omega_\alpha^{\mathrm{CK}}$.)

Fact 4.17: (See Koepke [32]) A real x is ITRM-computable relative to a real y if and only if $x \in L_{\omega_\omega^{\mathrm{CK},y}}[y]$.

Now we show our main result for ITRM computability by combining a local Cohen forcing argument and a lim-sup tree forcing argument.

Fact 4.18: (Folklore) Suppose that z is a $\mathbb{C}(r)$-generic real over $L_{\omega_n^{\mathrm{CK},r}+1}[r]$. Then, $\omega_n^{\mathrm{CK},z\oplus r} = \omega_n^{\mathrm{CK},r}$.

One can again use local Cohen forcing argument to show the following.

Lemma 4.19: *Suppose that $\mathbb{Q}$ is a lim-sup tree forcing with an ITRM-computable norm. There exists a real z such that*

(1) z is $\mathbb{Q}$-generic over $L_{\omega_n^{\mathrm{CK}}}$ for infinitely many $n \in \omega$.
(2) $\omega_\omega^{\mathrm{CK}} = \omega_\omega^{\mathrm{CK},z}$.

Proof: We start from the empty condition $p_0 = \mathbf{1}_{\mathbb{Q}}$. Suppose that a condition $p_n \in \mathbb{Q} \cap L_{\omega_\omega^{\mathrm{CK}}}$ is given. Choose $k(n) \in \omega$ such that $p_n \in L_{\omega_{k(n)}^{\mathrm{CK}}}$. In the next admissible rank (i.e., in $L_{\omega_{k(n)+1}^{\mathrm{CK}}}$), we construct a sequence $p_n = q_n^0 \geq_0 q_n^1 \geq_1 q_n^2 \geq_2 \cdots$ such that q_n^e decides the eth sentence over $L_{\omega_{k(n)}^{\mathrm{CK}}}$. By fusion argument, one can find a **nor**-perfect tree $q_n \in L_{\omega_{k(n)+1}^{\mathrm{CK}}}$ below p_n such that for any $x \in [q_n]$, x is $\mathbb{Q}$-generic over $L^{\mathrm{CK}}_{\omega_{k(n)}}$.

Then, we can next find a **nor**-perfect tree $p_{n+1} \subseteq q_n$ in $L_{\omega_{k(n)+n+1}^{\mathrm{CK}}}$ such that for any $x \in [p_{n+1}]$, x is $\mathbb{C}(q_n)$-generic over $L_{\omega_{k(n)+n}^{\mathrm{CK}}}$, where $\mathbb{C}(q_n)$ is "local" Cohen forcing inside the closed subspace $[q_n] \subseteq \omega^\omega$. We may assume that the length of the stem of p_{n+1} is greater than n. Since $q_n \in L_{\omega_{k(n)}^{\mathrm{CK}}}$, for every $1 \leq i < n$, $L_{\omega_{k(n)+i}^{\mathrm{CK}}}$ is admissible in any $x \in [p_{n+1}]$ by Fact 4.18. Then, for such $x \in [p_{n+1}]$, there are at least n many admissible-in-x ordinals below $L_{\omega_\omega^{\mathrm{CK}}}$. Therefore, if $z \in \bigcap_n [p_n]$, then there are ω many admissible-in-z ordinals below $L_{\omega_\omega^{\mathrm{CK}}}$. This implies $\omega_\omega^{\mathrm{CK},z} = \omega_\omega^{\mathrm{CK},z}$. □

Corollary 4.20: *There is a real z such that z is ITRM-unbounded, i.o. ITRM-traceable, ITRM-Laver traceable (hence, not ITRM-infinitely often equal), and that $\mathfrak{G}_{\mathrm{ITRM}}(z) = 1/2$.*

Proof: For $\mathbb{Q} = \mathbb{PT}$, suppose that z is a real as in Lemma 4.19. We only show that z is ITRM-Laver-traceable. If $g \in \omega^\omega$ is ITRM-computable in

z, then $g \in L_{\omega_\omega^{\mathrm{CK},z}}[z] = L_{\omega_\omega^{\mathrm{CK}}}[z]$ by Lemma 4.19. In particular, there is n such that $g \in L_{\omega_n^{\mathrm{CK}}}[z]$. Suppose that g is ITRM-bounded, say, there is $h \in L_{\omega_k^{\mathrm{CK}}}$ such that $g(n) < h(n)$ for all $n \in \omega$ So, we can choose some $m \geq n, k$ such that z is $\mathbb{PT}$-generic over $L_{\omega_m^{\mathrm{CK}}}$. Then, by Lemma 4.15, there is a $\Sigma_1(L_{\omega_m^{\mathrm{CK}}})$-slalom $(T_n)_{n\in\omega}$ such that $g \in T_n$ for all $n \in \omega$. Consequently, every ITRM-bounded function $g \leq_{\mathrm{ITRM}} z$ is traced by a ITRM-computable slalom $(T_n)_{n\in\omega}$, that is, z is ITRM-Laver traceable. □

4.7. *Recursively hyper-inaccessibles*

Our results in Section 4 can be applied to Kleene's higher type computability theory, where recall that the 2-envelope of any normal type 2 functional forms a Spector pointclass (see Kechris [24]). We first consider a hierarchy of Spector pointclasses generated by a normal type 3 functional sJ called *Gandy's superjump operator* (see Hinman [19] for instance), where $\mathrm{sJ}(F) : \omega \times \omega^\omega \to 2$ is defined as follows for any type 2 functional F:

$$\mathrm{sJ}(F)(e,x) = \begin{cases} 1 & \text{if } \Phi_e^F(x)\downarrow, \\ 0 & \text{if } \Phi_e^F(x)\uparrow. \end{cases}$$

Here, Φ_e^F is the eth computation relative to the functional F in the sense of Kleene's finite type computability. Define E_0 to be the Turing jump operator, and define E_{n+1} to be the superjump $\mathrm{sJ}(\mathrm{E}_n)$ of E_n. Note that E_1 is essentially equivalent to the hyperjump operator (hence, a transfinite iteration of hyperjumps is computable in E_1). We can introduce the E_n-reducibility $\leq_{\mathrm{E}_n}$ via the Spector pointclass ${}_2\mathrm{env}(\mathrm{E}_n)$, where ${}_2\mathrm{env}(\mathrm{E}_n)$ denotes the 2-envelop of the normal type 2 functional E_n, that is, the collection of all sets A of reals such that A is semi-computable in E_n. Clearly, $\leq_{\mathrm{E}_0}$ is equivalent to the hyperarithmetical reducibility $\leq_{\mathrm{h}}$. The structure of the E_n-degrees has been studied by Shinoda [49, 50].

Shinoda [49] showed that for every $n \in \omega$, the Sacks forcing over $L_{\omega_1^{\mathrm{E}_n}}$ preserves the least n-recursively inaccessible ordinal $\omega_1^{\mathrm{E}_n}$ under the nth normal type 2 hypercomputation E_n, by using the Σ_1-fusion property and local Cohen forcing. By the same argument, we show that any sufficiently generic with respect to a lim-sup tree forcing also preserves $\omega_1^{\mathrm{E}_n}$. First note that if z is a (local) Cohen-generic real over $L_{\omega_1^{\mathrm{E}_n}}$, then, $\omega_1^{\mathrm{E}_n,z} = \omega_1^{\mathrm{E}_n}$ (Shinoda [49] (Theorem 3.9)). The following two lemmas are modifications of Lemma 4.6 and Theorem 4.7 in Shinoda [49], respectively.

Lemma 4.21: *Suppose that κ is n-recursively inaccessible, and p is a* **nor**-*perfect tree in L_κ. Then, there exists a* **nor**-*perfect tree $p^* \subseteq p$ in L_{κ^+} such* *that for any $x \in [p^*]$, x is p-Cohen over L_κ and κ is n-recursively-in-x inaccessible, where κ^+ is the first admissible ordinal larger than κ.*

Lemma 4.22: *Suppose that $\mathbb{Q}$ is a lim-sup tree forcing. For every $n \in \omega$, there exists a $\mathbb{Q}$-generic real x over $L_{\omega_1^{\mathrm{E}_n}}$ such that $\omega_1^{\mathrm{E}_n,x} = \omega_1^{\mathrm{E}_n}$.*

As a consequence of the above lemmas combined with Sections 4.3 and 4.4, we can see the following:

Corollary 4.23: *For every $n \in \omega$, there is a real z such that z is E_n-unbounded, i.o. E_n-traceable, E_n-Laver traceable (hence, not E_n-infinitely often equal), and $\mathfrak{G}_{\mathrm{E}_n}(z) = 1/2$.*

4.8. *Reflecting spector pointclasses*

In this subsection, we deal with the following kind of pointclasses indroduced by Kechris [25] (Definition 1.11). A pointclass Γ is *nice* if Γ contains Δ^0_1 and closed under computable substitutions, and $\vee, \wedge, \exists^\omega$, Γ is ω-parametrized and scaled, and Γ contains Π^0_1 unless $\Gamma = \Sigma^0_1$. For instance, the pointclass Σ^0_n for each $n \geq 1$ is nice. A Spector pointclass Γ is $\Game$-*generated* if there is a nice pointclass Γ' such that $\Gamma = \Game\Gamma'$. Moreover, a Spector pointclass Γ is *reflecting* if for any oracle α and $\Gamma(\alpha)$ set $R \subseteq 2^\omega$, we have

$$\exists A \in \Gamma(\alpha)\, R(A) \Rightarrow \exists B \in \Delta(\alpha)\, R(B).$$

Kechris [25] pointed out that $\Game\Gamma'$ is reflecting for every nice pointclass $\Gamma' \supseteq \Pi^0_2$. Basically, we follow the argument in Kechris [22]. Let $A \subseteq \omega^\omega$. We say that p *forces* A (written as $p \Vdash \dot{r}_{gen} \in A$ or $p \Vdash A$) if

$$(\forall p_0 \leq p)(\exists p_1 \leq p_0)(\forall p_2 \leq p_1)(\exists p_3 \leq p_2) \dots \lim_n p_n \in A.$$

Here, $\lim p_n$ is the unique element contained in $\bigcap_n [p_n]$ if $\bigcap_n [p_n]$ is a singleton.

(1) $p \Vdash A$ and $A \subseteq B$ implies $p \Vdash B$.
(2) $p \Vdash A$ and $q \leq p$ implies $q \Vdash A$.

We say that $A \subseteq \omega^\omega$ is $\mathbb{P}$-*measurable* (or it has *the Baire property with respect to* $\mathbb{P}$) if there is a $\mathbb{P}$-open set $U \subseteq \omega^\omega$ (i.e., a set of the form $\bigcup_{p \in W} [p]$ for some $W \subseteq \mathbb{P}$) such that $\mathbf{1} \Vdash A = U$ (i.e., $\mathbf{1} \Vdash \dot{r}_{gen} \notin A \Delta U$).

Proposition 4.24: *(Kechris [22]) Suppose that A is $\mathbb{P}$-measurable.*

(1) *(Quasi-completeness)* $p \Vdash A$ *or there exists* $q \leq p$ *such that* $q \Vdash \neg A$.
(2) *(Forcing = Truth) For all sufficiently* $\mathbb{P}$*-generic reals* $r \in \omega^\omega$, $r \in A$ *if and only if there exists* $p \in \mathbb{P}$ *such that* $r \in [p]$ *and* $p \Vdash A$.

Proposition 4.25: *(Kechris [22]) Suppose that* A *and* $\{A_i\}_{i\in\omega}$ *are* $\mathbb{P}$*-measurable.*

(1) $p \Vdash \neg A$ *if and only if* $q \nVdash A$ *for every* $q \leq p$.
(2) $p \Vdash \bigcap_i A_i$ *if and only if* $p \Vdash A_i$ *for all* $i \in \omega$.
(3) $p \Vdash \bigcup_i A_i$ *if and only if for every* $q \leq p$*, there are* $r \leq q$ *and* $i \in \omega$ *such that* $r \Vdash A_i$.

The problem is to estimate the complexity of $p \Vdash A$. Most pointclasses Γ which we will consider in this section satisfy the following condition:

$$\Gamma = \Game\Gamma' \text{ is a } \Game\text{-generated reflecting Spector pointclass} \quad \text{and } \sigma\mathbf{\Gamma}'\text{-determinacy holds,} \tag{4.3}$$

where $\sigma\mathbf{\Gamma}'$ is the smallest σ-algebra including $\mathbf{\Gamma}'$. The Borel determinacy implies that the Spector pointclass $\Game\Sigma^0_n$ for each $n \geq 3$ satisfies (4.3). We use the notation $\mathbb{P}$ for a Γ-definable strongly arboreal forcing, and $\mathbb{Q}$ for the following

$$\mathbb{Q} \text{ is a lim-sup tree forcing with } \Delta\text{-trees and } \Delta\text{-norms.} \tag{4.4}$$

Note that such a forcing $\mathbb{Q}$ must be Γ-definable forcing.

Lemma 4.26: *(See also Kechris [22] (Theorem 5.2.2)) Suppose that* Γ *is a Spector pointclass satisfying (4.3), and* $\mathbb{P}$ *is a* Γ*-forcing. Then, every* $\mathbf{\Gamma}$ *set is* $\mathbb{P}$*-measurable.*

Proof: [Sketch of Proof] As in Kechris [22] (Theorem 5.2.2), it suffices to show that the Banach-Mazur game (with respect to the topology induced by the forcing $\mathbb{P}$) with the winning set $A \cap F$ is determined for any $A \in \mathbf{\Gamma}$ and a closed set F. Note that the game on forcing conditions can be identified with a game on ω by using the oracle $D = \{\langle i, j\rangle \in \omega^2 : \pi(i), \pi(j) \in \mathbb{P} \;\&\; \pi(i) \leq \pi(j)\}$, where $\pi :\subseteq \omega \twoheadrightarrow \Delta$ is a partial surjection. Since $A \in \Game\Gamma'$, by absorbing the Γ' game for A into the Banach-Mazur game, the whole game can also be viewed as a $\mathbf{\Gamma}'$ game on ω with parameter D. Hence, we have the desired conclusion by $\mathbf{\Gamma}'$-determinacy. $\square$

Now, we see that the forcing relation of any lim-sup tree forcings $\mathbb{Q}$ is characterized in the context of local Cohen forcing.

Lemma 4.27: *(See also Kechris [25] (Theorem1.14)) Suppose that Γ is a Spector pointclass satisfying (4.3), and $\mathbb{Q}$ is a forcing satisfying (4.4). Then, the following are equivalent:*

$$(\exists q \leq p)\ q \Vdash_{\mathbb{Q}} (\dot{r}_{gen}, n) \in A, \tag{4.5}$$

$$(\exists r \leq p)\ [r \in \mathbb{Q}\ \&\ \mathbf{1} \Vdash_{\mathbb{C}(r)} (\dot{r}_{gen}, n) \in A]. \tag{4.6}$$

Proof: [Sketch of Proof] We follow the argument in Kechris [25] (Theorem 1.14). We first see that (4.5) implies (4.6). Let $D = \{\langle i, j\rangle \in \omega^2 : \pi(i), \pi(j) \in \mathbb{Q}\ \&\ \pi(i) \leq \pi(j)\}$. As before, there is a $\Gamma'(D)$-game G such that q forces $(\dot{r}_{gen}, n) \in A$ for some $q \leq p$ if and only if $\exists$-player wins the game G. Then, there is a $\Delta(D)$ winning strategy τ for $\exists$-player in the game associated with $q \Vdash_{\mathbb{M}} (\dot{r}_{gen}, n) \in A$ for some $q \leq p$.

We construct an $\omega^{<\omega}$-indexed set $\{r_\sigma\}_{\sigma \in \omega^{<\omega}}$ of **nor**-perfect subtrees of q according to the strategy τ. Let $r_{\langle\rangle} = q$, and suppose that we have already constructed a **nor**-perfect tree $r_\sigma \in \mathbb{Q}$. Let ρ_σ be the first node of the **nor**-perfect tree r_σ such that $\mathbf{nor}(\rho_\sigma, \mathrm{succ}_{r_\sigma}(\rho_\sigma)) > |\sigma|$. Then, $\forall$-player defines $r^*_{\sigma^\frown\langle n\rangle}$ to be $r_\sigma \upharpoonright \rho_\sigma{}^\frown\langle s_n\rangle$, where s_n is the nth immediate successor node of ρ_σ in the tree r_σ. Then, the winning strategy τ for $\exists$-player returns a **nor**-perfect tree $r_{\sigma^\frown\langle n\rangle}$ from $r^*(\sigma^\frown\langle n\rangle)$. Clearly, $\{\rho_\sigma : \sigma \in \omega^{<\omega}\}$ generates a $\Delta(D)$ **nor**-perfect tree $r \leq p$ such that $[r] \subseteq \{x : (x, n) \in A\}$, since τ is a $\Delta(D)$-winning strategy.

Now, we proceeds local Cohen forcing $\mathbb{C}(r)$ inside **nor**-perfect trees r. Clearly, the empty condition forces $(\dot{r}_{gen}, n) \in A$ inside r. Then,

$$(\exists r \in \Delta(D))\ [r \leq p\ \&\ r \in \mathbb{Q}\ \&\ \mathbf{1} \Vdash_{\mathbb{C}(r)} (\dot{r}_{gen}, n) \in A]$$

is represented by a Γ formula with a parameter n and D, since $\mathbb{C}(r)$ is set-forcing. Since Γ is reflecting, we can replace D with a Δ set. Consequently, the condition (4.5) implies the condition (4.6).

Conversely, if (4.6) holds, by using a Δ winning strategy associated with $\Vdash_{\mathbb{C}(r)} (\dot{r}_{gen}, n) \in A$, we may find a Δ **nor**-perfect tree q inside the **nor**-perfect tree r such that $[q] \subseteq \{x : (x, n) \in A\}$ by the similar argument as above. Clearly, q forces $(\dot{r}_{gen}, n) \in A$. □

As a consequence, for every Γ set A,

$$\{(p, n) \in \omega^\omega \times \omega : (\exists q \leq p)\ q \Vdash_{\mathbb{Q}} (\dot{r}_{gen}, n) \in A\}$$

is in Γ, since by the definability of (local) Cohen forcing. Therefore, by quasi-completeness,

$$\{(i, n) \in \omega^2 : p_i \nVdash_{\mathbb{P}} (\dot{r}_{gen}, n) \notin A\}$$

is in Γ, since this is equivalent to the consequence of the previous lemma provided every $\mathbf{\Gamma}$ set is $\mathbb{P}$-measurable.

Lemma 4.28: *(Kechris [22] (Theorem 4.4.1)) Let Γ be a Spector pointclass, and $\mathbb{P}$ be a Γ-forcing notion. Moreover, suppose that every $\mathbf{\Gamma}$ subset of ω^ω is $\mathbb{P}$-measurable. For every Γ set $A \subseteq \omega^\omega \times \omega$, if*

$$\{(i,n) \in \omega^2 : p_i \not\Vdash_{\mathbb{P}} (\dot{r}_{gen}, n) \notin A\}$$

is in Γ, then for all sufficiently $\mathbb{P}$-generic x, we have $\lambda_\Gamma^x = \lambda_\Gamma$.

Lemma 4.29: *(See also Kechris [25] (Lemma 1.14)) Suppose that Γ is a Spector pointclass satisfying (4.3), and $\mathbb{Q}$ is a forcing satisfying (4.4). Then, every sufficiently $\mathbb{Q}$-generic real z satisfies $\lambda_\Gamma^z = \lambda_\Gamma$.*

By combining the results from Section 4 and Lemma 4.29 (with Kechris [25] (Lemma 1.5)), we have the following:

Lemma 4.30: *Suppose that Γ is a Spector pointclass satisfying (4.3), and $\mathbb{Q}$ is a forcing satisfying (4.4). Then,*

(1) $\mathbb{Q}$ has the Γ-continuous reading of names.
(2) $\mathbb{Q}$ adds no weakly Γ-meager engulfing real.

Lemma 4.31: *Suppose that Γ is a Spector pointclass satisfying (4.3), and $\mathbb{PT}(\Delta)$ is the rational perfect set forcing with Δ-trees. Then,*

(1) $\mathbb{PT}(\Delta)$ adds a Δ-unbounded real.
(2) $\mathbb{PT}(\Delta)$ has the Γ-Laver property.
(3) For any $\mathbb{PT}(\Delta)$-generic real z, we have $\mathfrak{G}_\Delta(z) = 1/2$.

Corollary 4.32: *Suppose that Γ is a Spector pointclass satisfying (4.3), and $\mathbb{PT}(\Delta)$ is the rational perfect set forcing with Δ-trees. Then, for every $\mathbb{PT}(\Delta)$-generic real z, z is Δ-unbounded, i.o. Δ-traceable, Δ-Laver traceable (hence, not Δ-infinitely often equal), and $\mathfrak{G}_\Delta(z) = 1/2$.*

5. Open Questions

In this section, we propose several open problems. First we focus on the separation problem of $[\mathsf{Cov}\mathcal{N}]_\Delta$ and $[\mathsf{Non}\mathcal{M}]_\Delta$, that is, the problem asking the existence of a real which is not i.o. Δ-traceable, and Δ-computes no Δ-random real. Since i.o. Δ-tt-traceability is equivalent to Hausdorff h-nullness for all Δ-gauge functions h (see [28]), there must be an extremely long hierarchy of Hausdorff measures (i.e., a hierarchy based on growing rates of Kolmogorov complexity) between $[\mathsf{Cov}\mathcal{N}]_\Delta$ and $[\mathsf{Non}\mathcal{M}]_\Delta$. However, as shown in Lemma 4.4, any real added by a lim-sup tree forcing is captured by a ground model set which is Hausdorff h-null for all gauge functions h in the ground model. Therefore, to separate these two notions, we need to develop a new technique other than lim-sup tree forcing.

Problem 5.1: *Does there exist a forcing over a companion model* $\mathbf{M}$ *(e.g.,* $\mathbf{M} = L_{\omega_1^{\mathrm{CK}}}$*) which satisfies the* Σ_1*-fusion property, but adds a weakly meager engulfing real?*

The second problem is whether the lim-sup forcing argument is generally applicable to all Spector pointclasses. At least, as observed in this chapter, we know that the lim-sup forcing argument works at pointclasses $\Game\Sigma^0_1$, $\Game(D_{<\omega}\Sigma^0_1)$, and $\Game\Sigma^0_n$ for $n \geq 3$. However, our work does not cover various important pointclasses; e.g., any pointclasses in the $\mathcal{R}$-hierarchy (the hierarchy between $\Game\Sigma^0_2$ and $\Game\Delta^0_3$; see [8]). Note that the ordinal for $\Game\Sigma^0_2$ (or equivalently $\mathrm{Ind}(\Sigma^1_1)$) is characterized as the least Σ^1_1-reflecting ordinal by Richter-Aczel [43]. More generally, the ordinal for $\Game(D_\alpha\Sigma^0_2)$ (where $D_\alpha\Sigma^0_2$ is the αth level of the difference hierarchy starting from the pointclass Σ^0_2) is characterized by Lubersky [34] in the context of reflecting ordinal. Beyond the $\mathcal{R}$-hierarchy, Welch [55] showed that Sacks forcing is available at the level of infinite-time Turing machine (ITTM) computability. It is natural to ask whether any lim-sup forcing works at these natural pointclasses. For instance:

Problem 5.2: *Let* λ^x *be the supremum of infinite-time Turing machine writable ordinals relative to* x *(see [56]). Suppose that* $\mathbb{Q}$ *is a lim-sup tree forcing over* L_λ*. Then, for every* $\mathbb{Q}$*-generic real* z *over* L_λ*, do we have* $\lambda^z = \lambda$*?*

The problem can be reduced to the problem asking whether one can define such ordinals (Σ^1_1-reflecting ordinals, n-gap reflecting ordinals, writable ordinals, etc.) in the context of admissibility. For instance, recursively inaccessible ordinals are defined as the admissible limits of admissible ordinals; therefore, preservation of admissible ordinals implies preservation of recursively inaccessible ordinals, too (indeed, this is the essence in the proofs of results in Section 4.7). Hiroshi Fujita taught the author that, shortly after he obtained his master's degree under Juichi Shinoda in 1989, Shinoda had given him a similar problem.

Next, we devote our attention to lim-inf forcing. To carry out the fusion argument with Laver trees, the standard method requires us to decide the truth-value a given Σ_1-sentence (to define a ranking function). Of course, this is impossible if our model is the companion of a Spector pointclass. Therefore, we conjecture that Laver forcing is unavailable over the companion of any Spector pointclass. At least, as mentioned in Section 2.3, we know that Laver forcing is not available at the level of arithmetical and hyperarithmetical degrees because (hyper)arithmetical dominant computes all (hyper)arithmetical reals. We may ask whether unavailability of Laver forcing over any companion model can be witnessed in this way, that is, if a function $f \in \omega^\omega$ dominates all Δ-functions, then does there exist $x \leq_\Delta f$ such that x computes all Δ-reals? As a first step, one can ask whether if a function $f \in \omega^\omega$ dominates all $\mathcal{O}$-hyperarithmetic functions, then f compute all $\mathcal{O}$-hyperarithmetical reals; however this problem turns out to be negative by using Solovay's characterization [51] saying that the *computably encodable sets* are exactly the hyperarithmetical sets.

More precisely, given a reducibility notion $\leq_r$, we say that $A \in \omega^\omega$ is $\leq_r$*-encodable* if for every $X \in [\omega]^\omega$ there is $Y \in [X]^\omega$ such that $A \leq_r Y$. We also say that $A \in \omega^\omega$ admits a $\leq_r$*-modulus* if there is $f \in \omega^\omega$ such that, for any $g \in \omega^\omega$ dominating f, $A \leq_r g$ holds. Obviously, if A admits a $\leq_r$-modulus, then it is $\leq_r$-encodable.

Fact 5.3: (See Solovay [51]) Let A be a real.

(1) A is computably encodable if and only if A admits a computable modulus if and only if A is hyperarithmetical.
(2) If A is arithmetically encodable then $A \in L_{\omega_1^{\mathrm{E}_1}}$ (that is, $A \leq_{\mathrm{E}_1} \emptyset$).
(3) A is hyperarithmetically encodable if and only if A admits a hyperarithmetical modulus if and only if $A \in L_{\sigma_1^1}$ where σ_1^1 is the least Σ_1^1-reflecting ordinal (that is, $A \leq_\Delta \emptyset$ for $\Gamma = \Game\Sigma_2^0$ as mentioned above).

Fact 5.3 implies the following:

Corollary 5.4: *Let Γ be a countable pointclass.*

(1) If $\Delta \supsetneq \Delta_1^1$ then there are a Δ-dominant f and a Δ real x such that x is not computable in f.
(2) If Δ contains a real which is not E_1-computable, then there are a Δ-dominant f and a Δ real x such that x is not arithmetical in f.
(3) If Δ contains a real which is not $\Game\Sigma_2^0$-computable, then there are a Δ-dominant f and a Δ real x such that x is not hyperarithmetical in f.

Day-Greenberg-Turetsky (in private communication) pointed out that if f dominates all $\mathcal{O}$-computable functions, all $\mathcal{O}$-computable reals are *arithmetical* in f. Indeed, their argument shows that if f dominates all $\mathcal{O}^n$-computable functions, all $\mathcal{O}^n$-computable reals are arithmetical in f. In particular, if a function f dominates all ITRM-computable functions, then all ITRM-computable reals are arithmetical in f. As in Fact 2.6, this shows that

$$[\mathsf{Add}\mathcal{N}]_{\mathrm{ITRM}} \equiv_{\mathrm{ITRMw}} [\mathsf{Add}\mathcal{M}]_{\mathrm{ITRM}} \equiv_{\mathrm{ITRMw}} [\mathsf{B}]_{\mathrm{ITRM}}.$$

Fact 5.3 (2) leads us to the conjecture that Day-Greenberg-Turetsky's observation can be extended to E_1-computability. We also note that the least n-recursively inaccessible ordinal $\omega_1^{\mathrm{E}_n}$ is much smaller than the least Σ^1_1-reflecting ordinal σ^1_1, while the ITTM ordinal λ is far larger than σ^1_1. Therefore, for instance, one can naturally ask the following:

Problem 5.5:

(1) If a function $f \in \omega^\omega$ dominates all E_1-computable functions, then are all E_1-computable reals arithmetical in f?

(2) For $n > 1$, if a function $f \in \omega^\omega$ dominates all E_n-computable functions, then are all E_n-computable reals hyperarithmetical in f?

(3) If a function $f \in \omega^\omega$ dominates all $\Game\Sigma^0_2$-computable functions, then are all $\Game\Sigma^0_2$-computable reals hyperarithmetical in f?

(4) If a function $f \in \omega^\omega$ dominates all ITTM-computable functions, then are all ITTM-computable reals ITRM-computable in f?

We also point out the limit of our approach. We write $\Gamma' \ll \Gamma$ if there is a Δ-enumeration of all total $\Delta'(x)$-functions on ω for any Δ-real x. Our strategy to show the Δ-analog of the property (2.8) is to find a pointclass $\Gamma' \ll \Gamma$ (which has a good closure property) such that, if f is a Δ-dominant, then every Δ-real is Δ'-reducible to f. By using the notion from Q-theory, Kechris [26] proved an analog of Solovay's characterization of encodable sets. For the basic terminology in Q-theory, we refer the reader to Kechris-Martin-Solovay [27].

Fact 5.6: (Kechris [26]) Assume Projective Determinacy. For $n > 0$, the Q_{2n+1}-encodable reals are exactly the Q_{2n+1}-reals.

This implies that, if $\Delta \supsetneq Q_{2n+1}$, some Δ-real is not Q_{2n+1}-reducible to some Δ-dominant. Therefore, for instance, if $\Delta \supsetneq Q_3$ is close to Q_3 (that is, $\Gamma' \ll \Gamma$ implies $\Delta' \subseteq Q_3$), this fact limits us to use the above mentioned strategy.

The fourth problem is about infinite often equality. The reason of the emphasis on the notion of infinite often equality in this article lies in previous works on forcing theory. It is known that two iterations of any forcing adding an infinitely often equal real always add a Cohen real (see [2] (Lemma 2.4.8)). Fremlin's so-called "half-a-Cohen-real problem" asks whether there is a forcing adding an infinitely often equal real over the ground model without adding a Cohen real. Zapletal [58] recently solved the problem by showing that forcing with nonzero-dimensional subcompacta of a Henderson compactum (a hereditarily infinite dimensional compactum) adds an infinitely often equal real, but no Cohen real (Zapletal also pointed out that if X is uncountable dimensional compactum, forcing with uncountable-dimensional subcompacta of X satisfies Fremlin's condition). It is very unusual that infinite dimensionality is applied in set theory and mathematical logic in such an essential way (except for set-theoretic topology). Therefore, the role of infinite dimensionality in set theory and logic is yet to be understood.

As another recent use of infinite dimensional topology in mathematical logic, Kihara-Pauly [30] observed that a non-total degree (in the sense of computability theory) can be thought of as effective genericity with respect to countable-dimensional subsets of Hilbert cube. Based on this observation, the authors [30] found a connection between computability theory and infinite dimensional topology, and used the connection to solve some problems in descriptive set theory and Banach space theory. We are interested in whether there is a hidden connection between these two recent discoveries [30, 58]. To develop an understanding of the relationship between these works, it seems reasonable to study the effective version of Zapletal's forcing. This is the reason why we consider the following problem as an important one:

Problem 5.7: *Does there exist a Δ^1_1-infinitely often equal real $x \in 2^\omega$ such that no $y \leq_h x$ is Δ^1_1-generic?*

Zapletal's proof heavily uses the result of R. Pol and Zakrzewski [42] (Theorem 5.1) to show that Zapletal's uncountable-dimensional forcing has the 1-1 or constant property (Spector-Sacks' minimality condition). However, it seems that the proof of Pol-Zakrzewski [42] (Theorem 5.1) requires a stronger separation axiom than KP.

Acknowledgment

The author was partially supported by Grant-in-Aid for JSPS fellows. The author is grateful to André Nies for proposing the main problem in this paper when the author visited him in New Zealand in November 2013. This work has been announced in CCR 2014 at the Institute for Mathematical Sciences of the National University of Singapore, and André Nies encouraged the author to publish this article. The author is also grateful to Jörg Brendle, Hiroshi Fujita, Noam Greenberg, Rupert Hölzl, Daisuke Ikegami, Masaru Kada, Hiroaki Minami, Benoit Monin, Takako Nemoto, Philipp Schlicht, Paul Shafer, and Kazuyuki Tanaka for helpful comments and valuable discussions. Finally, the author would like to thank the anonymous referee for valuable comments and suggestions; especially for pointing out that Corollary 5.4 follows from Fact 5.3.

References

1. Uri Andrews, Mingzhong Cai, David Diamondstone, Carl Jockusch, and Steffen Lempp. Asymptotic density, computable traceability and 1-randomness. *Fund. Math.*, 234(1):41–53, 2016.
2. Tomek Bartoszyński and Haim Judah. *Set Theory: On the Structure of the Real Line.* A K Peters, Ltd., Wellesley, MA, 1995.
3. Jon Barwise. *Admissible Sets and Structures.* Springer-Verlag, Berlin-New York, 1975. An Approach to Definability Theory, Perspectives in Mathematical Logic.
4. L. Bienvenu, N. Greenberg, and B. Monin. Continuous higher randomness. preprint.
5. Andreas Blass. Combinatorial cardinal characteristics of the continuum. In *Handbook of Set Theory. Vols. 1, 2, 3*, pages 395–489. Springer, Dordrecht, 2010.
6. Jörg Brendle, Andrew Brooke-Taylor, Keng Meng Ng, and André Nies. An analogy between cardinal characteristics and highness properties of oracles. In *Proceedings of the 13th Asian Logic Conference*, pages 1–28. World Sci. Publ., Hackensack, NJ, 2015.
7. John P. Burgess. Classical hierarchies from a modern standpoint. I. C-sets. *Fund. Math.*, 115(2):81–95, 1983.
8. John P. Burgess. Classical hierarchies from a modern standpoint. II. R-sets. *Fund. Math.*, 115(2):97–105, 1983.
9. Merlin Carl and Philipp Schlicht. Infinite computations with random oracles. to appear in Notre Dame Journal of Formal Logic.
10. C. T. Chong, Andre Nies, and Liang Yu. Lowness of higher randomness notions. *Israel J. Math.*, 166:39–60, 2008.
11. C. T. Chong and Liang Yu. Measure-theoretic applications of higher Demuth's theorem. *Trans. Amer. Math. Soc.*, 368:8249–8265, 2016.

12. C. T. Chong and Liang Yu. Randomness in the higher setting. *J. Symb. Log.*, 80(4):1131–1148, 2015.
13. Samuel Coskey, Tamás Mátrai, and Juris Steprāns. Borel Tukey morphisms and combinatorial cardinal invariants of the continuum. *Fund. Math.*, 223(1):29–48, 2013.
14. Rodney G. Downey and Denis R. Hirschfeldt. *Algorithmic Randomness and Complexity*. Theory and Applications of Computability. Springer, New York, 2010.
15. Herbert B. Enderton and Hilary Putnam. A note on the hyperarithmetical hierarchy. *J. Symb. Log.*, 35(3):429–430, 1970.
16. Jens Erik Fenstad. *General Recursion Theory: An Axiomatic Approach*. Perspectives in Mathematical Logic. Springer-Verlag, Berlin-New York, 1980.
17. Johanna N. Y. Franklin and Frank Stephan. Schnorr trivial sets and truth-table reducibility. *J. Symbolic Logic*, 75(2):501–521, 2010.
18. Martin Goldstern and Saharon Shelah. Many simple cardinal invariants. *Archive for Mathematical Logic*, 32(3):203–221, 1993.
19. Peter G. Hinman. *Recursion Theoretic Hierarchies*. Springer-Verlag, Berlin-New York, 1978. Perspectives in Mathematical Logic.
20. Denis R. Hirschfeldt, Jr. Carl G. Jockusch, Timothy McNicholl, and Paul E. Schupp. Asymptotic density and the coarse computability bound. *Computability*, 5(1):13–27, 2016.
21. Alexander S. Kechris. Measure and category in effective descriptive set theory. *Ann. Math. Logic*, 5:337–384, 1972/73.
22. Alexander S. Kechris. Forcing in analysis. In Gert H. Muller and Dana S. Scott, editors, *Higher Set Theory*, volume 669 of *Lecture Notes in Mathematics*, pages 277–302, 1978.
23. Alexander S. Kechris. On Spector classes. In *Cabal Seminar 76–77 (Proc. Caltech-UCLA Logic Sem., 1976–77)*, volume 689 of *Lecture Notes in Math.*, pages 245–277. Springer, Berlin, 1978.
24. Alexander S. Kechris. Spector second order classes and reflection. In *Generalized Recursion Theory, II (Proc. Second Sympos., Univ. Oslo, Oslo, 1977)*, volume 94 of *Stud. Logic Foundations Math.*, pages 147–183. North-Holland, Amsterdam-New York, 1978.
25. Alexander S. Kechris. Forcing with Δ perfect trees and minimal Δ-degrees. *Journal of Symbolic Logic*, 46(4):803–816, 1981.
26. Alexander S. Kechris. Effective Ramsey theorems in the projective hierarchy. In *Proceedings of the Herbrand symposium (Marseilles, 1981)*, volume 107 of *Stud. Logic Found. Math.*, pages 179–187. North-Holland, Amsterdam, 1982.
27. Alexander S. Kechris, Donald A. Martin, and Robert M. Solovay. Introduction to Q-theory. In *Cabal seminar 79–81*, volume 1019 of *Lecture Notes in Math.*, pages 199–282. Springer, Berlin, 1983.
28. Takayuki Kihara and Kenshi Miyabe. Null-additivity in the theory of algorithmic randomness. in preparation.
29. Takayuki Kihara and Kenshi Miyabe. Unified characterizations of lowness properties via Kolmogorov complexity. *Arch. Math. Logic*, 54(3-4):329–358, 2015.

30. Takayuki Kihara and Arno Pauly. Point degree spectra of represented spaces. submitted.
31. Bjørn Kjos-Hanssen, André Nies, Frank Stephan, and Liang Yu. Higher Kurtz randomness. *Ann. Pure Appl. Logic*, 161(10):1280–1290, 2010.
32. Peter Koepke. Ordinal computability. In *Mathematical theory and computational practice*, volume 5635 of *Lecture Notes in Comput. Sci.*, pages 280–289. Springer, Berlin, 2009.
33. Peter Koepke and Russell Miller. An enhanced theory of infinite time register machines. In Arnold Beckmann, Costas Dimitracopoulos, and Benedikt Löwe, editors, *Logic and Theory of Algorithms, 4th Conference on Computability in Europe, CiE 2008, Athens, Greece, June 15-20, 2008, Proceedings*, volume 5028 of *Lecture Notes in Computer Science*, pages 306–315. Springer, 2008.
34. Robert S. Lubarsky. μ-definable sets of integers. *J. Symbolic Logic*, 58(1):291–313, 1993.
35. A. R. D. Mathias. Provident sets and rudimentary set forcing. *Fund. Math.*, 230(2):99–148, 2015.
36. Benoit Monin. Higher randomness and forcing with closed sets. In *31st International Symposium on Theoretical Aspects of Computer Science*, volume 25 of *LIPIcs. Leibniz Int. Proc. Inform.*, pages 566–577. Schloss Dagstuhl. Leibniz-Zent. Inform., Wadern, 2014.
37. Benoit Monin and André Nies. A unifying approach to the Gamma question. In *30th Annual ACM/IEEE Symposium on Logic in Computer Science, LICS 2015, Kyoto, Japan, July 6-10, 2015*, pages 585–596. IEEE, 2015.
38. Yiannis N. Moschovakis. *Elementary Induction on Abstract Structures.* North-Holland Publishing Co., Amsterdam-London; American Elsevier Publishing Co., Inc., New York, 1974. Studies in Logic and the Foundations of Mathematics, Vol. 77.
39. Yiannis N. Moschovakis. *Descriptive Set Theory*, volume 155 of *Mathematical Surveys and Monographs.* American Mathematical Society, Providence, RI, second edition, 2009.
40. André Nies. *Computability and Randomness*, volume 51 of *Oxford Logic Guides.* Oxford University Press, Oxford, 2009.
41. Janusz Pawlikowski and Ireneusz Recław. Parametrized Cichoń's diagram and small sets. *Fund. Math.*, 147(2):135–155, 1995.
42. Roman Pol and Piotr Zakrzewski. On Borel mappings and σ-ideals generated by closed sets. *Advances in Mathematics*, 231:651–663, 2012.
43. Wayne Richter and Peter Aczel. Inductive definitions and reflecting properties of admissible ordinals. In I.E. Fenstad and P.G. Hinman, editors, *Generalized Recursion Theory*, pages 301–381. North Holland, Amsterdam, 1974.
44. A. Rosłanowski and S. Shelah. Norms on possibilities. i: Forcing with trees and creatures. *Mem. Amer. Math. Soc.*, 141(671), 1999.
45. Nicholas Rupprecht. *Effective Correspondents to Cardinal Characteristics in Cichon's Diagram.* PhD thesis, University of Michigan, 2010.
46. Nicholas Rupprecht. Relativized Schnorr tests with universal behavior. *Arch. Math. Logic*, 49(5):555–570, 2010.
47. Gerald E. Sacks. Measure-theoretic uniformity in recursion theory and set

theory. *Trans. Amer. Math. Soc.*, 142:381–420, 1969.
48. Gerald E. Sacks. *Higher Recursion Theory.* Perspectives in Mathematical Logic. Springer-Verlag, Berlin, 1990.
49. Juichi Shinoda. On the upper semi-lattice of J_a^S-degrees. *Nagoya Math. J.*, 80:75–106, 1980.
50. Juichi Shinoda. Countable J_a^S-admissible ordinals. *Nagoya Math. J.*, 99:1–10, 1985.
51. Robert M. Solovay. Hyperarithmetically encodable sets. *Trans. Amer. Math. Soc.*, 239:99–122, 1978.
52. Jacques Stern. Some measure theoretic results in effective descriptive set theory. *Israel J. Math.*, 20(2):97–110, 1975.
53. Kazuyuki Tanaka. Weak axioms of determinacy and subsystems of analysis I: Δ_2^0 games. *Mathematical Logic Quarterly*, 36(6):481–491, 1990.
54. Klaus Weihrauch. *Computable Analysis: An introduction.* Texts in Theoretical Computer Science. An EATCS Series. Springer-Verlag, Berlin, 2000.
55. P. D. Welch. *Friedman's Trick: Minimality Arguments in the Infinite Time Turing Degrees*, volume 259, pages 425–436. Cambridge University Press, 1999.
56. Philip D. Welch. The length of infinite time Turing machine computations. *J. Symb. Log.*, 65(3):1193–1203, 2000.
57. Jindřich Zapletal. *Forcing Idealized*, volume 174 of *Cambridge Tracts in Mathematics.* Cambridge University Press, Cambridge, 2008.
58. Jindřich Zapletal. Dimension theory and forcing. *Topology Appl.*, 167:31–35, 2014.
59. Andrzej Zarach. Generic extension of admissible sets. In Marian Srebrny Wiktor Marek and Andrzej Zarach, editors, *Set Theory and Hierarchy Theory: A Memorial Tribute to Andrzej Mostowski*, volume 537 of *Lecture Notes in Mathematics*, pages 321–333. Springer-Verlag, 1976.

ACKERMANNIAN GOODSTEIN PRINCIPLES FOR FIRST ORDER PEANO ARITHMETIC

Andreas Weiermann
Department of Mathematics
Ghent University
Krijgslaan 281 S22
9000 Ghent
Belgium
Andreas.Weiermann@UGent.be

We introduce a canonical Goodstein principle which is defined with respect to a natural variant of the Ackermann function. Since, as shown recently by Arai, Wainer and Weiermann, the l-level Goodstein principles (which are also defined with respect to a natural variant of the Ackermann function) for $l \leq \omega$ exhaust eventually the provably total functions of ATR_0 (a theory having the proof-theoretic ordinal Γ_0) it is in light of the hierarchy comparison theorem for the slow and fast growing hierarchy somewhat surprising that in the context of this paper the strength of the Goodstein principles can be classified in terms of transfinite induction for ordinals not exceeding ε_0.

1. Introduction

Inspired by Gentzen's constructive consistency proof for PA (which uses only transfinite induction up to ε_0 for bounded formulas as a non finitistic proof-principle [see, for example, [6] for a modern exposition]) Goodstein devised in [11] a number-theoretic principle which provides a very elegant reformulation of the combinatorial well-foundedness of ε_0. The Goodstein principle can nowadays be considered as the easiest and most direct example for providing a mathematical assertion which is true but – by results of Kirby and Paris from [13] – not provable in first order Peano arithmetic PA. This principle attracted the interest of quite a few researchers and several variants of Goodstein's principle have been devised until now. See, for example, [1, 2, 4, 9, 10, 21] for more details.

In fact Goodstein devised even more general principles regarding l-level Goodstein principles in [12] and indicated that they will increase in strength. These l-level Goodstein principles are defined with respect to a natural ternary variant of the Ackermann function which starts with the base function $A_0(k, b) = k^b$. The choice of the starting function is very natural since it builts on Goodstein's notion of k normal form which is defined in terms of plus, times and iterated exponentiations.

In [9] a first attempt has been made to classify the l-level Goodstein sequences and extensions thereof. Later that path has been extended to Goodstein principles which led in strength up to the proof-theoretic ordinal of $(\Pi^1_1 - \mathrm{CA})_0$ in [21]. The basic idea in these papers was to define certain sets of terms for number-theoretic functions and to define the Goodstein sequences relative to the term notations. The advantages of terms over numbers consisted in overcoming the problems with ambiguity in normal form representations for numbers.

T. Arai suggested to reconsider the Goodstein principles with regard to the normal form problem and this turned out to be a fruitful idea. During the workshop "Sets and Computations" Arai discussed his ideas with the current author and recently Arai, Wainer and Weiermann have been able to show in [3] that the extension of the Goodstein principle based on normal forms for numbers lead in strength eventually up to Γ_0, the proof-theoretic ordinal of ATR_0.

In this paper, we take a fresh look at the Goodstein sequences and we study a variant of the Goodstein principle which is defined with respect to a natural ternary variant of the Ackermann function which starts with the base function $A_0(k, b) = b + 1$.

This switch in the base function does not affect the non primitive recursiveness of the resulting Ackermann function and indeed it is a simple exercise to bound branches of one version of the Ackermann function by appropriate branches of the other. In view of the hierarchy comparison result (see, for example, [8] for further details) it is somewhat surprising that we are now able to show that the proof-theoretic strength of the new principle will collapse and it will not exceed in strength the ordinal ε_0. So in fact the new principle will have the same proof-theoretic strength as the classical Goodstein principle relative to plus, times and exponentiation.

We also discuss at the end certain restrictions of our principles which emerge when certain parameters are kept fixed. These will correspond in strength to the fragments of Peano arithmetic by routine modifications of this paper's approach.

We expect that it will be possible to show along the lines of this paper that an extension of Goodstein sequences relative to the extended Grzegorczyk hierarchy $(A_\alpha)_{\alpha<\varepsilon_0}$ will lead to a canonical independence result for $\mathrm{ACA}_0+(\Pi^1_1-\mathrm{CA})^-$ and ID_1. (Details might nevertheless become somewhat cumbersome.)

Our exposition reveals the notion of a canonical k normal form for a given natural number m. Quite interestingly the natural less than relation on normal forms for natural numbers models precisely the natural less than relation on the associated transfinite ordinals. This leads to a canonical representation of ordinals via a Π^1_2-style direct limit and we hope being able to exploit this relation elsewhere.

2. Basic Definitions

For the rest of the paper, let k denote a positive integer, i.e. $k > 0$.

Definition 2.1:

(1) $A_0(k,b) = b+1$.
(2) $A_{a+1}(k,0) = A_a(k,\cdot)^k(1)$.
(3) $A_{a+1}(k,b+1) = A_a(k,\cdot)^k(A_{a+1}(k,b))$.

In this definition the upper index refers to the number of function iterations. As usual the function $a,k,b \mapsto A_a(k,b)$ is not primitive recursive whereas the function $k,b \mapsto A_a(k,b)$ is primitive recursive for every fixed a.

This paper's approach will also work in case of $A_0(k,b) = b+k$ or $A_0(k,b) = b\cdot k$. The case $A_0(k,b) = k^b$ behaves fundamentally different and has been treated in [3]. We have not yet investigated other choices of the start function and this might be a topic of interest for further investigations with respect to possible phase transitions. See, for example, [15] for related results.

In the sequel we often write $A_a(b)$ for $A_a(k,b)$ when k is fixed. It is well known that the function A (and even the restriction of A when $k>0$ is fixed) is not primitive recursive. The classical binary Ackermann function appears namely in the case that $k=1$ is fixed in advance.

The following lemma regarding natural monotonicity properties of the function A can be routinely proved from the definitions.

Lemma 2.2:

(1) $b < A_a(k,b)$.

(2) If $a < a'$ then $A_a(k,b) < A_{a'}(k,b)$.
(3) If $b < b'$ then $A_a(k,b) < A_a(k,b')$.
(4) If $a > 0$ and $k < k'$ then $A_a(k,b) < A_a(k',b)$.

Definition 2.3: Let a natural number m be given. We are going to define the k normal form of m by a sandwiching procedure.

If $m = 0$ then m is its own k normal form and we write $m =_{k-NF} 0$. Assume now that $m > 0$.

First determine the unique a such that $A_a(0) \leq m < A_{a+1}(0)$ and let $a_0 := a$.

Next determine the unique b_0 such that $A_{a_0}(b_0) \leq m < A_{a_0}(b_0+1)$. To this situation we associate the k-pattern $\langle k, a_0, b_0 \rangle$. (When k is clear from the context we speak in the sequel just from patterns instead of k-patterns.)

If $A_{a_0}(b_0) = m$, then by definition this is the k normal form of m with pattern $\langle k, a_0, b_0 \rangle$ and we write

$$m =_{k-NF} A_{a_0}(b_0).$$

Assume recursively that we have arrived at a situation

$$A_{a_r}(b_r) \leq m < A_{a_r}(b_r+1)$$

and that we have assigned to this situation a pattern $\langle k, a_0, b_0, \ldots, a_r, b_r \rangle$.

If $A_{a_r}(b_r) = m$ then by definition this is the k normal form of m with pattern $\langle k, a_0, b_0, \ldots, a_r, b_r \rangle$ and we write $m =_{k-NF} A_{a_r}(b_r)$.

Otherwise we are in the situation that $A_{a_r}(b_r) < m < A_{a_r}(b_r+1)$.

The case $a_r = 0$ is impossible and so we may assume $a_r > 0$.

If $A_{a_r-1}(A_{a_r}(b_r)) \leq m < A_{a_r}(b_r+1)$ let $a_{r+1} := a_r - 1$. Then there exists a unique b_{r+1} such that $A_{a_{r+1}}(b_{r+1}) \leq m < A_{a_{r+1}}(b_{r+1}+1)$ such that $b_{r+1} \geq A_{a_r}(b_r)$. To this situation we assign the extended pattern $\langle k, a_0, b_0, \ldots, a_{r+1}, b_{r+1} \rangle$ and we can iterate.

If $A_{a_r}(b_r) < m < A_{a_r-1}(A_{a_r}(b_r))$ then there exists a minimal a^* such that $A_{a_r}(b_r) \leq m < A_{a^*}(A_{a_r}(b_r))$.

Assume first that $a^* = 0$. Then $m = A_{a_r}(b_r)$ but this has already been excluded.

Assume therefore that $a^* > 0$. Then $A_{a^*-1}(A_{a_r}(b_r)) \leq m < A_{a^*}(A_{a_r}(b_r))$. Let $a_{r+1} := a^* - 1$. Then there exists a unique b_{r+1} such that $A_{a_{r+1}}(b_{r+1}) \leq m < A_{a_{r+1}}(b_{r+1}+1)$ such that $b_{r+1} \geq A_{a_r}(b_r)$. To this situation we assign the extended pattern $\langle k, a_0, b_0, \ldots, a_{r+1}, b_{r+1} \rangle$ and we can iterate.

The underlying idea is to compare numbers via the lexicographic ordering on the patterns associated to their normal form. The sandwiching procedure is set up so that the following lemma is immediate by construction.

Lemma 2.4: *Let the pattern $\langle k, a_0, b_0, \dots, a_r, b_r\rangle$ be assigned to m. Then of course $m =_{k-NF} A_{a_r}(b_r)$. Moreover the following assertions hold.*

(1) $a_0 > \cdots > a_r$ *and* $b_l \geq A_{a_{l-1}}(b_{l-1})$ *holds for* $1 \leq l \leq r$.
(2) $m < A_{a_l+1}(A_{a_{l-1}}(b_{l-1}))$ *and for* $1 \leq l \leq r$ *either* $b_l < A_{a_l}^{k-1}(A_{a_{l-1}}(b_{l-1}))$ *holds if* $a_l = a_{l-1} - 1$, *or* $b_l < A_{a_{l-1}-1}(A_{a_{l-1}}(b_{l-1}))$ *holds if* $a_l < a_{l-1} - 1$.
(3) All numbers $A_{a_l}(b_l)$ *are in* k *normal form for* $0 \leq l \leq r$.
(4) All the numbers $A_{a_l}^n(A_{a_{l-1}}(b_{l-1}))$ *are in* k *normal form for* $1 \leq l \leq r$ *and* $1 \leq n < k$.
(5) All the numbers $A_{a_l}(b)$ *where* $1 \leq l \leq r$ *and* $A_{a_{l-1}}(b_{l-1}) \leq b \leq b_l$ *are in* k *normal form.*
(6) Let $b_r = A_e(d)$ *with* $e \leq a_r$ *and* $a_r > 0$. *Then* $A_{a_r}(k, b_r - 1)$ *is in* k *normal form. (Actually* $\langle k, a_0, b_0, \dots, a_r, b_r - 1\rangle$ *is the* k *pattern for the number* $A_{a_r}(k, b_r - 1)$.*)*

Definition 2.5: If $m =_{k-NF} A_a(k, b)$ then $m[k \leftarrow k+1] := A_{a[k\leftarrow k+1]}(k+1, b[k \leftarrow k+1])$.

The Goodstein sequences of this paper's approach are now defined as follows. Let us start with a given number m. Let $m_0 := m$ and define recursively $m_{l+1} := m_l[l+2 \leftarrow l+3] - 1$ if $m_l > 0$. For convenience we also put $m_{l+1} = 0$ if $m_l = 0$. Since our Goodstein sequences start with base 2 we assume from now on that $k \geq 2$.

Theorem 2.6:

(1) $\mathrm{PA} + \mathrm{TI}(\varepsilon_0) \vdash \forall m \exists l m_l = 0$.
(2) $\mathrm{PA} \not\vdash \forall m \exists l m_l = 0$.

For proving the first assertion of Theorem 1 we will later need specific ordinal notations for the ordinals less than ε_0. For technical reasons we use the ordinal notation system for Γ_0 where all terms involving the ordinal addition are deleted. The resulting system is then of order type ε_0.

Definition 2.7: Let T be the least set of terms such that

(1) $0 \in T$
(2) $\alpha, \beta \in T \Rightarrow \overline{\varphi}\alpha\beta \in T$.

Let $<$ be defined recursively on T as follows.

(1) $\alpha \neq 0 \Rightarrow 0 < \alpha$.
(2) $\overline{\varphi}\alpha\beta < \overline{\varphi}\gamma\delta$ holds if and only if one of the following cases holds
   (a) $\alpha < \gamma$ and $\beta < \overline{\varphi}\gamma\delta$.
   (b) $\alpha = \gamma$ and $\beta < \delta$.
   (c) $\gamma < \alpha$ and $\overline{\varphi}\alpha\beta \leq \delta$.

We often write $\overline{\varphi}_\alpha\beta$ for $\overline{\varphi}\alpha\beta$ and we use $\alpha+p$ as abbreviation for $\overline{\varphi}_0(\cdot)^p(\alpha)$. It is well known that the order type of $\langle T, < \rangle$ is equal to ε_0, that $\overline{\varphi}10 = \omega$, that $\overline{\varphi}_2 0 = \omega^\omega$ and that $\overline{\varphi}_{\overline{\varphi}10}0 = \omega^{\omega^\omega}$. To prove these claims one could for example employ the order preserving mapping $ord : T \to \varepsilon_0$ which is recursively defined by $ord(0) := 0$, $ord(\overline{\varphi}0\beta) := ord(\beta)+1$ and $ord(\overline{\varphi}\alpha\beta) := \omega^{\omega^{ord(\alpha)}} \cdot (ord(\beta)+1)$ for $\alpha > 0$. (A more precise embedding which leads to an order isomorphism can be extracted from Touzet's article [16].) By iterating the $\overline{\varphi}$ function in the first argument in a suitable way one can then built up all finite towers of ω.

We now introduce canonical fundamental sequences for the elements of T. These will provide a technical device in proving the second assertion of Theorem 1, namely the independence of the Goodstein principle.

Let $Fix_\alpha := \{\overline{\varphi}\gamma\delta : \gamma > \alpha\}$. The underlying rationale is that an element $\delta \in Fix_\alpha$ would be a fixed point of the version of $\overline{\varphi}$ which allows for fixed points. For us the definition is purely technical.

Definition 2.8: Recursive definition of $\alpha[x]$ for $\alpha \in T$.

(1) $0[x] := 0$.
(2) $(\alpha+1)[x] = \alpha$.
(3) $(\overline{\varphi}(\alpha+1)0)[x] := \overline{\varphi}_\alpha^x(1)$.
(4) $(\overline{\varphi}(\lambda)0)[x] := \overline{\varphi}(\lambda[x])(1)$ if λ is a limit.
(5) $(\overline{\varphi}(\alpha+1)(\beta+1))[x] := \overline{\varphi}_\alpha^x(\overline{\varphi}(\alpha+1)\beta)$.
(6) $(\overline{\varphi}(\lambda)(\beta+1))[x] := \overline{\varphi}(\lambda[x])(\overline{\varphi}\lambda\beta)$ if λ is a limit.
(7) $(\overline{\varphi}\alpha\lambda)[x] := \overline{\varphi}\alpha(\lambda[x])$ if λ is a limit and $\lambda \notin Fix_\alpha$ and $\alpha > 0$.
(8) $(\overline{\varphi}(\alpha+1)\lambda)[x] := \overline{\varphi}_\alpha^x\lambda$ if λ is a limit and $\lambda \in Fix_{\alpha+1}$.
(9) $(\overline{\varphi}\alpha\lambda)[x] := \overline{\varphi}(\alpha[x])\lambda$ if α and λ are limits and $\lambda \in Fix_\alpha$.

3. Proving Termination for the Goodstein Process

We are going to show that the Goodstein sequences terminate by providing an ordinal interpretation into the ordinal notation system T.

Definition 3.1: Let $\psi_k(0) := 0$ and let $\psi_k(m) := \overline{\varphi}_{\psi_k a}(\psi_k b)$ if $m =_{k-NF} A_a(k,b)$.

(In the sequel we often use $\psi_k m$ as abbreviation for $\psi_k(m)$.)

We aim at showing termination of the sequence m_k using the functions ψ_{k+2}. For this we need of course the strict monotonicity of ψ_k. Moreover we need that k normal forms are transferred to $k+1$ normal forms when the base is changed from k to $k+1$. This property is somewhat difficult to prove and we need to show that a base change preserves monotonicity. Although this is intuitively clear a proof seems to require to work along the definition of k normal forms. The required monotonicity will be proved by using (via a suitable induction hypothesis) the monotonicity of the k patterns involved.

Lemma 3.2: *If $m < n$ then $\psi_k(m) < \psi_k(n)$.*

Proof: By induction on n with subsidiary induction on m. As before we write $A_a(b)$ for $A_a(k,b)$ since k remains fixed during the proof.

Let us start with a preliminary observation. Let c_p, d_p be a couple occurring in the k-pattern $\langle k, c_0, d_0, \ldots, c_s, d_s\rangle$ for n. Assume that

$$A_{c_p}(d_p) < n < A_{c_p}(d_p+1).$$

Then

$$(*) \quad \overline{\varphi}(\psi_k c_p)(\psi_k d_p) < \psi_k n.$$

This is seen as follows. Let $p < s$. Then $c_p \leq A_{c_p}(d_p) \leq d_s < A_{c_s}(d_s) = n$ and $c_s < c_p$. The induction hypothesis yields $\overline{\varphi}(\psi_k c_p)(\psi_k d_p) = \psi_k(A_{c_p}(d_p)) \leq \psi_k(d_s)$ and $\psi_k c_s < \psi_k c_p$. Hence we obtain $\overline{\varphi}(\psi_k c_p)(\psi_k d_p) < \overline{\varphi}(\psi_k c_s)(\psi_k d_s)$ by $\beta < \overline{\varphi}\alpha\beta$. This finishes the proof of the starting remark.

Now let $A_a(0) \leq m < A_{a+1}(0)$ and $A_c(0) \leq n < A_{c+1}(0)$. Note that $A_a(0), A_{a+1}(0), A_c(0)$ and $A_{c+1}(0)$ are in k normal form. The inequality $m < n$ yields $a \leq c$.

Case 1. Assume $a < c$.

Case 1.1. Assume first that $A_c(0) < n$. Assume that $A_c(d) \leq n < A_c(d+1)$. Then $(*)$ yields $\psi_k n \geq \overline{\varphi}(\psi_k c)(\psi_k d) \geq \overline{\varphi}(\psi_k c)0$. The induction hypothesis yields $\psi_k m < \psi_k A_c(0) = \overline{\varphi}(\psi_k c)0$ hence $\psi_k m < \psi_k n$.

Case 1.2. Now assume that $A_c(0) = n$.

If $a+1 < c$ then the induction hypothesis yields $\psi_k m < \overline{\varphi}(\psi_k(a+1))0 < \overline{\varphi}(\psi_k c)0 = \psi_k n$. If $A_a(0) = m$ then the induction hypothesis applied to $a < c$ yields $\psi_k a < \psi_k c$ hence

$$\psi_k m = \overline{\varphi}(\psi_k a)0 < \overline{\varphi}(\psi_k c)0 = \psi_k n$$

(since $0 < \overline{\varphi}(\psi_k c)0$).

So we may assume that $a+1=c$ and

$$A_a(0) < m < A_{a+1}(0) = n.$$

Let $m =_{k-NF} A_{a_r}(b_r)$. Then $a_r < c = a+1$ and $b_r < m < n$. The induction hypothesis yields $\psi_k(a_r) < \psi_k(c)$ and moreover $\psi_k b_r < \overline{\varphi}(\psi_k c)0$ and hence $\psi_k m = \overline{\varphi}(\psi_k a_r)(\psi_k b_r) < \overline{\varphi}(\psi_k c)0 = \psi_k n$.

Case 2. Assume that $a = c$.

Moreover, assume that $A_a(b) \leq m < A_a(b+1)$ and $A_a(d) \leq n < A_a(d+1)$. Then $b \leq d$. Note that according to Lemma 2.4 $A_a(e)$ is in k normal form for every e with $b \leq e \leq d$.

Case 2.1. Assume first that $b < d$. If $A_a(d) < n$ then the induction hypothesis yields $\psi_k m < \overline{\varphi}(\psi_k a)(\psi_k(b+1)) \leq \overline{\varphi}(\psi_k a)(\psi_k d) < \psi_k n$ by $(*)$. We thus may assume that $A_a(d) = n$. If $A_a(b) = m$ then the induction hypothesis yields $\psi_k b < \psi_k d$ and $\psi_k m < \psi_k n$. If $b+1 < d$ then $\psi_k m < \overline{\varphi}(\psi_k a)(\psi_k(b+1)) < \overline{\varphi}(\psi_k a)(\psi_k d)) = n$. We may thus assume that

$$A_a(b) < m < A_a(b+1) = A_a(d) = n.$$

Let $m =_{k-NF} A_{a_r}(b_r)$ be in k normal form. Then $a_r < a$ and $b_r < m$ and the induction hypothesis yields $\psi_k(b_r) < \overline{\varphi}(\psi_k a)(\psi_k(b+1))$ and hence $\psi_k m = \overline{\varphi}(\psi_k a_r)(\psi_k(b_r)) < \overline{\varphi}(\psi_k a)(\psi_k(b+1)) = \psi_k n$.

Case 2.2. Assume that $b = d$.

We arrive at the situation that $A_a(b) \leq m < n < A_a(b+1)$. By $(*)$ we may assume

$$A_a(b) < m < n < A_a(b+1).$$

Assume that a pattern $\langle k, a_0, b_0, \ldots, a_r, b_r\rangle$ for the k normal form for m and a pattern $\langle k, c_0, d_0, \ldots, c_s, d_s\rangle$ for the k normal form for n are given. Then $a_0 = a = c_0$ and $b_0 = b = d_0$. Let p be maximal such that

$$A_{a_p}(b_p) < m < n < A_{a_p}(b_p+1)$$

so that $a_l = c_l$ and $b_l = d_l$ for $0 \leq l \leq p$ in the patterns for m and n. The choice of p yields $p < s$, $p < r$ and $a_p > 0$.

For n there exists a $c_{p+1} < a_p$ such that

$$A_{c_{p+1}}(d_{p+1}) \leq n < A_{c_p+1}(d_{p+1}+1)$$

where $A_{c_{p+1}}(A_{a_p}(b_p)) \leq n < A_{c_{p+1}+1}(A_{a_p}(b_p))$. Note that the case $A_{a_p}(b_p) \leq n < A_0(A_{a_p}(b_p))$ does not occur.

For m there exists an $a_{p+1} < a_p$ such that

$$A_{a_{p+1}}(b_{p+1}) \leq m < A_{a_{p+1}}(b_{p+1}+1)$$

where $A_{a_{p+1}}(A_{a_p}(b_p)) \leq m < A_{a_{p+1}+1}(A_{a_p}(b_p))$. Note that the case $A_{a_p}(b_p) \leq m < A_0(A_{a_p}(b_p))$ does not occur.

We see that $a_{p+1} \leq c_{p+1}$ since $m < n$.

Case 2.2.1. Assume that $a_{p+1} = c_{p+1}$.

Then $A_{a_{p+1}}(b_{p+1}) \leq m < A_{a_{p+1}}(b_{p+1}+1)$ and $A_{a_{p+1}}(d_{p+1}) \leq n < A_{a_{p+1}}(d_{p+1}+1)$. The maximality of p yields that $b_{p+1} < d_{p+1}$. Then $A_{a_{p+1}}(b_{p+1}+1)$ is in k normal form.

If $A_{a_{p+1}}(d_{p+1}) < n$ then $(*)$ and the induction hypothesis yield $\psi_k m < \overline{\varphi}(\psi_k a_{p+1})(\psi_k(b_{p+1}+1)) \leq \overline{\varphi}(\psi_k a_{p+1})(\psi_k(d_{p+1})) < \psi_k n$. So assume that $A_{a_{p+1}}(b_{p+1}) \leq m < A_{a_{p+1}}(b_{p+1}+1) = n = A_{a_{p+1}}(d_{p+1})$.

If $A_{a_{p+1}}(b_{p+1}) = m$ then the induction hypothesis yields $\psi_k m < \psi_k n$. So assume that $A_{a_{p+1}}(b_{p+1}) < m$.

Let $m =_{k-NF} A_{a_r}(b_r)$. Then $a_r < a_{p+1}$ and $b_r < m < n = A_{a_{p+1}}(d_{p+1})$ together with the induction hypothesis yield $\psi_k(b_r) < \overline{\varphi}(\psi_k(a_{p+1}))(\psi_k(d_{p+1})) = \psi_k n$ and $\psi_k m = \overline{\varphi}(\psi_k(a_r))(\psi_k(b_r)) < \psi_k n$.

Case 2.2.2. Assume that $a_{p+1} < c_{p+1}$.

Let $m =_{k-NF} A_{a_r}(b_r)$ be in k normal form. Then $a_r \leq a_{p+1}$ and the induction hypothesis yields $\psi_k a_r \leq \psi_k(a_{p+1}) < \psi_k(c_{p+1})$. Moreover we have $b_r < m < A_{a_{p+1}+1}(A_{a_p}(b_p)) \leq A_{c_{p+1}}(A_{a_p}(b_p)) \leq A_{c_{p+1}}(d_{p+1})$ and the induction hypothesis yields $\psi_k(b_r) < \overline{\varphi}(\psi_k c_{p+1})(\psi_k(d_{p+1}))$ hence $\psi_k m < \overline{\varphi}(\psi_k c_{p+1})(\psi_k(d_{p+1}))$. If $n = A_{c_{p+1}}(d_{p+1})$ then $\psi_k m < \psi_k n$. If $n > A_{c_{p+1}}(d_{p+1})$ then $(*)$ yields $\psi_k n > \psi_k(A_{c_{p+1}}(d_{p+1})) = \overline{\varphi}(\psi_k c_{p+1})(\psi_k d_{p+1}) > \psi_k m$. □

Lemma 3.3: $m \leq m[k \leftarrow k+1]$. *Moreover if* $m =_{k-NF} A_c(k,d)$ *and* $c > 0$ *then* $m < A_{c[k\leftarrow k+1]}(k+1, d[k \leftarrow k+1])$.

Proof: An easy induction on m yields $m \leq m[k \leftarrow k+1]$. The remaining assertion then follows by assertion 4 of Lemma 2.2. □

We are going to prove that the base shift is an order preserving operation. This looks obvious since k-times iterations will be replaced by $k+1$-times iterations but the problem is that one carefully has to control the k normal forms of the numbers involved. We first prove a technical lemma which is needed at the crucial step in the proof of Lemma 3.5.

Lemma 3.4: *Assume that* $a < c$ *and that* $A_c(d)$ *is in* k *normal form, and further assume that* $b = A_a^{k-1}(A_{a+1}(A_c(d)-1))$. *Then there exists some* e *with* $A_c(d) \leq e \leq A_a(A_c(d))$ *and an* $l \leq A_c(d) \cdot k - 2$ *such that*

$$b = A_a^l(e)$$

and hence $(**) \quad b \leq A_a^{l+1}(A_c(d))$.

Moreover we have for $1 \leq l' \leq l$ *that* $A_a^{l'}(e)$ *is in* k *normal form.*

Proof: Of course we have $A_c(d) \leq b$. Since $a < c$ we see that $A_a(A_c(d))$ is in k normal form. An easy inductive argument yields

$$(\star) \quad b = A_a^{k-1}(A_a^k(A_{a+1}(A_c(d)-2)) = A_a^{k-1}(A_a^{k\cdot(q-1)}(A_{a+1}(A_c(d)-q))$$

for $1 \leq q \leq A_c(d)$.

Since $A_{a+1}(0) \leq A_c(d)$ there will exist a unique $q \leq A_c(d)$ such that

$$A_{a+1}(A_c(d)-q) \leq A_c(d) < A_{a+1}(A_c(d)-q+1) = A_a^k(A_{a+1}(A_c(d)-q)).$$

Then there exists a unique k' such that $0 \leq k' < k$ and

$$A_a^{k'}(A_{a+1}(A_c(d)-q)) \leq A_c(d) < A_a^{k'+1}(A_{a+1}(A_c(d)-q)).$$

By applying A_a to this inequality we obtain for $e := A_a^{k'+1}(A_{a+1}(A_c(d)-q))$ that

$$A_c(d) \leq e \leq A_a(A_c(d)).$$

Let $l := q \cdot k - k' - 2$. Then $(\star)$ and the definition of e yield $b = A_a^l(e)$. Moreover the choice of q yields $l \leq A_c(d) \cdot k - 2$ and $e \leq A_a(A_c(d))$.

Since $A_c(d) \leq e \leq A_a^{k-1}(A_{a+1}(A_c(d)-1)) = b < A_{a+1}(A_c(d)))$ we see by our sandwiching procedure that $A_a^{l'}(e)$ is in k normal form for every $l' \leq l$. □

Lemma 3.5: *If* $m < n$ *then* $m[k \leftarrow k+1] < n[k \leftarrow k+1]$.

Proof: By induction on n with subsidiary induction on m. The proof is somehow similar to the proof of Lemma 3.2. The estimates mirror the "direct limit" structure of the ordinal numbers in terms of their Ackermannian representations on the natural numbers. Whereas ordinals denoted by values of large branches of the $\overline{\varphi}$-function are closed under arbitrary iterated applications from lower function branches of $\overline{\varphi}$ in the new context numbers denoted by values of the higher branches of the Ackermann function are in a certain sense closed under k-times iterated applications of lower branches of the Ackermann function (applied to a smaller starting value).

We start with an introductory claim: Let $A_{c_p}(d_p) < n < A_{c_p}(d_p+1)$ hold for a couple c_p, d_p which occurs in the pattern $\langle k, c_0, d_0, \ldots, c_s, d_s \rangle$ for n. Then $(*) \quad A_{c_p[k\leftarrow k+1]}(k+1, d_p[k \leftarrow k+1]) < n[k \leftarrow k+1]$.

Proof of the claim. Assume that $n = A_{c_s}(d_s)$ where $c_s < c_p$. Then $d_s \geq A_{c_{s-1}}(d_{s-1}) \geq A_{c_p}(d_p)$. The induction hypothesis and assertion 3 of Lemma 2.4 yield $d_s[k \leftarrow k+1] \geq A_{c_{s-1}[k\leftarrow k+1]}(k+1, d_{s-1}[k \leftarrow k+1]) \geq A_{c_p[k\leftarrow k+1]}(k+1, d_p[k \leftarrow k+1])$. Moreover, assertion 1 of Lemma 2.2 yields $n[k \leftarrow k+1] = A_{c_s[k\leftarrow k+1]}(k+1, d_s[k \leftarrow k+1]) > d_s[k \leftarrow k+1]$ and the claim follows. Note that the claim models the sandwiching idea.

To prove the assertion of the lemma let

$$A_a(0) < m < A_{a+1}(0)$$

and

$$A_c(0) \leq n < A_{c+1}(0).$$

Then $m < n$ yields $a \leq c$.

Case 1. Assume that $a < c$.

If $A_c(0) < n$ then $A_{c_0}(d_0) \leq n$ where $c_0 = c$ and d_0 are from the k pattern $\langle k, c_0, d_0, \ldots, c_s, d_s\rangle$ for n and $m < A_{a+1}(0) \leq A_c(0) \leq A_{c_0}(d_0) \leq n$. Then $m[k \leftarrow k+1] < A_{c[k\leftarrow k+1]}(k+1, 0) \leq A_{c[k\leftarrow k+1]}(k+1, d_0[k \leftarrow k+1]) < n[k \leftarrow k+1]$ follows by induction hypothesis and (*).

Assume that $A_c(0) = n$. If $a+1 < c$ then the induction hypothesis yields $m[k \leftarrow k+1] < A_{(a+1)[k\leftarrow k+1]}(k+1, 0) < A_{c[k\leftarrow k+1]}(k+1, 0) = n[k \leftarrow k+1]$.

Assume therefore that $m < A_{a+1}(0) = n$.

According to Lemma 2.4 assume that $m =_{k-NF} A_{a_r}(b_r)$ where

$$b_r < A^{k-1}_{a_{r-1}-1}(A_{a_{r-1}}(b_{r-1})), \ldots, b_1 < A^{k-1}_{a_0-1}(A_{a_0}(b_0)),$$

and $b_0 < A_a^{k-1}(1)$. Then the induction hypothesis yields

$$\begin{aligned}
& m[k \leftarrow k+1] \\
&= A_{a_r[k\leftarrow k+1]}(k+1, b_r[k \leftarrow k+1]) \\
&< A_{a_r[k\leftarrow k+1]}(k+1, A^{k-1}_{(a_{r-1}-1)[k\leftarrow k+1]}(k+1, A_{a_{r-1}[k\leftarrow k+1]} \\
&\quad \times (k+1, b_{r-1}[k \leftarrow k+1]))) \\
&\leq A_{a_{r-1}[k\leftarrow k+1]}(k+1, b_{r-1}[k \leftarrow k+1]+1) \\
&\leq \cdots \leq A_{a[k\leftarrow k+1]}(k+1, b_0[k \leftarrow k+1]+1) \\
&\leq A_{a[k\leftarrow k+1]}(A^{k-1}_{a[k\leftarrow k+1]}(k+1, 1)) \\
&< A_{(a+1)[k\leftarrow k+1]}(k+1, 0) \\
&= n[k \leftarrow k+1].
\end{aligned}$$

Case 2. Assume that $a = c$ and

$$A_a(0) \leq m < n < A_{a+1}(0).$$

If $A_a(b_0) \leq m < A_a(b_0+1)$ and $A_a(d_0) \leq n < A_a(d_0+1)$ (where b_0 is from the k pattern $\langle k, a, b_0, \ldots, a_r, b_r \rangle$ for m and d_0 is from the k pattern $\langle k, a, d_0, \ldots, c_r, d_r \rangle$ for n) then $b_0 \leq d_0$. The inequality $(*)$ yields $A_{a[k \leftarrow k+1]}(k+1, d_0[k \leftarrow k+1]) \leq n[k \leftarrow k+1]$.

Case 2.1. Assume that $b_0 < d_0$.

If $b_0+1 < d_0$ then the induction hypothesis and $(*)$ yield $m[k \leftarrow k+1] < A_{a[k \leftarrow k+1]}(k+1, (b_0+1)[k \leftarrow k+1]) < A_{a[k \leftarrow k+1]}(k+1, d_0[k \leftarrow k+1]) \leq n[k \leftarrow k+1]$.

Assume that $b_0+1 = d_0$. If $A_a(d_0) < n$ then the induction hypothesis and $(*)$ yield $m[k \leftarrow k+1] < A_{a[k \leftarrow k+1]}(k+1, d_0[k \leftarrow k+1]) < n[k \leftarrow k+1]$. So we may assume $A_a(b_0+1) = n = A_{a_0}(b_0+1)$.

According to Lemma 2.4 assume that $m = A_{a_r}(b_r)$ where

$$b_r < A^{k-1}_{a_{r-1}-1}(A_{a_{r-1}}(b_{r-1})), \ldots,$$

and $b_1 < A^{k-1}_{a_0-1}(A_{a_0}(b_0))$. Then the induction hypothesis yields

$$\begin{aligned}
&m[k \leftarrow k+1] \\
&= A_{a_r[k \leftarrow k+1]}(k+1, b_r[k \leftarrow k+1]) \\
&< A_{a_r[k \leftarrow k+1]}(k+1, A^{k-1}_{(a_{r-1}-1)[k \leftarrow k+1]}(k+1, A_{a_{r-1}[k \leftarrow k+1]} \\
&\quad \times (k+1, b_{r-1}[k \leftarrow k+1]))) \\
&\leq A_{a_{r-1}[k \leftarrow k+1]}(k+1, b_{r-1}[k \leftarrow k+1]+1) \\
&\leq \cdots \leq A_{a_1[k \leftarrow k+1]}(k+1, b_1[k \leftarrow k+1]+1) \\
&\leq A_{a_1[k \leftarrow k+1]}(k+1, A^{k-1}_{(a_0-1)[k \leftarrow k+1]}(k+1, A_{a[k \leftarrow k+1]} \\
&\quad \times (k+1, b_0[k \leftarrow k+1]))) \\
&< A_{a[k \leftarrow k+1]}(k+1, b_0[k \leftarrow k+1]+1) \\
&\leq A_{a[k \leftarrow k+1]}(k+1, (b_0+1)[k \leftarrow k+1]) \\
&= n[k \leftarrow k+1].
\end{aligned}$$

Case 2.2. Assume that $a = c = a_0$, $b_0 = d_0$ and $A_{a_0}(b_0) \leq m < n < A_{a_0}(b_0+1)$.

Choose p maximal such that

$$A_{a_p}(b_p) \leq m < n < A_{a_p}(b_p+1).$$

Then $a_l = c_l$ and $b_l = d_l$ for $0 \leq l \leq p$. We may safely assume that $a_p > 0$. If $A_{a_p}(b_p) = m$ then $(*)$ yields $m[k \leftarrow k+1] < n[k \leftarrow k+1]$.

Now let us observe that $a_{p+1} \leq c_{p+1}$: If $A_{c_p-1}(A_{c_p}(d_p)) \leq n$ then $c_{p+1} = c_p - 1$. Since $a_{p+1} < a_p$ we see indeed that $a_{p+1} \leq c_{p+1}$. If $m < n <$

$A_{c_p-1}(A_{c_p}(d_p))$ then the least c such that $n < A_c(A_{c_p}(d_p))$ is as least as big as the least c' such that $m < A_{c'}(A_{c_p}(d_p))$, hence $a_{p+1} \leq c_{p+1}$.

Let us now work with the assumption that

$$A_{c_{p+1}}(d_{p+1}) \leq n < A_{c_{p+1}}(d_{p+1}+1)$$

and

$$A_{a_{p+1}}(b_{p+1}) \leq m < A_{a_{p+1}}(b_{p+1}+1)$$

where $0 \leq a_{p+1} \leq c_{p+1}$. By $(*)$ we may assume $0 < a_{p+1}$.

Case 2.2.1. Assume that $a_{p+1} = c_{p+1}$.

Then the maximality of p and $m < n$ yield $b_{p+1} < d_{p+1}$. If $A_{c_{p+1}}(d_{p+1}) < n$ then $(*)$ and the induction hypothesis yield

$$\begin{aligned}
&m[k \leftarrow k+1] \\
&< A_{a_{p+1}[k\leftarrow k+1]}(k+1,(b_{p+1}+1)[k \leftarrow k+1]) \\
&\leq A_{a_{p+1}[k\leftarrow k+1]}(k+1,(d_{p+1})[k \leftarrow k+1]) \\
&< n[k \leftarrow k+1].
\end{aligned}$$

Now consider the case $A_{c_{p+1}}(d_{p+1}) = n$.

If $b_{p+1}+1 < d_{p+1}$ then $m[k \leftarrow k+1] < A_{a_{p+1}}(k+1,(b_{p+1}+1)[k \leftarrow k+1]) < A_{a_{p+1}}(k+1,(d_{p+1})[k \leftarrow k+1]) = n[k \leftarrow k+1]$.

If $A_{a_{p+1}}(b_{p+1}) = m$ then the induction hypothesis yields $b_{p+1}[k \leftarrow k+1] < d_{p+1}[k \leftarrow k+1]$. Hence $m[k \leftarrow k+1] < n[k \leftarrow k+1]$. So we may assume $b_{p+1}+1 = d_p$ and $A_{a_{p+1}}(b_{p+1}) < m < A_{a_{p+1}}(b_{p+1}+1) = n$.

According to Lemma 2.4 assume that $m =_{k-NF} A_{a_r}(b_r)$ where $a_r < a_{p+1}$ and $b_r < A^{k-1}_{a_{r-1}-1}(A_{a_{r-1}}(b_{r-1})), \ldots$ and $b_1 < A^{k-1}_{a_0-1}(A_{a_0}(b_0))$. Then the induction hypothesis yields

$$\begin{aligned}
&m[k \leftarrow k+1] \\
&= A_{a_r[k\leftarrow k+1]}(k+1, b_r[k \leftarrow k+1]) \\
&< A_{a_r[k\leftarrow k+1]}(k+1, A^{k-1}_{(a_{r-1}-1)[k\leftarrow k+1]}(k+1, A_{a_{r-1}[k\leftarrow k+1]} \\
&\quad \times(k+1, b_{r-1}[k \leftarrow k+1]))) \\
&\leq A_{a_{r-1}[k\leftarrow k+1]}(k+1, b_{r-1}[k \leftarrow k+1]+1) \\
&\leq \cdots \leq A_{a_{p+1}[k\leftarrow k+1]}(k+1, b_{p+1}[k \leftarrow k+1]+1) \\
&\leq n[k \leftarrow k+1].
\end{aligned}$$

Case 2.2.2. Let us now consider the case $a_{p+1} < c_{p+1}$.

If $A_{c_{p+1}}(d_{p+1}) < n$ then $(*)$ yields $A_{c_{p+1}[k\leftarrow k+1]}(d_{p+1}[k \leftarrow k+1]) < n[k \leftarrow k+1]$. Moreover we have $m < A_{a_{p+1}+1}(A_{a_p}(b_p)) \leq A_{c_{p+1}}(A_{a_p}(b_p)) \leq$

$A_{c_{p+1}}(d_{p+1})$ and the induction hypothesis yields $m[k \leftarrow k+1] < A_{c_{p+1}[k\leftarrow k+1]}(d_{p+1}[k \leftarrow k+1]) < n[k \leftarrow k+1]$.

So we may assume

$$n =_{k-NF} A_{c_{p+1}}(d_{p+1}).$$

If $a_{p+1}+1 < c_{p+1}$ then $m < A_{a_{p+1}+1}(A_{a_p}(b_p)) < A_{c_{p+1}}(A_{a_p}(b_p)) \leq A_{c_{p+1}}(d_{p+1}) = n$ and the induction hypothesis yields

$$\begin{aligned}
&m[k \leftarrow k+1] \\
&< A_{(a_{p+1}+1)[k\leftarrow k+1]}(k+1, A_{a_p[k\leftarrow k+1]}(k+1, b_p[k \leftarrow k+1])) \\
&\leq A_{c_{p+1}[k\leftarrow k+1]}(k+1, A_{a_p[k\leftarrow k+1]}(k+1, b_p[k \leftarrow k+1])) \\
&\leq A_{c_{p+1}[k\leftarrow k+1]}(k+1, d_{p+1}[k \leftarrow k+1]) \\
&= n[k \leftarrow k+1].
\end{aligned}$$

So we may assume $a_{p+1}+1 = c_{p+1}$ and hence

$$m < n = A_{a_{p+1}+1}(d_{p+1}).$$

If $d_{p+1} > A_{c_p}(d_p)$ then $m < A_{a_{p+1}+1}(A_{c_p}(d_p)) < A_{a_{p+1}+1}(d_{p+1})$ and the induction hypothesis yields

$$\begin{aligned}
&m[k \leftarrow k+1]) \\
&< A_{(a_{p+1}+1)[k\leftarrow k+1]}(A_{c_p[k\leftarrow k+1]}(d_p[k \leftarrow k+1]))) \\
&< A_{(a_{p+1}+1)[k\leftarrow k+1]}(d_{p+1}[k \leftarrow k+1])) \\
&= n[k \leftarrow k+1]).
\end{aligned}$$

So we arrive at the situation $d_{p+1} = A_{c_p}(d_p)$ and

$$m < n = A_{a_{p+1}+1}(A_{c_p}(d_p)) = A_{a_{p+1}+1}(A_{a_p}(b_p)).$$

According to Lemma 2.4 assume that $m =_{k-NF} A_{a_r}(b_r)$ where $a_r < a_{r-1}$, $b_r < A^{k-1}_{a_{r-1}-1}(A_{a_{r-1}}(b_{r-1})), \ldots,$ and $b_1 < A^{k-1}_{a_0-1}(A_{a_0}(b_0))$.

Let us first assume that $r > p+1$.

Then the induction hypothesis yields

$$\begin{aligned}
&m[k \leftarrow k+1] \\
&= A_{a_r[k\leftarrow k+1]}(k+1, b_r[k \leftarrow k+1]) \\
&< A_{a_r[k\leftarrow k+1]}(k+1, A^{k-1}_{(a_{r-1}-1)[k\leftarrow k+1]}(k+1, A_{a_{r-1}[k\leftarrow k+1]} \\
&\quad \times(k+1, b_{r-1}[k \leftarrow k+1]))) \\
&\leq A_{a_{r-1}[k\leftarrow k+1]}(k+1, b_{r-1}[k \leftarrow k+1]+1) \\
&\leq \cdots
\end{aligned}$$

$$\leq A_{a_{p+2}[k\leftarrow k+1]}(k+1, A^{k-1}_{(a_{p+1}-1)[k\leftarrow k+1]}(k+1, A_{a_{p+1}[k\leftarrow k+1]} \times (k+1, b_{p+1}[k\leftarrow k+1]))).$$

We have $A_{a_{p+1}}(b_{p+1}) \leq m < A_{a_{p+1}}(b_{p+1}+1)$ and $m < n = A_{a_{p+1}+1}(A_{c_p}(d_p)) = A^k_{a_{p+1}}(A_{a_{p+1}+1}(A_{c_p}(d_p)-1))$. Then $b_{p+1} < A^{k-1}_{a_{p+1}}(A_{a_{p+1}+1}(A_{c_p}(d_p)-1)) =: b$. The induction hypothesis yields $(b_{p+1}+1)[k\leftarrow k+1] \leq b[k\leftarrow k+1]$. According to Lemma 3.4 we can write $b =_{k-NF} A^l_{a_{p+1}}(e)$ where $A_{c_p}(d_p) \leq e \leq A_{a_{p+1}}(A_{c_p}(d_p))$ and $l \leq A_{c_p}(d_p)\cdot k - 2$.

The induction hypothesis applied to $e \leq A_{a_{p+1}}(A_{c_p}(d_p))$ yields $e[k\leftarrow k+1] \leq A_{a_{p+1}[k\leftarrow k+1]}(A_{c_p[k\leftarrow k+1]}(k+1, d_p[k\leftarrow k+1]))$.

Therefore the induction hypothesis applied to $b =_{k-NF} A^l_{a_{p+1}}(e)$ yields

$$\begin{aligned} & b[k\leftarrow k+1] \\ &= A_{a_{p+1}[k\leftarrow k+1]}(k+1,\cdot)^l(e[k\leftarrow k+1]) \\ &\leq A^{l+1}_{a_{p+1}[k\leftarrow k+1]}(A_{c_p[k\leftarrow k+1]}(k+1, d_p[k\leftarrow k+1])). \end{aligned}$$

By Lemma 3.3 we have $l+2 \leq A_{c_p}(d_p)\cdot k \leq A_{c_p[k\leftarrow k+1]}(k+1, d_p[k\leftarrow k+1])\cdot k$. Hence

$$A_{a_{p+1}[k\leftarrow k+1]}(k+1, b[k\leftarrow k+1]) \leq A^{A_{c_p[k\leftarrow k+1]}(k+1, d_p[k\leftarrow k+1])\cdot k}_{a_{p+1}[k\leftarrow k+1]}(A_{c_p[k\leftarrow k+1]}(k+1, d_p[k\leftarrow k+1]))).$$

The induction hypothesis yields

$$\begin{aligned} & m[k\leftarrow k+1] \\ &< A_{a_{p+2}[k\leftarrow k+1]}(A^{k-1}_{a_{p+1}[k\leftarrow k+1]-1}(k+1, A_{a_{p+1}[k\leftarrow k+1]} \\ &\quad \times (k+1, b_{p+1}[k\leftarrow k+1]))) \\ &\leq A_{a_{p+1}[k\leftarrow k+1]}(k+1, (b_{p+1}+1)[k\leftarrow k+1]) \\ &\leq A_{a_{p+1}[k\leftarrow k+1]}(k+1, b[k\leftarrow k+1]) \\ &\leq A_{a_{p+1}[k\leftarrow k+1]}{}^{k\cdot A_{c_p[k\leftarrow k+1]}(k+1, d_p[k\leftarrow k+1])}(k+1, A_{c_p[k\leftarrow k+1]} \\ &\quad \times (k+1, d_p[k\leftarrow k+1])) \\ &\leq A_{a_{p+1}[k\leftarrow k+1]}{}^{(k+1)\cdot A_{c_p[k\leftarrow k+1]}(k+1, d_p[k\leftarrow k+1])}(k+1,\cdot) \\ &\quad \times (A_{(a_{p+1}+1)[k\leftarrow k+1]}(k+1, 0)) \\ &\leq A_{(a_{p+1}+1)[k\leftarrow k+1]}(k+1, A_{c_p[k\leftarrow k+1]}(d_p[k\leftarrow k+1])) \\ &= n[k\leftarrow k+1] \end{aligned}$$

since in general $A_{a+1}(A_c(d)) = A_a^{k\cdot A_c(d)}(A_{a+1}(0))$ and $b \leq A^b_a(0)$. Note that we used here crucially the occurrence of the factor $k+1$ in the exponent.

The extra 1 allows for a majorization of $A_{c_p[k\leftarrow k+1]}(k+1, d_p[k \leftarrow k+1])$ in the argument in the last sequence of inequations.

For the remaining case $r = p+1$ we simply have

$$m[k \leftarrow k+1] = A_{a_{p+1}[k\leftarrow k+1]}(b_{p+1}[k \leftarrow k+1])$$

and we obtain similarly as before

$$\begin{aligned} &m[k \leftarrow k+1] \\ &= A_{a_{p+1}[k\leftarrow k+1]}(k+1, (b_{p+1})[k \leftarrow k+1]) \\ &< A_{a_{p+1}[k\leftarrow k+1]}(k+1, (b_{p+1}+1)[k \leftarrow k+1]) \\ &\le A_{a_{p+1}[k\leftarrow k+1]}(k+1, b[k \leftarrow k+1]) \\ &\le \cdots \\ &\le n[k \leftarrow k+1]. \end{aligned}$$

□

Lemma 3.6: *If $m =_{k-NF} A_a(k, b)$, then $m[k \leftarrow k+1] =_{k+1-NF} A_{a[k\leftarrow k+1]}(k+1, b[k \leftarrow k+1])$ is in $k+1$ normal form.*

Proof: Let $\langle k, a_0, b_0, \ldots, a_r, b_r\rangle$ be the pattern for m. Let $\langle k+1, a'_0, b'_0, \ldots, a'_{r'}, b'_{r'}\rangle$ be the pattern for $m[k \leftarrow k+1]$. We prove that $r = r'$ and that $a_p[k \leftarrow k+1] = a'_p$ and that $b_p[k \leftarrow k+1] = b'_p$ for $0 \le p \le r$. This assertion is clearly in accordance with the idea of sandwiching. Moreover this assertion is in accordance with the just proven monotonicity of the operation $m \mapsto m[k \leftarrow k+1]$ which we apply tacitly in the proof of this lemma.

We have $A_a(k, 0) \le m < A_{a+1}(k, 0)$. Then according to Lemma 2

$$m =_{k-NF} A_{a_r}(b_r)$$

where $a_l > a_{l+1}$ for $l < r$ and

$$b_r < A^{k-1}_{a_{r-1}-1}(A_{a_{r-1}}(b_{r-1})), \ldots, b_1 < A^{k-1}_{a_0-1}(A_{a_0}(b_0)),$$

$a_0 = a$, and $b_0 < A^{k-1}_a(1)$.

Then

$$\begin{aligned} &A_{a[k\leftarrow k+1]}(k+1, 0) \\ &\le m[k \leftarrow k+1] \\ &< A_{a_r[k\leftarrow k+1]}(k+1, A^{k-1}_{(a_{r-1}-1)[k\leftarrow k+1]}(k+1, A_{a_{r-1}[k\leftarrow k+1]} \\ &\quad \times (k+1, b_{r-1}[k \leftarrow k+1])) \\ &< \cdots \le A_{a[k\leftarrow k+1]+1}(k+1, 0). \end{aligned}$$

This shows that $a'_0 = a[k \leftarrow k+1] = a_0[k \leftarrow k+1]$.

Assume that $A_a(k,b) \leq m < A_a(k,b+1)$ with $b_0 = b$.

If $A_a(k,b) = m$ then $m[k \leftarrow k+1] = A_{a[k \leftarrow k+1]}(k+1, b[k \leftarrow k+1]) < A_{a[k \leftarrow k+1]}(k+1, b[k \leftarrow k+1]+1)$ and $b'_0 = b_0[k \leftarrow k+1]$.

Otherwise we have $A_{a_1}(k,b_1) \leq m < A_{a_1}(k,b_1+1)$ with $a_1 < a$ and $b_1 < A^{k-1}_{a_0-1}(k, A_{a_0}(k,b))$. Then $A_{a_1[k \leftarrow k+1]}(k+1, b_1[k \leftarrow k+1]) \leq m[k \leftarrow k+1] < A_{a_1[k \leftarrow k+1]}(k+1, A^{k-1}_{(a_0-1)[k \leftarrow k+1]}(k+1, A_{a_0[k \leftarrow k+1]}(k+1, b[k \leftarrow k+1]))) < A_{a_0[k \leftarrow k+1]}(k+1, b[k \leftarrow k+1]+1)$. This proves $b'_0 = b[k \leftarrow k+1]$.

Assume now that $a_0[k \leftarrow k+1] = a'_0, \ldots, b_{p-2}[k \leftarrow k+1] = b'_{p-2}, a_{p-1}[k \leftarrow k+1] = a'_{p-1}$ and $b_{p-1}[k \leftarrow k+1] = b'_{p-1}$. Assume further that $A_{a_{p-1}}(k, b_{p-1}) \leq m < A_{a_{p-1}}(k, b_{p-1}+1)$. If $A_{a_{p-1}}(k, b_{p-1}) = m$ then $m[k \leftarrow k+1] =_{k+1-NF} A_{a_{p-1}[k \leftarrow k+1]}(k+1, b_{p-1}[k \leftarrow k+1])$. Assume that $A_{a_{p-1}}(k, b_{p-1}) < m < A_{a_{p-1}}(k, b_{p-1}+1)$. Then $a_{p-1} > 0$.

Case 1. Assume that

$$A_{a_{p-1}-1}(A_{a_{p-1}}(k, b_{p-1})) \leq m < A^k_{a_{p-1}-1}(A_{a_{p-1}}(k, b_{p-1})).$$

Then $a_p = a_{p-1} - 1$. Moreover, $A_{a_p}(b_p) \leq m < A_{a_p}(b_p + 1)$ where $b_p + 1 \leq A^{k-1}_{a_p}(A_{a_{p-1}}(k, b_{p-1}))$.

According to Lemma 2.4 assume that $m =_{k-NF} A_{a_r}(b_r)$ with $b_r < A_{a_{r-1}}(b_{r-1})$, etc.

These inequalities yield as in the proof of Lemma 3.5

$$\begin{aligned}
&A_{a_p[k \leftarrow k+1]}(k+1, A_{a_{p-1}[k \leftarrow k+1]}(k+1, b_{p-1}[k \leftarrow k+1])) \\
&\leq m[k \leftarrow k+1] \\
&= A_{a_r[k \leftarrow k+1]}(k+1, b_r[k \leftarrow k+1]) \\
&< A_{a_r[k \leftarrow k+1]}(k+1, (b_r+1)[k \leftarrow k+1]) \\
&\leq A_{a_p[k \leftarrow k+1]}(k+1, (b_p+1)[k \leftarrow k+1])) \\
&\leq A^k_{a_p[k \leftarrow k+1]}(k+1, \cdot)(A_{a_{p-1}[k \leftarrow k+1]}(k+1, b_{p-1})[k \leftarrow k+1]) \\
&< A_{a_p[k \leftarrow k+1]+1}(k+1, A_{a_{p-1}[k \leftarrow k+1]}(k+1, b_{p-1})[k \leftarrow k+1]).
\end{aligned}$$

This yields $a'_p = a_p[k \leftarrow k+1]$.

Case 2. Assume that $A_{a_{p-1}}(k, b_{p-1}) \leq m < A_{a_{p-1}-1}(A_{a_{p-1}}(k, b_{p-1}))$. Then $A_{a_p}(A_{a_{p-1}}(k, b_{p-1})) \leq m < A_{a_p+1}(A_{a_{p-1}}(k, b_{p-1}))$ where $a_p + 1 \leq a_{p-1} - 1$.

Moreover, since a_p, b_p is part of the k pattern for m we see $m < A_{a_p}(b_p + 1)$ where according to assertion 2 of Lemma 2.4

$$b_p < b := A^{k-1}_{a_p}(A_{a_p+1}(A_{a_{p-1}}(k, b_{p-1}) - 1)).$$

Lemma 3.4 yields the existence of numbers e and l such that $b =_{k-NF} A^l_{a_p}(e)$ with $A_{a_{p-1}}(b_{p-1}) \leq e \leq A_{a_p}(A_{a_{p-1}}(b_{p-1}))$ and $l \leq A_{a_{p-1}}(b_{p-1}) \cdot k - 2 \leq A_{a_{p-1}[k \leftarrow k+1]}(k+1, b_{p-1}[k \leftarrow k+1]) \cdot k - 2$.

Lemma 3.5 applied to $e \leq A_{a_p}(A_{a_{p-1}}(b_{p-1}))$ yields

$$e[k \leftarrow k+1] \leq A_{a_p[k \leftarrow k+1]}(k+1, A_{a_{p-1}[k \leftarrow k+1]}(k+1, b_{p-1}[k \leftarrow k+1])).$$

In addition, Lemma 3.5 and inequation (**) from Lemma 3.4 yield

$$\begin{aligned} &(b_p + 1)[k \leftarrow k+1] \\ &\leq b[k \leftarrow k+1] \\ &\leq A^{l+1}_{a_p[k \leftarrow k+1]}(k+1, \cdot)(A_{a_{p-1}[k \leftarrow k+1]}(k+1, b_{p-1}[k \leftarrow k+1])). \end{aligned}$$

Moreover, according to Lemma 2.4 assume that $m =_{k-NF} A_{a_r}(b_r)$ with $b_r < A_{a_{r-1}}(b_{r-1})$, etc.

These inequalities yield as in the proof of Lemma 6

$$\begin{aligned} &m[k \leftarrow k+1] \\ &= A_{a_r[k \leftarrow k+1]}(k+1, b_r[k \leftarrow k+1]) \\ &< A_{a_r[k \leftarrow k+1]}(k+1, (b_r+1)[k \leftarrow k+1]) \\ &\leq A_{a_p[k \leftarrow k+1]}(k+1, (b_p+1)[k \leftarrow k+1])) \\ &\leq A_{a_p[k \leftarrow k+1]}(k+1, b[k \leftarrow k+1])) \\ &= A_{a_p[k \leftarrow k+1]}(k+1, A^{l+1}_{a_p[k \leftarrow k+1]}(k+1, A_{a_{p-1}[k \leftarrow k+1]}(k+1, b_{p-1}[k \leftarrow k+1]))) \\ &\leq A^{(k+1) \cdot A_{a_{p-1}[k \leftarrow k+1]}(k+1, b_{p-1}[k \leftarrow k+1])}_{a_p[k \leftarrow k+1]}(k+1, \cdot)(A_{a_p[k \leftarrow k+1]+1}(k+1, 0)) \\ &= A_{a_p[k \leftarrow k+1]+1}(k+1, A_{a_{p-1}[k \leftarrow k+1]}(k+1, b_{p-1}[k \leftarrow k+1])). \end{aligned}$$

This estimate makes again critical use of the factor $k+1$ in the exponent.

Moreover $A_{a_p[k \leftarrow k+1]}(k+1, A_{a_{p-1}[k \leftarrow k+1]}(k+1, b_{p-1}[k \leftarrow k+1])) \leq m[k \leftarrow k+1]$. This yields $a'_p = a_p[k \leftarrow k+1]$.

We have in Case 1 and Case 2 that $A_{a_p}(k, b_p) \leq m < A_{a_p}(k, b_p + 1) = A^k_{a_p - 1}(A_{a_p}(k, b_p))$ so that we obtain as before

$$\begin{aligned} &A_{a_p[k \leftarrow k+1]}(k+1, b_p[k \leftarrow k+1]) \\ &\leq m[k \leftarrow k+1] \\ &< A^k_{(a_p-1)[k \leftarrow k+1]}(k+1, \cdot)(A_{a_p[k \leftarrow k+1]}(k, b_p[k \leftarrow k+1])) \\ &\leq A_{a_p[k \leftarrow k+1]}(k+1, b_p[k \leftarrow k+1] + 1). \end{aligned}$$

This yields $b'_p = b_p[k \leftarrow k+1]$ and we see $a_0[k \leftarrow k+1] = a'_0, \ldots, b_{p-1}[k \leftarrow k+1] = b'_{p-1}, a_p[k \leftarrow k+1] = a'_p$ and $b_p[k \leftarrow k+1] = b'_p$ for all $p \leq r$. We have $m = A_{a_r}(k, b_r)$ and hence $m[k \leftarrow k+1] =$

$A_{a_r[k\leftarrow k+1]}(k+1, b_r[k \leftarrow k+1])$ so that finally $r = r'$ and $m[k \leftarrow k+1] =_{k+1-NF} A_{a_r[k\leftarrow k+1]}(k+1, b_r[k \leftarrow k+1])$. □

Lemma 3.7: $\psi_k m = \psi_{k+1}(m[k \leftarrow k+1])$.

Proof: By induction on m using Lemma 3.6. Indeed, let $m =_{k-NF} A_a(k,b)$. Then $m[k \leftarrow k+1]$ has the $k+1$ normal form $A_{a[k\leftarrow k+1]}(k+1, b[k \leftarrow k+1])$ and the induction hypothesis yields

$$\begin{aligned}&\psi_{k+1}(m[k \leftarrow k+1])\\&= \overline{\varphi}_{\psi_{k+1}(a[k\leftarrow k+1])}(\psi_{k+1}(b[k \leftarrow k+1])\\&= \overline{\varphi}_{\psi_k(a)}(\psi_k(b))\\&= \psi_k(m).\end{aligned}$$

□

Proof of Assertion 1 of Theorem 1. The results which have been obtained in the meantime suffice to prove the first assertion of Theorem 1 using Lemma 3.2 and Lemma 3.7. Indeed, let us argue informally in the formal system under consideration. Let $o(m,l) := \psi_{l+2}(m_l)$. Then $m_l > 0$ yields

$$\begin{aligned}o(m, l+1) &= \psi_{l+3}(m_l[l+2 \leftarrow l+3] - 1)\\&< \psi_{l+3}(m_l[l+2 \leftarrow l+3])\\&= \psi_{l+2}(m_l)\\&= o(m,l).\end{aligned}$$

So, arguing in PA+TI(ε_0), an infinite Goodstein sequence with positive entries would yield a primitive recursive descending chain of ordinals below ε_0 which contradicts TI(ε_0).

4. Proving the PA Independence of the Goodstein Principle

To prove the second assertion of Theorem 1 we are going to analyze the Goodstein sequences in terms of canonical fundamental sequences. The basic idea is to show that $o(m,k)$ is not smaller than the ordinal $o(m,0)[0][1]\dots[k-1]$ which emerges by stepping down the canonical fundamental sequences with a linear speed up. For this technical verification we use the Bachmann property of our system of fundamental sequences.

By classical results on the Hardy functions the lengths of such chains is bounded from below by $H_{o(m,0)}$. Termination of the Goodstein process then would yield the totality of the function H_{ε_0} and so termination of the

Goodstein process cannot be proved in PA by standard results from proof theory (confer, e.g. [4], [10] or [20]).

The details are as follows. Let $\succ_k$ denote the stepping down relation with respect to argument k along the fundamental sequences. Thus $\succ_k$ is the transitive closure of $\{(\lambda, \delta) : \delta = \lambda[k]\}$. Let $\succeq_k$ be the reflexive closure of $\succ_k$. It is well known that for $\lambda > \beta > \lambda[k]$ one has $\beta \succeq_1 \lambda[k]$. This is the so called Bachmann property. This property is of key importance for proving unprovability in our context. (See, for example, [5], [19] and [17] for more details.) It is further well known that $\alpha \succeq_k \beta$ yields $\alpha \succeq_{k+1} \beta$.

Lemma 4.1: *Assume that* $\alpha > 0$ *and* $\beta > 0$.

(1) $\overline{\varphi}^k_{\alpha[k]}(1) \geq (\overline{\varphi}\alpha 0)[k]$.
(2) $\overline{\varphi}^k_{\alpha[k]}(\overline{\varphi}\alpha(\beta[k])) \geq (\overline{\varphi}\alpha\beta)[k]$ *if* $\beta \notin Fix_\alpha$.
(3) $\overline{\varphi}^k_{\alpha[k]}\beta \geq (\overline{\varphi}\alpha\beta)[k]$ *if* $\beta \in Fix_\alpha$.

Proof: The assertions follow easily by inspecting the definition of the fundamental sequences. □

Lemma 4.2: $\psi_{k+1}(c[k \leftarrow k+1] - 1) \succeq_1 \psi_k(c)[k]$.

Proof: By induction on c.

We write $A_a(b)$ for $A_a(k, b)$ and $B_a(b)$ for $A_a(k+1, b)$. It suffices to show

$$\psi_{k+1}(c[k \leftarrow k+1] - 1) \geq \psi_k(c)[k].$$

The refined inequality follows from the Bachmann property since Lemma 3 and Lemma 8 yield

$$\psi_k(c) = \psi_{k+1}(c[k \leftarrow k+1]) > \psi_{k+1}(c[k \leftarrow k+1] - 1).$$

Case 1. $c =_{k-NF} A_0(k, b)$. Then $\psi_{k+1}(c[k \leftarrow k+1] - 1) = \psi_{k+1}(b[k \leftarrow k+1] + 1 - 1) = \psi_{k+1}(b[k \leftarrow k+1]) = \psi_k(c)[k]$.

Case 2. $c =_{k-NF} A_a(k, 0)$ where $a > 0$. Note that $B^l_{a[k \leftarrow k+1]-1}(1)$ is in $k+1$ normal form for $1 \leq l \leq k$ (since at most k but not $k+1$ iterations are involved). Then Lemma 3.2, assertion 1 of Lemma 4.1 and the induction hypothesis yield

$$\begin{aligned}
&\psi_{k+1}(c[k \leftarrow k+1]-1) \\
&= \psi_{k+1}(B_{a[k\leftarrow k+1]}(0)-1) \\
&= \psi_{k+1}(B^{k+1}_{a[k\leftarrow k+1]-1}(1)-1) \\
&\geq \psi_{k+1}(B^{k}_{a[k\leftarrow k+1]-1}(1)) \\
&\geq \overline{\varphi}^{k}_{\psi_{k+1}(a[k\leftarrow k+1]-1)}(1) \\
&\geq \overline{\varphi}^{k}_{\psi_k(a)[k]}(1) \\
&\geq (\overline{\varphi}_{\psi_k(a)}(0))[k] \\
&= \psi_k(c)[k].
\end{aligned}$$

Case 3. $c =_{k-NF} A_a(k,b)$ where $a > 0$, $b > 0$ and b has not the k normal form $A_e(d)$ with $e > a$. Then by Lemma 3.2 $\psi_k b \notin Fix_{\psi_k a}$ and thus $\overline{\varphi}^{k}_{\psi_k(a)[k]}(\overline{\varphi}_{\psi_k(a)}((\psi_k b)[k])) \geq (\overline{\varphi}_{\psi_k(a)}(\psi_k b))[k]$ according to assertion 2 of Lemma 4.1.

Let $f := A_a(k, b-1)$. Then $f =_{k-NF} A_a(k, b-1)$. This follows from assertion 6 of Lemma 2.4.

Lemma 3.6 yields that $B^{l}_{a[k\leftarrow k+1]-1}(B_{a[k\leftarrow k+1]}(b[k \leftarrow k+1]-1))$ is in $k+1$ normal form for $1 \leq l \leq k$ (since at most k but not $k+1$ iterations are involved). Then Lemma 3.2 and the induction hypothesis yield

$$\begin{aligned}
&\psi_{k+1}(c[k \leftarrow k+1]-1) \\
&= \psi_{k+1}(B_{a[k\leftarrow k+1]}(b[k \leftarrow k+1])-1) \\
&= \psi_{k+1}(B^{k+1}_{a[k\leftarrow k+1]-1}(B_{a[k\leftarrow k+1]}(b[k \leftarrow k+1]-1)-1) \\
&\geq \psi_{k+1}(B^{k}_{a[k\leftarrow k+1]-1}(B_{a[k\leftarrow k+1]}(b[k \leftarrow k+1]-1)) \\
&\geq \overline{\varphi}^{k}_{\psi_{k+1}(a[k\leftarrow k+1]-1)}(\overline{\varphi}_{\psi_{k+1}(a[k\leftarrow k+1])}(\psi_{k+1}(b[k \leftarrow k+1]-1)) \\
&\geq \overline{\varphi}^{k}_{\psi_k(a)[k]}(\overline{\varphi}_{\psi_k(a)}((\psi_k b)[k])) \\
&\geq (\overline{\varphi}_{\psi_k(a)}(\psi_k b))[k] \\
&= \psi_k(c)[k].
\end{aligned}$$

Case 4. $c =_{k-NF} A_a(k,b)$ where $a > 0$, $b > 0$ and b has the $k+1$ normal form $A_e(d)$ with $e > a$. Note that in this case $\psi_k b \in Fix_{\psi_k a}$ so that $\overline{\varphi}^{k}_{\psi_k(a)[k]}(\psi_k(b))) \geq (\overline{\varphi}_{\psi_k(a)}(\psi_k b))[k]$ according to assertion 3 of Lemma 4.1.

Moreover note that according to Lemma 3.6 $B^{l}_{a[k\leftarrow k+1]-1}(b[k \leftarrow k+1])$ is in $k+1$ normal form for $1 \leq l \leq k$. Therefore Lemma 3.2 and the induction hypothesis yield

$$\begin{aligned}
&\psi_{k+1}(c[k \leftarrow k+1]-1)\\
&= \psi_{k+1}(B_{a[k\leftarrow k+1]}(b[k \leftarrow k+1])-1)\\
&= \psi_{k+1}(B^{k+1}_{a[k\leftarrow k+1]-1}(B_{a[k\leftarrow k+1]}(b[k \leftarrow k+1]-1)-1))\\
&\geq \psi_{k+1}(B^{k}_{a[k\leftarrow k+1]-1}(B_{a[k\leftarrow k+1]}(b[k \leftarrow k+1]-1)))\\
&\geq \psi_{k+1}(B^{k}_{a[k\leftarrow k+1]-1}(b[k \leftarrow k+1]))\\
&\geq \overline{\varphi}^{k}_{\psi_{k+1}(a[k\leftarrow k+1]-1)}(\psi_{k+1}(b[k \leftarrow k+1]))\\
&\geq \overline{\varphi}^{k}_{\psi_k(a)[k]}(\psi_k(b)))\\
&\geq (\overline{\varphi}_{\psi_k(a)}(\psi_k b))[k]\\
&= \psi_k(c)[k].
\end{aligned}$$

□

Recall that $o(m,l) := \psi_{l+2}(m_l)$.

Lemma 4.3: $o(m,k) \succeq_{k+1} (o(m,0))[2]\ldots[k+1]$.

Proof: By induction on k. The assertion holds for $k=0$. Assume now that

$$o(m,k) \succeq_{k+1} (o(m,0))[2]\ldots[k+1].$$

Then $o(m,k) \succeq_{k+2} (o(m,0))[2]\ldots[k+1]$ and hence

$$o(m,k)[k+2] \succeq_{k+2} (o(m,0))[2]\ldots[k+1][k+2].$$

Moreover Lemma 4.2 yields

$$o(m,k+1) \succeq_1 o(m,k)[k+2],$$

hence

$$o(m,k+1) \succeq_{k+2} o(m,k)[k+2].$$

Putting things together we arrive at

$$o(m,k+1) \succeq_{k+2} (o(m,0))[2]\ldots[k+1][k+2].$$

□

Proof of Assertion 2 of Theorem 1. By recursion on n let us define $\gamma_0 := 0$ and $\gamma_{n+1} := \overline{\varphi}_{\gamma_n}(0)$. Moreover let $a_0 := 0$ and $a_{n+1} := A_{a_n}(2,0)$. Recall that $o(a_m,l) = \psi_{l+2}((a_m)_l)$. Then $o(a_m,0) = \psi_2(a_m) = \gamma_m = \omega_{2m-3}$ for $m \geq 3$. Lemma 11 yields $o(a_m,0)[2]\ldots[k+1] \preceq_{k+1} o(a_m,k)$. A standard result from proof theory (confer, for example, [4], [10] or [20] for a proof) yields

$$\mathrm{PA} \nvdash \forall m \exists l \gamma_m[2]\ldots[l+1]=0.$$

Assume that

$$\mathrm{PA} \vdash \forall m \exists l m_l = 0.$$

The function $m \mapsto a_m$ is provably recursive in PA and therefore

$$\mathrm{PA} \vdash \forall m \exists l (a_m)_l = 0.$$

Then

$$\mathrm{PA} \vdash \forall m \exists l o(a_m, l) = 0.$$

But then by formalizing the proof of Lemma 11

$$\mathrm{PA} \vdash \forall m \exists l \gamma_m[2] \ldots [l+1] = 0.$$

Contradiction. □

5. Possible Variants of the Goodstein Sequences

For $m =_{k-NF} A_a(k,b)$ define $m[k \leftarrow k+1]' := A_a(k+1, b[k \leftarrow k+1]')$. Moreover put $m_0' := m$ and for $m_l' > 0$ $m_{l+1}' := m_l'[k \leftarrow k+1]' - 1$. If $m'_l = 0$ then as before put $m'_{l+1} = 0$. Let $\psi_k' m := \overline{\varphi}_a(\psi_k' b)$.

Theorem 5.1:

(1) $\mathrm{I}\Sigma_2 + TI(\omega^{\omega^\omega}) \vdash \forall m \exists l m_l' = 0$.
(2) $\mathrm{I}\Sigma_2 \nvdash \forall m \exists l m_l' = 0$.

Proof: By adapting the previous proofs using properties of ψ_k'. □

To obtain an independence result for $\mathrm{I}\Sigma_1$ one can refine the definition of k normal form as follows. If $m < A_1(k,0)$ then m is in k normal form and $m\{k \leftarrow k+1\} := m$. If $A_1(k,b) \leq m < A_1(b+1)$ then there exists a p such that $m = A_1(k,b) + p$ which is by definition the k normal form. Note that in this case $p < k$. We may then put $m\{k \leftarrow k+1\} := A_1(k+1, b\{k \leftarrow k+1\}) + p$. (In this case we have $p\{k \leftarrow k+1\} = p$.) Let $\tilde{m}_0 := m$ and for $\tilde{m}_l > 0$ $\tilde{m}_{l+1} := \tilde{m}_l\{k \leftarrow k+1\} - 1$. If $\tilde{m}_l = 0$ then as before put $\tilde{m}_{l+1} = 0$.

Theorem 5.2:

(1) $\mathrm{I}\Sigma_1 + TI(\omega^\omega) \vdash \forall m \exists l \tilde{m}_l = 0$.
(2) $\mathrm{I}\Sigma_1 \nvdash \forall m \exists l \tilde{m}_l = 0$.

Proof: By adapting the previous proofs. □

Another variant concerns the start function $A_0(k, b) := k \cdot b$. For the corresponding k normal form one then can find for $A_a(k, b) \leq m < A_0(k, A_a(k, b))$ a unique p such that $m = A_a(k, b) + p$ which is by definition then the k normal form.

By varying the previous approaches we obtain independence results for the fragments of PA. To this end let $\gamma_0 := \overline{\varphi}_1 0$ and $\gamma_{n+1} := \overline{\varphi}_{\gamma_n} 0$. Then $\gamma_m = \omega_{2m+1}$. Moreover let $\delta_1 := \overline{\varphi}_2 0$ and $\delta_{n+1} = \overline{\varphi}_{\delta_n} 0$. Then one easily verifies $\delta_m = \omega_{2m}$. Goodstein principles for the fragments can then be defined to match those ordinals exactly.

Another theme of research would concern possible phase transitions with respect to the Goodstein principles considered in this chapter. We assume that the results from [19] and methods from [14] can be adapted without much difficulty to yields results analogous to [14] and we expect an emergence of comparable threshold functions.

Acknowledgements

The author would like to thank Paul Shafer and the anonymous referee for their very insightful comments and many helpful suggestions.

References

1. V.M. Abrusci. Dilators, generalized Goodstein sequences, independence results: a survey. *Logic and combinatorics* (Arcata, Calif., 1985), 1–23, Contemp. Math., **65**, Amer. Math. Soc., Providence, RI, 1987.
2. V.M. Abrusci. Some uses of dilators in combinatorial problems. III. Independence results by means of decreasing F-sequences (F weakly finite dilator). *Arch. Math. Logic* **29** (1989), no. 2, 85–109.
3. T. Arai, S. Wainer and A. Weiermann. Generalized Goodstein sequences. In preparation.
4. W. Buchholz and S. Wainer. Provably computable functions and the fast growing hierarchy. *Logic and combinatorics* (Arcata, Calif., 1985), 179–198, Contemp. Math., **65**, Amer. Math. Soc., Providence, RI, 1987.
5. W. Buchholz, E.A. Cichon and A. Weiermann. A uniform approach to fundamental sequences and hierarchies. *Math. Logic Quart.* **40** (1994), no. 2, 273–286
6. W. Buchholz. Explaining Gentzen's consistency proof within infinitary proof theory. *Computational logic and proof theory* (Vienna, 1997), 4–17, Lecture Notes in Comput. Sci., 1289, Springer, Berlin, 1997.
7. E.A. Cichon. A short proof of two recently discovered independence results using recursion theoretic methods. *Proceedings of the AMS* **87** (1983), 704-706.

8. E.A. Cichon and S. S. Wainer. The slow-growing and the Grzegorczyk hierarchies. *J. Symbolic Logic* **48** (1983), no. 2, 399–408.
9. M. De Smet and A. Weiermann. Goodstein sequences for prominent ordinals up to the Bachmann-Howard ordinal. *Ann. Pure Appl. Logic* **163** (2012), no. 6, 669–680.
10. M. Fairtlough and S.S. Wainer. *Hierarchies of provably recursive functions.* Handbook of proof theory, 149–207, Stud. Logic Found. Math., **137**, North-Holland, Amsterdam, 1998.
11. R.L. Goodstein. On the restricted ordinal theorem. *J. Symbolic Logic* **9**, (1944). 33–41.
12. R.L. Goodstein. Transfinite ordinals in recursive number theory. *J. Symbolic Logic* **12**, (1947). 123–129.
13. L. Kirby and J. Paris. Accessible independence results for Peano arithmetic. *Bull. London Math. Soc.* **14** (1982), no. 4, 285–293.
14. F. Meskens and A. Weiermann. Classifying Phase Transition Thresholds for Goodstein Sequences and Hydra Games. *Gentzen's Centenary. The Quest for Consistency*, R. Kahle and M. Rathjen (eds.), Springer 2015. 455-478.
15. E. Omri and A. Weiermann. Classifying the phase transition threshold for Ackermannian functions. *Ann. Pure Appl. Logic* **158** (2009), no. 3, 156–162.
16. H. Touzet. A characterisation of multiply recursive functions with Higman's lemma. RTA '99 (Trento). *Inform. and Comput.* **178** (2002), no. 2, 534544.
17. A. Weiermann. A very slow growing hierarchy for Γ_0. *Logic Colloquium '99*, 182–199, Lect. Notes Log., **17**, Assoc. Symbol. Logic, Urbana, IL, 2004.
18. A. Weiermann. Phase transition thresholds for some Friedman-style independence results. *Math. Log. Q.* **53** (2007), no. 1, 418.
19. A. Weiermann. Some interesting connections between the slow growing hierarchy and the Ackermann function. *J. Symbolic Logic* **66** (2001), no. 2, 609628.
20. A. Weiermann. Classifying the provably total functions of PA. *Bull. Symbolic Logic* **12** (2006), no. 2, 177–190.
21. A. Weiermann and G. Wilken. Goodstein sequences for prominent ordinals up to the ordinal of $\Pi^1_1 - \mathrm{CA}_0$. *Ann. Pure Appl. Logic* **164** (2013), no. 12, 1493–1506.

TRACKING CHAINS REVISITED

Gunnar Wilken[a]
Structural Cellular Biology Unit
Okinawa Institute of Science and Technology
1919-1 Tancha, Onna-son, 904-0495 Okinawa, Japan
wilken@oist.jp

The structure $\mathcal{C}_2 := (1^\infty, \le, \le_1, \le_2)$, introduced and first analyzed in [5], is shown to be elementary recursive. Here, 1^∞ denotes the proof-theoretic ordinal of the fragment Π^1_1-CA_0 of second order number theory, or equivalently the set theory $\mathrm{KP}\ell_0$, which axiomatizes limits of models of Kripke-Platek set theory with infinity. The partial orderings $\le_1$ and $\le_2$ denote the relations of Σ_1- and Σ_2-elementary substructure, respectively. In a subsequent article [10], we will show that the structure $\mathcal{C}_2$ comprises the core of the structure $\mathcal{R}_2$ of pure elementary patterns of resemblance of order 2. In [5], the stage has been set by showing that the least ordinal containing a cover of each pure pattern of order 2 is 1^∞. However, it is not obvious from [5] that $\mathcal{C}_2$ is an elementary recursive structure. This is shown here through a considerable disentanglement in the description of connectivity components of $\le_1$ and $\le_2$. The key to and starting point of our analysis is the apparatus of ordinal arithmetic developed in [7] and in Section 5 of [4], which was enhanced in [5], specifically for the analysis of $\mathcal{C}_2$.

1. Introduction

Let $\mathcal{R}_2 = (\mathrm{Ord}; \le, \le_1, \le_2)$ be the structure of ordinals with standard linear ordering $\le$ and partial orderings $\le_1$ and $\le_2$, simultaneously defined by induction on β in

$$\alpha \le_i \beta :\Leftrightarrow (\alpha; \le, \le_1, \le_2) \preceq_{\Sigma_i} (\beta; \le, \le_1, \le_2)$$

[a]The author would like to acknowledge the Institute for Mathematical Sciences of the National University of Singapore for its partial support of this work during the "Interactions" week of the workshop *Sets and Computations* in April 2015.

where $\preceq_{\Sigma_i}$ is the usual notion of Σ_i-elementary substructure (without bounded quantification), see [1, 2] for fundamentals and groundwork on elementary patterns of resemblance. Pure patterns of order 2 are the finite isomorphism types of $\mathcal{R}_2$. The *core* of $\mathcal{R}_2$ consists of the union of *isominimal realizations* of these patterns within $\mathcal{R}_2$, where a finite substructure of $\mathcal{R}_2$ is called isominimal, if it is pointwise minimal (with respect to increasing enumerations) among all substructures of $\mathcal{R}_2$ isomorphic to it, and where an isominimal substructure of $\mathcal{R}_2$ realizes a pattern P, if it is isomorphic to P. It is a basic observation, cf. [2], that the class of pure patterns of order 2 is contained in the class $\mathcal{RF}_2$ of *respecting forests of order* 2: finite structures P over the language $(\leq_0, \leq_1, \leq_2)$ where $\leq_0$ is a linear ordering and $\leq_1, \leq_2$ are forests such that $\leq_2 \subseteq \leq_1 \subseteq \leq_0$ and $\leq_{i+1}$ *respects* $\leq_i$, i.e. $p \leq_i q \leq_i r \;\&\; p \leq_{i+1} r$ implies $p \leq_{i+1} q$ for all $p, q, r \in P$, for $i = 0, 1$.

In [5] we showed that every pattern has a cover below 1^∞, the least such ordinal. Here, an order isomorphism (embedding) is a cover (covering, respectively) if it maintains the relations $\leq_1$ and $\leq_2$. The ordinal of $\mathrm{KP}\ell_0$ is therefore least such that there exist arbitrarily long finite $\leq_2$-chains. Moreover, by determination of enumeration functions of (relativized) connectivity components of $\leq_1$ and $\leq_2$ we were able to describe these relations in terms of classical ordinal notations. The central observation in connection with this is that every ordinal below 1^∞ is the greatest element in a $\leq_1$-chain in which $\leq_1$- and $\leq_2$-chains alternate. We called such chains *tracking chains* as they provide all $\leq_2$-predecessors and the greatest $\leq_1$-predecessors insofar as they exist.

In the present article, we will review and slightly extend the ordinal arithmetical toolkit and then verify through a disentangling reformulation, that [5] in fact yields an elementary recursive characterization of the restriction of $\mathcal{R}_2$ to the structure $\mathcal{C}_2 = (1^\infty; \leq, \leq_1, \leq_2)$. It is not obvious from [5] that $\mathcal{C}_2$ is an elementary recursive structure since several proofs there make use of transfinite induction up to 1^∞, which allowed for a somewhat shorter argumentation there. We will summarize the results in [5] to a sufficient and convenient degree. As a byproduct, [5] will become considerably more accessible. We will prove the equivalence of the arithmetical descriptions of $\mathcal{C}_2$ given in [5] and here. Note that the equivalence of this elementary recursive characterization with the original structure based on elementary substructurehood is proven in Section 7 of [5], using full transfinite induction up to the ordinal of $\mathrm{KP}\ell_0$. In this article, we rely on this result and henceforth identify $\mathcal{C}_2$ with its arithmetical characterization given in [5] and further illuminated in Section 5 of the present article, where we also show that the finite isomorphism types of the arithmetical $\mathcal{C}_2$ are respecting forests of order 2, without relying on semantical characterization of the arithmetical $\mathcal{C}_2$.

With these preparations out of the way we will be able to provide, in a subsequent article [10], an algorithm that assigns an isominimal realization within $\mathcal{C}_2$ to each respecting forest of order 2, thereby showing that each such respecting forest is in fact (up to isomorphism) a pure pattern of order 2. The approach is to formulate the corresponding theorem flexibly so that isominimal realizations above certain relativizing tracking chains are considered. There we will also define an elementary recursive function that assigns descriptive patterns $P(\alpha)$ to ordinals $\alpha \in 1^\infty$. A descriptive pattern for an ordinal α in the above sense is a pattern, the isominimal realization of which contains α. Descriptive patterns will be given in a way that makes a canonical choice for normal forms, since in contrast to the situation in $\mathcal{R}_1^+$, cf. [9, 4], there is no unique notion of normal form in $\mathcal{R}_2$. The chosen normal forms will be of least possible cardinality.

The mutual order isomorphisms between hull and pattern notations that will be given in [10] enable classification of a new independence result for $\mathrm{KP}\ell_0$: We will demonstrate that the result by Carlson in [3], according to which the collection of respecting forests of order 2 is well-quasi ordered with respect to coverings, cannot be proven in $\mathrm{KP}\ell_0$ or, equivalently, in the restriction $\Pi^1_1-\mathrm{CA}_0$ of second order number theory to Π^1_1-comprehension and set induction. On the other hand, we know that transfinite induction up to the ordinal 1^∞ of $\mathrm{KP}\ell_0$ suffices to show that every pattern is covered [5].

This article therefore delivers the first part of an in depth treatment of insights and results presented in a lecture during the "Interactions" week of the workshop *Sets and Computations*, held at the Institute for Mathematical Sciences of the National University of Singapore in April 2015.

2. Preliminaries

The reader is assumed to be familiar with basics of ordinal arithmetic (see e.g. [6]) and the ordinal arithmetical tools developed in [7] and Section 5 of [4]. See the index at the end of [7] for quick access to its terminology. Section 2 of [9] (2.1–2.3) provides a summary of results from [7]. As mentioned before, we will build upon [5], the central concepts and results of which will be reviewed here and in the next section. For detailed reference see also the index of [5].

2.1. *Basics*

Here we recall terminology already used in [5] (Section 2) for the reader's convenience. Let $\mathbb{P}$ denote the class of additive principal numbers, i.e. nonzero ordinals that are closed under ordinal addition, that is the image of ordinal exponentiation to base ω. Let $\mathbb{L}$ denote the class of limits of additive principal numbers, i.e. the limit points of $\mathbb{P}$, and let $\mathbb{M}$ denote the class of multiplicative principal numbers, i.e. nonzero ordinals closed under ordinal multiplication. By $\mathbb{E}$ we denote the class of epsilon numbers, i.e. the fixed-points of ω-exponentiation.

We write $\alpha =_{\text{ANF}} \alpha_1 + \cdots + \alpha_n$ if $\alpha_1, \ldots, \alpha_n \in \mathbb{P}$ such that $\alpha_1 \geq \cdots \geq \alpha_n$, which is called the representation of α in additive normal form, and $\alpha =_{\text{NF}} \beta + \gamma$ if the expansion of β into its additive normal form (ANF) in the sum $\beta + \gamma$ syntactically results in the additive normal form of α. The Cantor normal form representation of an ordinal α is given by $\alpha =_{\text{CNF}} \omega^{\alpha_1} + \cdots + \omega^{\alpha_n}$ where $\alpha_1 \geq \cdots \geq \alpha_n$ with $\alpha > \alpha_1$ unless $\alpha \in \mathbb{E}$. For $\alpha =_{\text{ANF}} \alpha_1 + \cdots + \alpha_n$, we define $\text{mc}(\alpha) := \alpha_1$ and $\text{end}(\alpha) := \alpha_n$. We set $\text{end}(0) := 0$. Given ordinals α, β with $\alpha \leq \beta$ we write $-\alpha + \beta$ for the unique γ such that $\alpha + \gamma = \beta$. As usual let $\alpha \dot{-} \beta$ be 0 if $\beta \geq \alpha$, γ if $\beta < \alpha$ and there exists the minimal γ s.t. $\alpha = \gamma + \beta$, and α otherwise.

For $\alpha \in \text{Ord}$ we denote the least multiplicative principal number greater than α by $\alpha^{\mathbb{M}}$. Notice that if $\alpha \in \mathbb{P}$, $\alpha > 1$, say $\alpha = \omega^{\alpha'}$, we have $\alpha^{\mathbb{M}} = \alpha^{\omega} = \omega^{\alpha' \cdot \omega}$. For $\alpha \in \mathbb{P}$ we use the following notations for *multiplicative normal form*:

(1) $\alpha =_{\text{NF}} \eta \cdot \xi$ if and only if $\xi = \omega^{\xi_0} \in \mathbb{M}$ (i.e. $\xi_0 \in \{0\} \cup \mathbb{P}$) and either $\eta = 1$ or $\eta = \omega^{\eta_1 + \cdots + \eta_n}$ such that $\eta_1 + \cdots + \eta_n + \xi_0$ is in additive normal form. When ambiguity is unlikely, we sometimes allow ξ to be of a form $\omega^{\xi_1 + \cdots + \xi_m}$ such that $\eta_1 + \cdots + \eta_n + \xi_1 + \cdots + \xi_m$ is in additive normal form.

(2) $\alpha =_{\text{MNF}} \alpha_1 \cdot \ldots \cdot \alpha_k$ if and only if $\alpha_1, \ldots, \alpha_k$ is the unique decreasing sequence of multiplicative principal numbers, the product of which is equal to α.

For $\alpha \in \mathbb{P}$, $\alpha =_{\text{MNF}} \alpha_1 \cdot \ldots \cdot \alpha_k$, we write $\text{mf}(\alpha)$ for α_1 and $\text{lf}(\alpha)$ for α_k. Note that if $\alpha \in \mathbb{P} - \mathbb{M}$ then $\text{lf}(\alpha) \in \mathbb{M}^{>1}$ and $\alpha =_{\text{NF}} \bar{\alpha} \cdot \text{lf}(\alpha)$ where the definition of $\bar{\alpha}$ given in [7] for limits of additive principal numbers is extended to ordinals α of a form $\alpha = \omega^{\alpha'+1}$ by $\bar{\alpha} := \omega^{\alpha'}$, see Section 5 of [4].

Given $\alpha, \beta \in \mathbb{P}$ with $\alpha \leq \beta$ we write $(1/\alpha) \cdot \beta$ for the uniquely determined ordinal $\gamma \leq \beta$ such that $\alpha \cdot \gamma = \beta$. Note that with the representations

$\alpha = \omega^{\alpha'}$ and $\beta = \omega^{\beta'}$ we have

$$(1/\alpha) \cdot \beta = \omega^{-\alpha'+\beta'}.$$

For any α of a form $\omega^{\alpha'}$ we write $\log(\alpha)$ for α', and we set $\log(0) := 0$. For an arbitrary ordinal β we write $\operatorname{logend}(\beta)$ for $\log(\operatorname{end}(\beta))$.

2.2. *Relativized notation systems* T^τ

Settings of relativization are given by ordinals from $\mathbb{E}_1 := \{1\} \cup \mathbb{E}$ and frequently indicated by Greek letters, preferably σ or τ. Clearly, in this context $\tau = 1$ denotes the trivial setting of relativization. For a setting τ of relativization we define $\tau^\infty := \mathrm{T}^\tau \cap \Omega_1$ where T^τ is defined as in [7] and reviewed in Section 2.2 of [9]. T^τ is the closure of parameters below τ under addition and the stepwise, injective, and fixed-point free collapsing functions ϑ_k the domain of which is $\mathrm{T}^\tau \cap \Omega_{k+2}$, where $\vartheta^\tau := \vartheta_0$ is relativized to τ. As in [5], most considerations will be confined to the segment 1^∞. Translation between different settings of relativization, see Section 6 of [7], is effective on the term syntax and enjoys convenient invariance properties regarding the operators described below, as was verified in [7], [4], and [5]. We therefore omit the purely technical details here.

2.3. *Refined localization*

The notion of τ-localization (Definition 2.11 of [9]) and its refinement to τ-fine-localization by iteration of the operator $\bar{\cdot}$, see Definitions 5.1 and 5.5 of [4], continue to be essential as they locate ordinals in terms of closure properties (fixed-point level and limit point thinning). These notions are effectively derived from the term syntax. We refer to Subsection 2.3 of [9] and Section 5 of [4] for a complete picture of these concepts. In the present article, the operator $\bar{\cdot}$ is mostly used to decompose ordinals that are not multiplicative principal, i.e. if $\alpha =_{\mathrm{NF}} \eta \cdot \xi$ where $\eta > 1$ and $\xi \in \mathbb{M}$, then $\bar{\alpha} = \eta$. The notion of τ-localization enhanced with multiplicative decomposition turns out to be the appropriate tool for the purposes of the present article, whereas general τ-fine-localization will re-enter the picture through the notion of closedness in a subsequent article [10].

2.4. *Operators related to connectivity components*

The function log (logend) is described in T^τ-notation in Lemma 2.13 of [9], and for $\beta = \vartheta^\tau(\eta) \in \mathrm{T}^\tau$ where $\eta < \Omega_1$ we have

$$\log((1/\tau)\cdot\beta) = \begin{cases} \eta+1 & \text{if } \eta = \varepsilon + k \text{ where } \varepsilon \in \mathbb{E}^{>\tau}, k < \omega \\ \eta & \text{otherwise.} \end{cases} \tag{2.1}$$

The foregoing distinction reflects the property of ϑ-functions to omit fixed points.

The operators $\iota_{\tau,\alpha}$ indicating the fixed-point level, ζ^τ_α displaying the degree of limit point thinning, and their combination λ^τ_α measuring closure properties of ordinals $\alpha \in \mathrm{T}^\tau$ are as in Definitions 2.14 and 2.18 of [9], which also reviews the notion of base transformation $\pi_{\sigma,\tau}$ and its smooth interaction with these operators.

The operator λ^τ_α already played a central role in the analysis of $\mathcal{R}^+_1$-patterns as it displays the number of $\le_1$-connectivity components that are $\le_1$-connected to the component with index α in a setting of relativization $\tau \in \mathbb{E}_1$, cf. Lemma 2.31 part (a) of [9]. It turns out that λ^τ_α plays a similar role in $\mathcal{R}_2$, see below.

In order to avoid excessive repetition of formal definitions from [5] we continue to describe operators and functions introduced for analysis of $\mathcal{C}_2$ in [5] in terms of their meaning in the context of $\mathcal{C}_2$. Those (relativized) $\le_1$-components, the enumeration index of which is an epsilon number, give rise to infinite $\le_1$-chains, along which new $\le_2$-components arise. Omitting from these $\le_1$-chains those elements that have a $\le_2$-predecessor in the chain and enumerating the remaining elements, we obtain the so-called ν-functions, see Definition 4.4 of [5], and the μ-operator provides the length of such enumerations up to the *final newly arising* $\le_2$*-component*, cf. the remark before Definition 4.4 of [5]. This *terminal point* on a main line, at which the largest newly arising $\le_2$-component originates, is crucial for understanding the structure $\mathcal{C}_2$. Note that in general the terminal point has an infinite increasing continuation in the $\le_1$-chain under consideration, leading to $\le_2$-components which have isomorphic copies below, i.e. which are *not* new. Recall Convention 2.9 of [9].

Definition 2.1: (3.4 of [5]**)** Let $\tau \in \mathbb{E}_1$ and $\alpha \in (\tau, \tau^\infty) \cap \mathbb{E}$, say $\alpha = \vartheta^\tau(\Delta + \eta)$ where $\eta < \Omega_1$ and $\Delta = \Omega_1 \cdot (\lambda + k)$ such that $\lambda \in \{0\} \cup \mathrm{Lim}$ and $k < \omega$. We define

$$\mu^\tau_\alpha := \omega^{\iota_{\tau,\alpha}(\lambda)+\chi^\alpha(\iota_{\tau,\alpha}(\lambda))+k}.$$

The χ-indicator occurring above is given in Definition 3.1 of [5] and indicates whether the maximum $\le_2$-component starting from an ordinal on the infinite $\le_1$-chain under consideration itself $\le_1$-reconnects to that chain which we called a *main line.* The question remains which $\le_1$-component starting from such a point on a main line is the largest that is also $\le_2$-connected to it. This is answered by the ϱ-operator:

Definition 2.2: (3.9 of [5]) Let $\alpha \in \mathbb{E}$, $\beta < \alpha^\infty$, and $\lambda \in \{0\} \cup \mathrm{Lim}$, $k < \omega$ be such that $\mathrm{logend}(\beta) = \lambda + k$. We define

$$\varrho^\alpha_\beta := \alpha \cdot (\lambda + k \dot{-} \chi^\alpha(\lambda)).$$

Now, the terminal point on a main line, given as, say, $\nu^{\boldsymbol{\tau},\alpha}_{\mu^\tau_\alpha}$ with a setting of relativization $\boldsymbol{\tau} = (\tau_1, \ldots, \tau_n)$ that will be discussed later and $\tau = \tau_n$, $\alpha \in \mathbb{E}^{>\tau}$, connects to λ^τ_α-many $\le_1$-components. The following lemma is a direct consequence of the respective definitions.

Lemma 2.3: *(3.12 of* [5]*) Let $\tau \in \mathbb{E}_1$ and $\alpha = \vartheta^\tau(\Delta + \eta) \in (\tau, \tau^\infty) \cap \mathbb{E}$. Then we have*

(1) $\iota_{\tau,\alpha}(\Delta) = \varrho^\alpha_{\mu^\tau_\alpha}$ and hence $\lambda^\tau_\alpha = \varrho^\alpha_{\mu^\tau_\alpha} + \zeta^\tau_\alpha$.
(2) $\varrho^\alpha_\beta \le \lambda^\tau_\alpha$ for every $\beta \le \mu^\tau_\alpha$. For $\beta < \mu^\tau_\alpha$ such that[b] *$\chi^\alpha(\beta) = 0$ we even have $\varrho^\alpha_\beta + \alpha \le \lambda^\tau_\alpha$.*
(3) If $\mu^\tau_\alpha < \alpha$ we have $\mu^\tau_\alpha < \alpha \le \lambda^\tau_\alpha < \alpha^2$, while otherwise

$$\max\left((\mu^\tau_\alpha + 1) \cap \mathbb{E}\right) = \max\left((\lambda^\tau_\alpha + 1) \cap \mathbb{E}\right).$$

(4) If $\lambda^\tau_\alpha \in \mathbb{E}^{>\alpha}$ we have $\mu^\tau_\alpha = \lambda^\tau_\alpha \cdot \omega$ in case of $\chi^\alpha(\lambda^\tau_\alpha) = 1$ and $\mu^\tau_\alpha = \lambda^\tau_\alpha$ otherwise.

Notice that we have $\mu^\tau_\alpha = \iota_{\tau,\alpha}(\Delta) = \mathrm{mc}(\lambda^\tau_\alpha)$ whenever $\mu^\tau_\alpha \in \mathbb{E}^{>\alpha}$.

3. Tracking Sequences and Their Evaluation

3.1. *Maximal and minimal μ-coverings*

The following sets of sequences are crucial for the description of settings of relativization, which in turn is the key to understanding the structure of connectivity components in $\mathcal{C}_2$.

---

[b] This condition is missing in [5]. However, that inequality was only applied under this condition, cf. Def. 5.1 and L. 5.7 of [5].

Definition 3.1: **(4.2 of** [5]**)** Let $\tau \in \mathbb{E}_1$. A nonempty sequence $(\alpha_1, \ldots, \alpha_n)$ of ordinals in the interval $[\tau, \tau^\infty)$ is called a τ-tracking sequence if

(1) $(\alpha_1, \ldots, \alpha_{n-1})$ is either empty or a strictly increasing sequence of epsilon numbers greater than τ.
(2) $\alpha_n \in \mathbb{P}$, $\alpha_n > 1$ if $n > 1$.
(3) $\alpha_{i+1} \le \mu^\tau_{\alpha_i}$ for every $i \in \{1, \ldots, n-1\}$.

By TS^τ we denote the set of all τ-tracking sequences. Instead of TS^1 we also write TS.

According to Lemma 3.5 of [5] the length of a tracking sequence is bounded in terms of the largest index of ϑ-functions in the term representation of the first element of the sequence.

Definition 3.2: Let $\tau \in \mathbb{E}_1$, $\alpha \in \mathbb{E} \cap (\tau, \tau^\infty)$, and $\beta \in \mathbb{P} \cap (\alpha, \alpha^\infty)$. A sequence $(\alpha_0, \ldots, \alpha_{n+1})$ where $\alpha_0 = \alpha$, $\alpha_{n+1} = \beta$, $(\alpha_1, \ldots, \alpha_{n+1}) \in \mathrm{TS}^\alpha$, and $\alpha < \alpha_1 \le \mu^\tau_\alpha$ is called a *μ-covering from α to β.*

Lemma 3.3: *Any μ-covering from α to β is a subsequence of the α-localization of β.*

Proof: Let $(\alpha_0, \ldots, \alpha_{n+1})$ be a μ-covering from α to β. Stepping down from α_{n+1} to α_0, Lemmas 3.5 of [5] and 4.9, 6.5 of [7] apply to show that the α-localization of β is the successive concatenation of the α_i-localization of α_{i+1} for $i = 1, \ldots, n$, modulo translation between the respective settings of relativization. □

Definition 3.4: Let $\tau \in \mathbb{E}_1$.

(1) For $\alpha \in \mathbb{P} \cap (\tau, \tau^\infty)$ we define $\text{max-cov}^\tau(\alpha)$ to be the longest subsequence $(\alpha_1, \ldots, \alpha_{n+1})$ of the τ-localization of α which satisfies $\tau < \alpha_1$, $\alpha_{n+1} = \alpha$, and which is *μ-covered*, i.e. which satisfies $\alpha_{i+1} \le \mu^\tau_{\alpha_i}$ for $i = 1, \ldots, n$.
(2) For $\alpha \in \mathbb{E} \cap (\tau, \tau^\infty)$ and $\beta \in \mathbb{P} \cap (\alpha, \alpha^\infty)$ we denote the shortest subsequence $(\beta_0, \beta_1, \ldots, \beta_{n+1})$ of the α-localization of β which is a μ-covering from α to β by $\text{min-cov}^\alpha(\beta)$, if such sequence exists.

We recall the notion of the tracking sequence of an ordinal, for greater clarity only for multiplicative principals at this stage.

Definition 3.5: **(cf. 3.13 of** [5]**)** Let $\tau \in \mathbb{E}_1$ and $\alpha \in \mathbb{M} \cap (\tau, \tau^\infty)$ with τ-localization $\tau = \alpha_0, \ldots, \alpha_n = \alpha$. The tracking sequence of α above τ, $\mathrm{ts}^\tau(\alpha)$, is defined as follows. If there exists the largest index $i \in \{1, \ldots, n-1\}$ such that $\alpha \le \mu^\tau_{\alpha_i}$, then

$$\mathrm{ts}^\tau(\alpha) := \mathrm{ts}^\tau(\alpha_i)^\frown(\alpha),$$

otherwise $\mathrm{ts}^\tau(\alpha) := (\alpha)$.

Definition 3.6: Let $\tau \in \mathbb{E}_1$, $\alpha \in \mathbb{E} \cap (\tau, \tau^\infty)$, $\beta \in \mathbb{P} \cap (\alpha, \alpha^\infty)$, and let $\alpha = \alpha_0, \ldots, \alpha_{n+1} = \beta$ be the α-localization of β. If there exists the least index $i \in \{0, \ldots, n\}$ such that $\alpha_i < \beta \le \mu^\tau_{\alpha_i}$, then

$$\mathrm{mts}^\alpha(\beta) := \mathrm{mts}^\alpha(\alpha_i)^\frown(\beta),$$

otherwise $\mathrm{mts}^\alpha(\beta) := (\alpha)$.

Note that $\mathrm{mts}^\alpha(\beta)$ reaches β if and only if it is a μ-covering from α to β.

Lemma 3.7: *Fix* $\tau \in \mathbb{E}_1$.

(1) For $\alpha \in \mathbb{P} \cap (\tau, \tau^\infty)$ *let* $\text{max-cov}^\tau(\alpha) = (\alpha_1, \ldots, \alpha_{n+1}) = \boldsymbol{\alpha}$. *If* $\alpha_1 < \alpha$ *then* $\boldsymbol{\alpha}$ *is a* μ*-covering from* α_1 *to* α *and* $\mathrm{mts}^{\alpha_1}(\alpha) \subseteq \boldsymbol{\alpha}$.

(2) If $\alpha \in \mathbb{M} \cap (\tau, \tau^\infty)$ *then* $\text{max-cov}^\tau(\alpha) = \mathrm{ts}^\tau(\alpha)$.

(3) Let $\alpha \in \mathbb{E} \cap (\tau, \tau^\infty)$ *and* $\beta \in \mathbb{P} \cap (\alpha, \alpha^\infty)$. *Then* $\text{min-cov}^\alpha(\alpha)$ *exists if and only if* $\mathrm{mts}^\alpha(\beta)$ *is a* μ*-covering from* α *to* β*, in which case these sequences are equal, characterizing the lexicographically maximal* μ*-covering from* α *to* β.

Proof: These are immediate consequences of the definitions. □

Recall Definition 3.16 from [5], which for $\tau \in \mathbb{E}_1$, $\alpha \in \mathbb{E} \cap (\tau, \tau^\infty)$ defines $\widehat{\alpha}$ to be the minimal $\gamma \in \mathbb{M}^{>\alpha}$ such that $\mathrm{ts}^\alpha(\gamma) = (\gamma)$ and $\mu^\tau_\alpha < \gamma$.

Lemma 3.8: *Let* $\tau \in \mathbb{E}_1$, $\alpha \in \mathbb{E} \cap (\tau, \tau^\infty)$, *and* $\beta \in \mathbb{M} \cap (\alpha, \alpha^\infty)$. *Then* $\mathrm{mts}^\alpha(\beta)$ *is a* μ*-covering from* α *to* β *if and only if* $\beta < \widehat{\alpha}$. *This holds if and only if for* $\mathrm{ts}^\alpha(\beta) = (\beta_1, \ldots, \beta_m)$ *we have* $\beta_1 \le \mu^\tau_\alpha$.

Proof: Suppose first that $\mathrm{mts}^\alpha(\beta)$ is a μ-covering from α to β. Then α is an element of the τ-localization of β, and modulo term translation we obtain the τ-localization of β by concatenating the τ-localization of α with the α-localization of β. By Lemma 3.7 we therefore have $\mathrm{mts}^\alpha(\beta) \subseteq (\alpha)^\frown\mathrm{ts}^\alpha(\beta)$ where $\beta_1 \le \mu_\alpha$. Let $\gamma \in \mathbb{M} \cap (\alpha, \beta]$ be given. Then by Lemma 3.15 of [5] we have

$$(\gamma_1, \ldots, \gamma_k) := \mathrm{ts}^\alpha(\gamma) \le_{\mathrm{lex}} \mathrm{ts}^\alpha(\beta),$$

so $\gamma_1 \le \beta_1 \le \mu_\alpha$, and hence $\beta < \widehat{\alpha}$.

Toward proving the converse, suppose that $\beta < \widehat{\alpha}$. We have $\mathrm{ts}^{\alpha}(\beta_1) = (\beta_1)$, so $\beta_1 \le \mu_\alpha$ since $\beta_1 < \widehat{\alpha}$. This implies that $\mathrm{mts}^{\alpha}(\beta)$ reaches β as a subsequence of $(\alpha)^\frown \mathrm{ts}^{\alpha}(\beta)$. □

Definition 3.9: (4.3 of [5]) Let $\tau \in \mathbb{E}_1$. A sequence $\boldsymbol{\alpha}$ of ordinals below τ^∞ is a τ-reference sequence if

(1) $\boldsymbol{\alpha} = ()$ or
(2) $\boldsymbol{\alpha} = (\alpha_1, \ldots, \alpha_n) \in \mathrm{TS}^\tau$ such that $\alpha_n \in \mathbb{E}^{>\alpha_{n-1}}$ (where $\alpha_0 := \tau$).

We denote the set of τ-reference sequences by RS^τ. In case of $\tau = 1$ we simply write RS and call its elements reference sequences.

Definition 3.10: (c.f. 4.9 of [5]) For $\gamma \in \mathbb{M} \cap 1^\infty$ and $\varepsilon \in \mathbb{E} \cap 1^\infty$ let $\mathrm{sk}_\gamma(\varepsilon)$ be the maximal sequence $\delta_1, \ldots, \delta_l$ such that (setting $\delta_0 := 1$)

(1) $\delta_1 = \varepsilon$ and
(2) if $i \in \{1, \ldots, l-1\}$ & $\delta_i \in \mathbb{E}^{>\delta_{i-1}}$ & $\gamma \le \mu_{\delta_i}$, then $\delta_{i+1} = \overline{\mu_{\delta_i} \cdot \gamma}$.

Remark ([5]). Lemma 3.5 of [5] guarantees that the above definition terminates. We have $(\delta_1, \ldots, \delta_{l-1}) \in \mathrm{RS}$ and $(\delta_1, \ldots, \delta_l) \in \mathrm{TS}$. Notice that $\gamma \le \delta_i$ for $i = 2, \ldots, l$.

Definition 3.11: Let $\boldsymbol{\alpha}^\frown\beta \in \mathrm{RS}$ and $\gamma \in \mathbb{M}$.

(1) If $\gamma \in (\beta, \widehat{\beta})$, let $\mathrm{mts}^{\beta}(\gamma) = \boldsymbol{\eta}^\frown(\varepsilon, \gamma)$ and define

$$\mathrm{h}_\gamma(\boldsymbol{\alpha}^\frown\beta) := \boldsymbol{\alpha}^\frown\boldsymbol{\eta}^\frown \mathrm{sk}_\gamma(\varepsilon).$$

(2) If $\gamma \in (1, \beta]$ and $\gamma \le \mu_\beta$ then

$$\mathrm{h}_\gamma(\boldsymbol{\alpha}^\frown\beta) := \boldsymbol{\alpha}^\frown \mathrm{sk}_\gamma(\beta).$$

(3) If $\gamma \in (1, \beta]$ and $\gamma > \mu_\beta$ then

$$\mathrm{h}_\gamma(\boldsymbol{\alpha}^\frown\beta) := \boldsymbol{\alpha}^\frown\beta.$$

Remark. In 1. let $\mathrm{ts}^{\beta}(\gamma) =: (\gamma_1, \ldots, \gamma_m)$. Then we have $\gamma_1 \le \mu_\beta$ according to Lemma 3.17 of [5], so that $\mathrm{mts}^{\beta}(\gamma)$ reaches γ, $\beta \le \varepsilon < \gamma \le \mu_\varepsilon$ and $\beta < \mu_\beta$. In 2. we have $\mathrm{sk}_\gamma(\beta) = (\beta, \overline{\mu_\beta \cdot \gamma})$ with $\gamma \le \overline{\mu_\beta \cdot \gamma} \le \beta$ in case of $\mu_\beta \le \beta$. In 3. we have $\mu_\beta < \gamma \le \beta$ and hence $\mathrm{sk}_\gamma(\beta) = (\beta)$.

Lemma 3.12: *Let* $\boldsymbol{\alpha}^\frown\beta \in \mathrm{RS}$ *and* $\gamma \in \mathbb{M}$. *Then* $\mathrm{h}_\gamma(\boldsymbol{\alpha}^\frown\beta)$ *is of a form* $\boldsymbol{\alpha}^\frown\boldsymbol{\eta}^\frown \mathrm{sk}_\gamma(\varepsilon)$ *where* $\boldsymbol{\eta} = (\eta_1, \ldots, \eta_r)$, $r \ge 0$, $\eta_1 = \beta$, $\eta_{r+1} := \varepsilon$, *and* $\mathrm{sk}_\gamma(\varepsilon) = (\delta_1, \ldots, \delta_{l+1})$, $l \ge 0$, *with* $\delta_1 = \varepsilon$. *We have*

$$\mathrm{lf}(\delta_{l+1}) \ge \gamma,$$

and for $\boldsymbol{\tau}^\frown\sigma \in \mathrm{TS}$*, where* $\boldsymbol{\tau} = (\tau_1, \ldots, \tau_s)$ *and* $\tau_{s+1} := \sigma$*, such that* $\boldsymbol{\alpha}^\frown\beta \subseteq \boldsymbol{\tau}^\frown\sigma$ *and* $\mathrm{h}_\gamma(\boldsymbol{\alpha}^\frown\beta) <_{\mathrm{lex}} \boldsymbol{\tau}^\frown\sigma$ *we either have*

(1) $\boldsymbol{\tau} = \boldsymbol{\alpha}^\frown\boldsymbol{\eta}^\frown\boldsymbol{\delta}_{\restriction i}$ *for some* $i \in \{1, \ldots, l+1\}$ *and* $\sigma =_{\mathrm{NF}} \delta_{i+1} \cdot \sigma'$ *for some* $\sigma' < \gamma$*, setting* $\delta_{l+2} := 1$*, or*

(2) $\boldsymbol{\tau}_{\restriction s_0} = \boldsymbol{\alpha}^\frown\boldsymbol{\eta}_{\restriction r_0}$ *for some* $r_0 \in [1, r]$ *and* $s_0 \leq s$ *such that* $\eta_{r_0+1} < \tau_{s_0+1}$*, in which case we have* $\mu_{\tau_j} < \gamma$ *for all* $j \in \{s_0+1, \ldots, s\}$*, and* $\mu_\sigma < \gamma$ *if* $\sigma \in \mathbb{E}^{>\tau_s}$*.*

Proof: Suppose first that $\boldsymbol{\tau}$ is a maximal initial segment $\boldsymbol{\alpha}^\frown\boldsymbol{\eta}^\frown\boldsymbol{\delta}_{\restriction i}$ for some $i \in \{1, \ldots, l+1\}$.

In the case $i = l+1$ we have $\delta_{l+1} \in \mathbb{E}^{>\delta_l}$ and $\mu_{\delta_{l+1}} < \gamma \leq \delta_{l+1}$, so $\sigma \leq \mu_{\delta_{l+1}} < \gamma$, and we also observe that $\boldsymbol{\tau}^\frown\sigma$ could not be extended further. Now suppose that $i \leq l$. Then $\delta_{i+1} < \sigma \leq \mu_{\delta_i}$, and since $\gamma \leq \delta_{i+1} = \overline{\mu_{\delta_i} \cdot \gamma} \leq \mu_{\delta_i}$, we obtain $\sigma = \delta_{i+1} \cdot \sigma'$ for some $\sigma' \in (1, \gamma)$.

Otherwise $\boldsymbol{\tau}$ must be of the form given in part 2 of the claim. This implies $\gamma \in \mathbb{M} \cap (\beta, \widehat{\beta})$ and $\mathrm{mts}^\beta(\gamma) = \boldsymbol{\eta}^\frown(\varepsilon, \gamma)$. Let us assume, toward contradiction, there existed a least $j \in \{s_0+1, \ldots, s+1\}$ such that $\tau_j \in \mathbb{E}^{>\tau_{j-1}}$ and $\gamma \leq \mu_{\tau_j}$. Then $\boldsymbol{\eta}_{\restriction r_0}{}^\frown(\tau_{s_0+1}, \ldots, \tau_j, \gamma)$ is a μ-covering from β to γ, hence by part 3 of Lemma 3.7 it must be lexicographically less than or equal to $\mathrm{mts}^\beta(\gamma)$ and therefore $\boldsymbol{\eta}_{\restriction r_0}{}^\frown(\tau_{s_0+1}, \ldots, \tau_j) \leq_{\mathrm{lex}} \boldsymbol{\eta}^\frown\varepsilon$: contradiction. □

Corollary 3.13: *For* $\boldsymbol{\alpha}^\frown\beta \in \mathrm{RS}$ *and* $\gamma, \delta \in \mathbb{M}$ *such that* $\delta \in (1, \gamma)$ *we have*

$$\mathrm{h}_\gamma(\boldsymbol{\alpha}^\frown\beta) \leq_{\mathrm{lex}} \mathrm{h}_\delta(\boldsymbol{\alpha}^\frown\beta).$$

3.2. *Evaluation*

Definition 3.14: Let $\boldsymbol{\alpha}^\frown\beta \in \mathrm{TS}$, where $\boldsymbol{\alpha} = (\alpha_1, \ldots, \alpha_n)$, $n \geq 0$, $\beta =_{\mathrm{MNF}} \beta_1 \cdot \ldots \cdot \beta_k$, and set $\alpha_0 := 1$, $\alpha_{n+1} := \beta$, $h := \mathrm{ht}_1(\alpha_1)+1$, and $\boldsymbol{\gamma}_i := \mathrm{ts}^{\alpha_{i-1}}(\alpha_i)$, $i = 1, \ldots, n$,

$$\boldsymbol{\gamma}_{n+1} := \begin{cases} (\beta) & \text{if } \beta \leq \alpha_n \\ \mathrm{ts}^{\alpha_n}(\beta_1)^\frown\beta_2 & \text{if } k > 1, \beta_1 \in \mathbb{E}^{>\alpha_n} \ \&\ \beta_2 \leq \mu_{\beta_1} \\ \mathrm{ts}^{\alpha_n}(\beta_1) & \text{otherwise,} \end{cases}$$

and write $\boldsymbol{\gamma}_i = (\gamma_{i,1}, \ldots, \gamma_{i,m_i})$, $i = 1, \ldots, n+1$. Then define

$$\mathrm{lSeq}(\boldsymbol{\alpha}^\frown\beta) := (m_1, \ldots, m_{n+1}) \in [h]^{\leq h}.$$

Let $\beta' := 1$ if $k = 1$ and $\beta' := \beta_2 \cdot \ldots \cdot \beta_k$ otherwise. We define $\mathrm{o}(\boldsymbol{\alpha}^\frown\beta)$ recursively in $\mathrm{lSeq}(\boldsymbol{\alpha}^\frown\beta)$, as well as auxiliary parameters $n_0(\boldsymbol{\alpha}^\frown\beta)$ and $\gamma(\boldsymbol{\alpha}^\frown\beta)$, which are set to 0 where not defined explicitly.

(1) $\mathrm{o}((1)) := 1$.
(2) If $\beta_1 \leq \alpha_n$, then $\mathrm{o}(\boldsymbol{\alpha}^\frown\beta) :=_{\mathrm{NF}} \mathrm{o}(\boldsymbol{\alpha}) \cdot \beta$.
(3) If $\beta_1 \in \mathbb{E}^{>\alpha_n}$, $k > 1$, and $\beta_2 \leq \mu_{\beta_1}$, then set $n_0(\boldsymbol{\alpha}^\frown\beta) := n+1$, $\gamma(\boldsymbol{\alpha}^\frown\beta) := \beta_1$, and define

$$\mathrm{o}(\boldsymbol{\alpha}^\frown\beta) :=_{\mathrm{NF}} \mathrm{o}(\mathrm{h}_{\beta_2}(\boldsymbol{\alpha}^\frown\beta_1)) \cdot \beta'.$$

(4) Otherwise. Then setting

$$n_0 := n_0(\boldsymbol{\alpha}^\frown\beta) := \max\left(\{i \in \{1,\dots,n+1\} \mid m_i > 1\} \cup \{0\}\right),$$

define

$$\mathrm{o}(\boldsymbol{\alpha}^\frown\beta) :=_{\mathrm{NF}} \begin{cases} \beta & \text{if } n_0 = 0 \\ \mathrm{o}(\mathrm{h}_{\beta_1}(\boldsymbol{\alpha}_{\restriction n_0 - 1}{}^\frown\gamma)) \cdot \beta & \text{if } n_0 > 0, \end{cases}$$

where $\gamma := \gamma(\boldsymbol{\alpha}^\frown\beta) := \gamma_{n_0, m_{n_0}-1}$.

Remark. As indicated in writing $=_{\mathrm{NF}}$ in the above definition, we obtain terms in multiplicative normal form denoting the values of o. The **fixed points of** o, i.e. those $\boldsymbol{\alpha}^\frown\beta$ that satisfy $\mathrm{o}(\boldsymbol{\alpha}^\frown\beta) = \beta$ are therefore characterized by 1. and 4. for $n_0 = 0$.

Recall Definition 3.13 of [5], extending Definition 3.5 to additive principal numbers that are not multiplicative principal ones.

Definition 3.15: (cf. 3.13 of [5]) Let $\tau \in \mathbb{E}_1$ and $\alpha \in [\tau, \tau^\infty) \cap \mathbb{P}$. The tracking sequence of α above τ, $\mathrm{ts}^\tau(\alpha)$, is defined as in Definition 3.5 if $\alpha \in \mathbb{M}^{>\tau}$, and otherwise recursively in the multiplicative decomposition of α as follows.

(1) If $\alpha \leq \tau^\omega$ then $\mathrm{ts}^\tau(\alpha) := (\alpha)$.
(2) Otherwise. Then $\bar{\alpha} \in [\tau, \alpha)$ and $\alpha =_{\mathrm{NF}} \bar{\alpha} \cdot \beta$ for some $\beta \in \mathbb{M}^{>1}$. Let $\mathrm{ts}^\tau(\bar{\alpha}) = (\alpha_1, \dots, \alpha_n)$ and set $\alpha_0 := \tau$.[c]

2.1. If $\alpha_n \in \mathbb{E}^{>\alpha_{n-1}}$ and $\beta \leq \mu^\tau_{\alpha_n}$ then $\mathrm{ts}^\tau(\alpha) := (\alpha_1, \dots, \alpha_n, \beta)$.
2.2. Otherwise. For $i \in \{1, \dots, n\}$ let $(\beta^i_1, \dots, \beta^i_{m_i})$ be $\mathrm{ts}^{\alpha_i}(\beta)$ provided $\beta > \alpha_i$, and set $m_i := 1$, $\beta^i_1 := \alpha_i \cdot \beta$ if $\beta \leq \alpha_i$. We first define the critical index

$$i_0(\alpha) = i_0 := \max\left(\{1\} \cup \{j \in \{2, \dots, n\} \mid \beta^j_1 \leq \mu^\tau_{\alpha_{j-1}}\}\right).$$

Then $\mathrm{ts}^\tau(\alpha) := (\alpha_1, \dots, \alpha_{i_0-1}, \beta^{i_0}_1, \dots, \beta^{i_0}_{m_{i_0}})$.

[c]As verified in part 2 of the lemma below we have $\beta \leq \alpha_n$.

Instead of $\mathrm{ts}^1(\alpha)$ we also simply write $\mathrm{ts}(\alpha)$.

Lemma 3.16: *If in the above definition, part 2.2, we have $\beta > \alpha_{i_0}$, then for all $j \in (i_0, \ldots, n]$ we have*

$$\beta_1^{i_0} \leq \alpha_j \leq \mu_{\alpha_{j-1}},$$

in particular $\beta_1^{i_0} \leq \mu_{\alpha_{i_0}}$.

Proof: Assume toward contradiction that there exists the maximal $j \in (i_0, \ldots, n]$ such that $\alpha_j < \beta_1^{i_0}$. Since $\beta_1^{i_0} \leq \beta \leq \alpha_n$ we have $j < n$ and obtain

$$\alpha_{i_0} < \alpha_j < \beta_1^{i_0} \leq \alpha_{j+1} \leq \mu_{\alpha_j},$$

so that $\alpha_j \in \mathrm{ts}^{\alpha_{i_0}}(\beta)$, contradicting the minimality of $\beta_1^{i_0}$ in $\mathrm{ts}^{\alpha_{i_0}}(\beta)$. □

Lemma 3.17: ***(3.14 of** [5]) Let $\tau \in \mathbb{E}_1$ and $\alpha \in [\tau, \tau^\infty) \cap \mathbb{P}$. Let further $(\alpha_1, \ldots, \alpha_n)$ be $\mathrm{ts}^\tau(\alpha)$, the tracking sequence of α above τ.*

(1) If $\alpha \in \mathbb{M}$ then $\alpha_n = \alpha$ and $\mathrm{ts}^\tau(\alpha_i) = (\alpha_1, \ldots, \alpha_i)$ for $i = 1, \ldots, n$.
(2) If $\alpha =_{\mathrm{NF}} \eta \cdot \xi \notin \mathbb{M}$ then $\alpha_n \in \mathbb{P} \cap [\xi, \alpha]$ and $\alpha_n =_{\mathrm{NF}} \overline{\alpha_n} \cdot \xi$.
(3) $(\alpha_1, \ldots, \alpha_{n-1})$ is either empty or a strictly increasing sequence of epsilon numbers in the interval (τ, α).
(4) For $1 \leq i \leq n-1$ we have $\alpha_{i+1} \leq \mu^\tau_{\alpha_i}$, and if $\alpha_i < \alpha_{i+1}$ then $(\alpha_1, \ldots, \alpha_{i+1})$ is a subsequence of the τ-localization of α_{i+1}.

Proof: The proof proceeds by straightforward induction along the definition of $\mathrm{ts}^\tau(\alpha)$, i.e. along the length of the τ-localization of multiplicative principal numbers and the number of factors in the multiplicative decomposition of additive principal numbers. In part 4 Lemma 6.5 of [7] and the previous remark apply. □

Lemma 3.18: ***(3.15 of** [5]) Let $\tau \in \mathbb{E}_1$ and $\alpha, \gamma \in [\tau, \tau^\infty) \cap \mathbb{P}$, $\alpha < \gamma$. Then we have*

$$\mathrm{ts}^\tau(\alpha) <_{\mathrm{lex}} \mathrm{ts}^\tau(\gamma).$$

Proof: The proof given in [5] is in fact an induction along the inductive definition of $\mathrm{ts}^\tau(\gamma)$ with a subsidiary induction along the inductive definition of $\mathrm{ts}^\tau(\alpha)$. □

Theorem 3.19: *For all $\alpha \in \mathbb{P} \cap 1^\infty$ we have*

$$\mathrm{o}(\mathrm{ts}(\alpha)) = \alpha.$$

Proof: The theorem is proved by induction along the inductive definition of $\mathrm{ts}(\alpha)$.

Case 1: $\alpha \in \mathbb{M}$. Then $\mathrm{lSeq}(\mathrm{ts}(\alpha)) = (1, \ldots, 1)$ and hence $\mathrm{o}(\mathrm{ts}(\alpha)) = \alpha$ immediately by definition.

Case 2: $\alpha =_{\mathrm{NF}} \bar{\alpha} \cdot \beta \in \mathbb{P} - \mathbb{M}$. Let $\mathrm{ts}(\bar{\alpha}) =: (\alpha_1, \ldots, \alpha_n)$ and $\alpha_0 := 1$. By the i.h. $\mathrm{o}(\boldsymbol{\alpha}) = \bar{\alpha}$. We have $\beta \leq \mathrm{lf}(\alpha_n) \leq \alpha_n$, $n \geq 1$.

Subcase 2.1: $\alpha_n \in \mathbb{E}^{>\alpha_{n-1}}$ & $\beta \leq \mu_{\alpha_n}$. Then $\mathrm{ts}(\alpha) = \boldsymbol{\alpha}^\frown\beta$, and since $\beta \in \mathbb{M}^{\leq \alpha_n}$ according to the definition of o we obtain $\mathrm{o}(\boldsymbol{\alpha}^\frown\beta) = \mathrm{o}(\boldsymbol{\alpha}) \cdot \beta = \bar{\alpha} \cdot \beta = \alpha$.

Subcase 2.2: Otherwise. Let $(\beta_1^i, \ldots, \beta_{m_i}^i)$ for $i = 1, \ldots, n$ as well as the index i_0 be defined as in case 2.2 of Definition 3.15, so that $\mathrm{ts}(\alpha) = (\alpha_1, \ldots, \alpha_{i_0-1}, \beta_1^{i_0}, \ldots, \beta_{m_{i_0}}^{i_0})$.

2.2.1: $i_0 = n$. Then we have $\mathrm{ts}(\alpha) = (\alpha_1, \ldots, \alpha_{n-1}, \alpha_n \cdot \beta)$, and using the i.h. we obtain $\mathrm{o}(\mathrm{ts}(\alpha)) = \mathrm{o}(\boldsymbol{\alpha}) \cdot \beta = \alpha$.

2.2.2: $i_0 < n$ and $\beta \leq \alpha_{i_0}$. Then we have $\alpha_{i_0} \in \mathbb{E}^{>\alpha_{i_0-1}}$, $\alpha_{i_0} \cdot \beta \leq \mu_{\alpha_{i_0-1}}$, and $\mathrm{ts}(\alpha) = (\alpha_1, \ldots, \alpha_{i_0-1}, \alpha_{i_0} \cdot \beta)$. It follows that for all $j \in (i_0, n]$ we have $\beta \leq \alpha_j$ and $\alpha_j \cdot \beta > \mu_{\alpha_{j-1}}$, hence $\beta \leq \mu_{\alpha_{i_0}}$ and thus

$$\mathrm{o}(\mathrm{ts}(\alpha)) = \mathrm{o}(\mathrm{h}_\beta(\boldsymbol{\alpha}_{\restriction_{i_0}})) \cdot \beta.$$

The sequence $\mathrm{h}_\beta(\boldsymbol{\alpha}_{\restriction_{i_0}})$ is of the form $\boldsymbol{\alpha}_{\restriction_{i_0-1}}{}^\frown\boldsymbol{\delta}$ where $\boldsymbol{\delta} := \mathrm{sk}_\beta(\alpha_{i_0})$. Since $\mu_{\alpha_{i_0}} < \alpha_{i_0+1} \cdot \beta$ we have $\overline{\mu_{\alpha_{i_0}} \cdot \beta} = \alpha_{i_0+1} = \delta_2$. In the case $i_0 + 1 = n$ we obtain $\boldsymbol{\delta} = (\alpha_{n-1}, \alpha_n)$, hence $\mathrm{h}_\beta(\boldsymbol{\alpha}_{\restriction_{i_0}}) = \boldsymbol{\alpha}$, otherwise we iterate the above argumentation to see that $\boldsymbol{\delta} = (\alpha_{i_0}, \ldots, \alpha_n)$. Hence $\mathrm{o}(\mathrm{ts}(\alpha)) = \mathrm{o}(\boldsymbol{\alpha}) \cdot \beta$ as desired.

2.2.3: $i_0 < n$ and $\alpha_{i_0} < \beta$. Then we have $\boldsymbol{\beta}^{i_0} = \mathrm{ts}^{\alpha_{i_0}}(\beta)$ with $\beta_1^{i_0} \leq \mu_{\alpha_{i_0-1}}$ if $i_0 > 1$. By Lemma 3.16 α_{i_0} is the immediate predecessor of $\beta_1^{i_0}$ in $\mathrm{ts}^{\alpha_{i_0-1}}(\beta)$. By definition of o we have $\mathrm{o}(\mathrm{ts}(\alpha)) = \mathrm{o}(\mathrm{h}_\beta(\boldsymbol{\alpha}_{\restriction_{i_0}})) \cdot \beta$ and therefore have to show that $\mathrm{h}_\beta(\boldsymbol{\alpha}_{\restriction_{i_0}}) = \boldsymbol{\alpha}$. We define

$$j_0 := \min\{j \in \{i_0, \ldots, n-1\} \mid \beta \leq \mu_{\alpha_j}\},$$

which exists, because $\beta \leq \alpha_n \leq \mu_{\alpha_{n-1}}$.

Claim: $\mathrm{mts}^{\alpha_{i_0}}(\beta) = (\alpha_{i_0}, \ldots, \alpha_{j_0}, \beta)$.
Proof: For every $j \in \{i_0, \ldots, j_0 - 1\}$ the minimality of j_0 implies $\beta > \mu_{\alpha_j} \geq \alpha_{j+1}$, and thus by the maximality of i_0 also $\beta_1^{j+1} > \mu_{\alpha_j}$. Moreover, we have

$$\beta_1^{j+1} \leq \alpha_{j+2} :$$

Assume otherwise and let j be maximal in $\{i_0, \dots, j_0 - 1\}$ such that $\beta_1^{j+1} > \alpha_{j+2}$. Since $\beta_1^{j+1} \le \beta \le \alpha_n$ we must have $j \le n-3$. But then $\alpha_{j+1} < \alpha_{j+2} < \beta_1^{j+1} \le \alpha_{j+3} \le \mu_{j+2}$ and hence $\alpha_{j+2} \in \mathrm{ts}^{\alpha_{j+1}}(\beta)$, contradicting the minimality of β_1^{j+1} in $\mathrm{ts}^{\alpha_{j+1}}(\beta)$. Therefore

$$\mu_{\alpha_j} < \beta_1^{j+1} \le \alpha_{j+2} \le \mu_{\alpha_{j+1}},$$

which concludes the proof of the claim. □

It remains to be shown that $\mathrm{sk}_\beta(\alpha_{j_0}) = (\alpha_{j_0}, \dots, \alpha_n)$, i.e. to successively check for $j = j_0, \dots, n-1$ that $\beta \le \alpha_{j+1} \le \mu_{\alpha_j}$ and $\alpha_{j+1} \cdot \beta > \mu_{\alpha_j}$, whence $\overline{\mu_{\alpha_j} \cdot \beta} = \alpha_{j+1}$. This concludes the verification of $\mathrm{h}_\beta(\boldsymbol{\alpha}_{\restriction i_0}) = \boldsymbol{\alpha}$ and consequently the proof of the theorem.

Theorem 3.20: *For all* $\boldsymbol{\alpha}^\frown\beta \in \mathrm{TS}$ *we have*

$$\mathrm{ts}(\mathrm{o}(\boldsymbol{\alpha}^\frown\beta)) = \boldsymbol{\alpha}^\frown\beta.$$

Proof: The theorem is proved by induction on $\mathrm{lSeq}(\boldsymbol{\alpha}^\frown\beta)$ along the ordering $(\mathrm{lSeq}, <_{\mathrm{lex}})$. Let $\beta =_{\mathrm{NF}} \beta_1 \cdot \ldots \cdot \beta_k$ and set $n_0 := n_0(\boldsymbol{\alpha}^\frown\beta)$, $\gamma := \gamma(\boldsymbol{\alpha}^\frown\beta)$ according to Definition 3.14, which provides us with an NF-representation of $\mathrm{o}(\boldsymbol{\alpha}^\frown\beta)$, where in the interesting cases the i.h. applies to the term $\mathrm{ts}\left(\overline{\mathrm{o}(\boldsymbol{\alpha}^\frown\beta)}\right)$.

Case 1: $n = 0$ and $\beta = 1$. Trivial.

Case 2: $1 < \beta_1 \le \alpha_n$. Then $\mathrm{o}(\boldsymbol{\alpha}^\frown\beta) =_{\mathrm{NF}} \mathrm{o}(\boldsymbol{\alpha}) \cdot \beta$, and it is straightforward to verify the claim from the i.h. applied to $\boldsymbol{\alpha}$ by inspecting case 2.1 of Definition 3.15.

Case 3: $k > 1$ with $\beta_1 \in \mathbb{E}^{>\alpha_n}$ and $\beta_2 \le \mu_{\beta_1}$. Then by definition $\mathrm{o}(\boldsymbol{\alpha}^\frown\beta) =_{\mathrm{NF}} \mathrm{o}(\mathrm{h}_{\beta_2}(\boldsymbol{\alpha}^\frown\beta_1)) \cdot \beta'$ where $\beta' = (1/\beta_1) \cdot \beta$. According to part 2 of Definition 3.11 we have

$$\mathrm{h}_{\beta_2}(\boldsymbol{\alpha}^\frown\beta_1) = \boldsymbol{\alpha}^\frown\boldsymbol{\delta},$$

where $\boldsymbol{\delta} = (\delta_1, \dots, \delta_{l+1}) := \mathrm{sk}_{\beta_2}(\beta_1)$. Assume first that $k = 2$. Then the maximality of the length of $\boldsymbol{\delta}$ excludes the possibility $\delta_{l+1} \in \mathbb{E}^{>\delta_l}$ & $\beta_2 \le \mu_{\delta_{l+1}}$. We have $\beta_2 \le \beta_1$ and $\beta \le \mu_{\alpha_n}$. For $j \in \{2, \dots, l+1\}$ we have $\beta_2 \le \delta_j = \overline{\mu_{\delta_{j-1}} \cdot \beta_2} \le \mu_{\delta_{j-1}}$ and $\delta_j \cdot \beta_2 = \mu_{\delta_{j-1}} \cdot \beta_2 > \mu_{\delta_{j-1}}$. This implies that $\mathrm{ts}(\mathrm{o}(\boldsymbol{\alpha}^\frown\beta)) = \boldsymbol{\alpha}^\frown\beta$. The claim now follows easily for $k > 2$ since $\beta \le \mu_{\alpha_n}$.

Case 4: Otherwise.

Subcase 4.1: $n_0 = 0$. Then we have $\mathrm{o}(\boldsymbol{\alpha}^\frown\beta) = \beta$, $\mathrm{ts}^{\alpha_{i-1}}(\alpha_i) = (\alpha_i)$ for $i = 1, \ldots, n$, and $\mathrm{ts}^{\alpha_n}(\beta_1) = (\beta_1)$, whence $\mathrm{ts}(\beta) = (\beta)$.

Subcase 4.2: $n_0 > 0$. Using the abbreviation $\boldsymbol{\alpha}' := \boldsymbol{\alpha}_{\restriction n_0 - 1}$ we then have $\mathrm{o}(\boldsymbol{\alpha}^\frown\beta) =_{\mathrm{NF}} \mathrm{o}(\mathrm{h}_{\beta_1}(\boldsymbol{\alpha}'^\frown\gamma)) \cdot \beta$. Setting $\mathrm{mts}^\gamma(\beta_1) =: (\gamma_1, \ldots, \gamma_{m+1})$ where $\gamma_1 = \gamma$, $\gamma_{m+1} = \beta_1$, and $\mathrm{sk}_{\beta_1}(\gamma_m) =: \boldsymbol{\delta} = (\delta_1, \ldots, \delta_{l+1})$ where $\delta_1 = \gamma_m$, we have

$$\mathrm{h}_{\beta_1}(\boldsymbol{\alpha}'^\frown\gamma) = \boldsymbol{\alpha}'^\frown\boldsymbol{\gamma}_{\restriction m-1}{}^\frown\boldsymbol{\delta},$$

which by the i.h. is the tracking sequence of $\overline{\mathrm{o}(\boldsymbol{\alpha}^\frown\beta_1)}$. Assuming first that $k = 1$, we now verify that the tracking sequence of $\mathrm{o}(\boldsymbol{\alpha}^\frown\beta)$ actually is $\boldsymbol{\alpha}^\frown\beta$, by checking that case 2.2 of Definition 3.15 applies, with n_0 playing the role of the critical index $i_0(\mathrm{o}(\boldsymbol{\alpha}^\frown\beta))$. Note first that the maximality of the length of $\boldsymbol{\delta}$ rules out the possibility $\delta_{l+1} \in \mathbb{E}^{>\delta_l}$ & $\beta_1 \le \mu_{\delta_{l+1}}$ and hence case 2.1 of Definition 3.15. According to the choice of n_0 and part 2 of Lemma 3.7 we have $\mathrm{ts}^\gamma(\beta_1) = \text{max-cov}^\gamma(\beta_1) = (\alpha_{n_0}, \ldots, \alpha_n, \beta_1)$ and of course $\alpha_{n_0} \le \mu_{\alpha_{n_0-1}}$. Thus n_0 qualifies for the critical index, once we show its maximality: Firstly, for any $i \in \{2, \ldots, m\}$ we have $\gamma_i < \beta_1$, and setting $\boldsymbol{\beta}^i := \mathrm{ts}^{\gamma_i}(\beta_1)$ the assumption $\beta^i_1 \le \mu_{\gamma_{i-1}}$ would imply that $\boldsymbol{\gamma}_{\restriction i-1}{}^\frown\boldsymbol{\beta}^i$ is a μ-covering from γ to β_1 such that

$$\mathrm{mts}^\gamma(\beta_1) <_{\mathrm{lex}} \boldsymbol{\gamma}_{\restriction i-1}{}^\frown\boldsymbol{\beta}^i,$$

contradicting part 3 of Lemma 3.7. Secondly, for any $j \in \{2, \ldots, l+1\}$ we have $\delta_j = \overline{\mu_{\delta_{j-1}} \cdot \beta_1}$, so $\delta_j \cdot \beta_1 = \mu_{\delta_{j-1}} \cdot \beta_1 > \mu_{\delta_{j-1}}$. These considerations entail

$$\mathrm{ts}(\boldsymbol{\alpha}^\frown\beta_1) = \boldsymbol{\alpha}'^\frown\mathrm{ts}^\gamma(\beta_1),$$

and it is easy now to verify the claim for arbitrary k, since $\beta \le \mu_{\alpha_n}$. □

Corollary 3.21: o *is strictly increasing with respect to the lexicographic ordering on* TS *and continuous in the last vector component.*

Proof: The first statement is immediate from Lemma 3.18 and Theorems 3.19 and 3.20. In order to verify continuity, let $\boldsymbol{\alpha} = (\alpha_1, \ldots, \alpha_n) \in \mathrm{RS}$ and $\beta \in \mathbb{L}^{\le\mu_{\alpha_n}}$ be given. For any $\gamma \in \mathbb{P} \cap \beta$, we have $\tilde{\gamma} := \mathrm{o}(\boldsymbol{\alpha}^\frown\gamma) < \mathrm{o}(\boldsymbol{\alpha}^\frown\beta) =: \tilde{\beta}$ and $\boldsymbol{\alpha}^\frown\gamma = \mathrm{ts}(\tilde{\gamma}) <_{\mathrm{lex}} \mathrm{ts}(\tilde{\beta}) = \boldsymbol{\alpha}^\frown\beta$. For given $\tilde{\delta} \in \mathbb{P} \cap (\tilde{\gamma}, \tilde{\beta})$ set $\boldsymbol{\delta} := \mathrm{ts}(\tilde{\delta})$, so that

$$\boldsymbol{\alpha}^\frown\gamma <_{\mathrm{lex}} \boldsymbol{\delta} <_{\mathrm{lex}} \boldsymbol{\alpha}^\frown\beta,$$

whence $\boldsymbol{\alpha} \subseteq \boldsymbol{\delta}$ is an initial segment. Writing $\boldsymbol{\delta} = \boldsymbol{\alpha}^\frown(\zeta_1, \ldots, \zeta_m)$ we obtain $\gamma < \zeta_1 < \beta$. For $\boldsymbol{\delta}' := \boldsymbol{\alpha}^\frown \zeta_1 \cdot \omega$ we then have $\tilde{\delta} < \mathrm{o}(\boldsymbol{\delta}') < \tilde{\beta}$. □

Remark. Theorems 3.19 and 3.20 establish Lemma 4.10 of [5] in a weak theory, adjusted to our redefinition of o. Its equivalence with the definition in [5] follows, since the definition of ts has not been modified. In the next section we will continue in this way in order to obtain suitable redefinitions of the κ- and ν-functions.

The following lemma will not be required in the sequel, but it has been included to further illuminate the approach.

Lemma 3.22: *Let $\boldsymbol{\alpha}^\frown\beta \in \mathrm{RS}$, $\gamma \in \mathbb{M} \cap (1, \widehat{\beta})$, and let $\boldsymbol{\tau}^\frown\sigma \in \mathrm{TS}$ be such that $\boldsymbol{\alpha}^\frown\beta \subseteq \boldsymbol{\tau}^\frown\sigma$ and $\mathrm{h}_\gamma(\boldsymbol{\alpha}^\frown\beta) <_{\mathrm{lex}} \boldsymbol{\tau}^\frown\sigma$. Then we have*

$$\mathrm{o}(\boldsymbol{\tau}^\frown\sigma) =_{\mathrm{NF}} \mathrm{o}(\mathrm{h}_\gamma(\boldsymbol{\alpha}^\frown\beta)) \cdot \delta$$

for some $\delta \in \mathbb{P} \cap (1, \gamma)$.

Proof: The lemma is proved by induction on $\mathrm{lSeq}(\boldsymbol{\tau}^\frown\sigma)$, using Lemma 3.12. In order to fix some notation, set $\boldsymbol{\tau} = (\tau_1, \ldots, \tau_s)$ and $\tau_{s+1} := \sigma$. Write $\mathrm{h}_\gamma(\boldsymbol{\alpha}^\frown\beta) = \boldsymbol{\alpha}^\frown\boldsymbol{\eta}^\frown\boldsymbol{\delta}$ where $\boldsymbol{\eta} = (\eta_1, \ldots, \eta_r)$ and $\boldsymbol{\delta} = (\delta_1, \ldots, \delta_{l+1}) = \mathrm{sk}_\gamma(\varepsilon)$, $\delta_1 = \varepsilon =: \eta_{r+1}$, and $\delta_{l+2} := 1$. Note that since $\boldsymbol{\alpha}^\frown\beta \in \mathrm{RS}$, we have $\varepsilon \in \mathbb{E}^{>\eta_r}$. According to Lemma 3.12 either one of the two following cases applies to $\boldsymbol{\tau}^\frown\sigma$.

Case 1: $\boldsymbol{\tau} = \boldsymbol{\alpha}^\frown\boldsymbol{\eta}^\frown\boldsymbol{\delta}_{\restriction i}$ for some $i \in \{1, \ldots, l+1\}$ and $\sigma =_{\mathrm{NF}} \delta_{i+1} \cdot \zeta$ for some $\zeta \in \mathbb{P} \cap (1, \gamma)$. Let $\zeta =_{\mathrm{MNF}} \zeta_1 \cdot \ldots \cdot \zeta_j$.

Subcase 1.1: $i = l+1$. Then $\mathrm{h}_\gamma(\boldsymbol{\alpha}^\frown\beta) = \boldsymbol{\tau}$, $\delta_{l+1} \in \mathbb{E}^{>\delta_l}$, and $\sigma \le \mu_{\delta_{l+1}} < \gamma \le \delta_{l+1}$. According to the definition of o we have $\mathrm{o}(\boldsymbol{\tau}^\frown\sigma) = \mathrm{o}(\mathrm{h}_\gamma(\boldsymbol{\alpha}^\frown\beta)) \cdot \sigma$.

Subcase 1.2: $i \le l$ where $\delta_{i+1} \in \mathbb{E}^{>\delta_i}$ and $\zeta_1 \le \mu_{\delta_{i+1}}$. By definition, $\mathrm{o}(\boldsymbol{\tau}^\frown\sigma) = \mathrm{o}(\mathrm{h}_{\zeta_1}(\boldsymbol{\tau}^\frown\delta_{i+1})) \cdot \zeta$. As $\boldsymbol{\tau}^\frown\delta_{i+1}$ is an initial segment of $\mathrm{h}_\gamma(\boldsymbol{\alpha}^\frown\beta)$ and $\zeta_1 < \gamma$, we have

$$\mathrm{h}_\gamma(\boldsymbol{\alpha}^\frown\beta) = \mathrm{h}_\gamma(\boldsymbol{\tau}^\frown\delta_{i+1}) \le_{\mathrm{lex}} \mathrm{h}_{\zeta_1}(\boldsymbol{\tau}^\frown\delta_{i+1})$$

by Corollary 3.13. The claim now follows by the i.h. (if necessary) applied to $\mathrm{h}_{\zeta_1}(\boldsymbol{\tau}^\frown\delta_{i+1})$.

Subcase 1.3: Otherwise. Here we must have $i = l$, since if $i < l$ it follows that $\delta_{i+1} \in \mathbb{E}^{>\delta_i}$ and $\zeta_1 < \gamma \le \overline{\mu_{\delta_{i+1}} \cdot \gamma} = \delta_{i+2} \le \mu_{\delta_{i+1}}$, which has been covered by the previous subcase. We therefore have $\mathrm{o}(\boldsymbol{\tau}^\frown\sigma) = \mathrm{o}(\mathrm{h}_\gamma(\boldsymbol{\alpha}^\frown\beta)) \cdot \zeta$.

Case 2: $\boldsymbol{\tau}_{\restriction_{s_0}} = \boldsymbol{\alpha}^\frown\boldsymbol{\eta}_{\restriction_{r_0}}$ for some $r_0 \in [1,r]$, $s_0 \le s$, $\eta_{r_0+1} < \tau_{s_0+1}$, whence according to Lemma 3.12 we have $\sigma \le \mu_{\tau_s} < \gamma$, $\mu_{\tau_j} < \gamma$ for $j = s_0+1,\dots,s$, and $\mu_\sigma < \gamma$ if $\sigma \in \mathbb{E}^{>\tau_s}$. Let $\sigma =_{\text{MNF}} \sigma_1 \cdot \ldots \cdot \sigma_k$ and $\sigma' := (1/\sigma_1)\cdot\sigma$.

In the case $s_0 < s$ we have $\mathrm{h}_\gamma(\boldsymbol{\alpha}^\frown\beta) <_{\text{lex}} \boldsymbol{\tau}$, and noting that $\sigma < \gamma$, the i.h. straightforwardly applies to $\boldsymbol{\tau}$. Let us therefore assume that $s_0 = s$, whence $\boldsymbol{\tau} = \boldsymbol{\alpha}^\frown\boldsymbol{\eta}_{\restriction_{r_0}}$ and $\eta_{r_0+1} < \sigma$. Then we have $\boldsymbol{\eta}^\frown\varepsilon <_{\text{lex}} \boldsymbol{\tau}^\frown\sigma$ and $\eta_{r_0} < \eta_{r_0+1} < \sigma \le \mu_{\eta_{r_0}} < \gamma$.

Subcase 2.1: $k > 1$ where $\sigma_1 \in \mathbb{E}^{>\eta_{r_0}}$ and $\sigma_2 \le \mu_{\sigma_1}$. Then $\mathrm{o}(\boldsymbol{\tau}^\frown\sigma) = \mathrm{o}(\mathrm{h}_{\sigma_2}(\boldsymbol{\tau}^\frown\sigma_1))\cdot\sigma'$ and $\gamma > \sigma_1 \ge \eta_{r_0+1} \in \mathbb{E}^{>\eta_{r_0}}$. In the case $\sigma_1 > \eta_{r_0+1}$ the i.h. applies to $\mathrm{h}_{\sigma_2}(\boldsymbol{\tau}^\frown\sigma_1)$. Now assume $\sigma_1 = \eta_{r_0+1}$. As in subcase 1.2 we obtain

$$\mathrm{h}_\gamma(\boldsymbol{\alpha}^\frown\beta) = \mathrm{h}_\gamma(\boldsymbol{\tau}^\frown\sigma_1) \le_{\text{lex}} \mathrm{h}_{\sigma_2}(\boldsymbol{\tau}^\frown\sigma_1),$$

and (if necessary) the i.h. applies to $\mathrm{h}_{\sigma_2}(\boldsymbol{\tau}^\frown\sigma_1)$.

Subcase 2.2: Otherwise, i.e. case 4 of Definition 3.14 applies. Let $n_0 := n_0(\boldsymbol{\tau}^\frown\sigma)$. We first assume that $k = 1$, i.e. $\sigma \in \mathbb{M}\cap(\eta_{r_0+1},\gamma)$.

2.2.1: $n_0 \le s$. We obtain $\mathrm{o}(\boldsymbol{\tau}) = \xi\cdot\tau_s$ and $\mathrm{o}(\boldsymbol{\tau}^\frown\sigma) = \xi\cdot\sigma$ where $\xi = 1$ if $n_0 = 0$ and $\xi = \mathrm{h}_{\tau_s}(\boldsymbol{\tau}_{\restriction_{n_0-1}}{}^\frown\gamma(\boldsymbol{\tau}))$. The $<, <_{\text{lex}}$-order isomorphism between TS and $\mathbb{P}^{<1^\infty}$ established by Lemma 3.18 and Theorems 3.19 and 3.20 yields

$$\mathrm{o}(\boldsymbol{\tau}) < \mathrm{o}(\mathrm{h}_\gamma(\boldsymbol{\alpha}^\frown\beta)) < \mathrm{o}(\boldsymbol{\tau}^\frown\sigma),$$

and hence the claim.

2.2.2: $n_0 = s+1$. We have $\sigma \in \mathbb{M}\cap(\eta_{r_0+1},\gamma)$. Let ζ be the immediate predecessor of σ in $\mathrm{ts}^{\eta_{r_0}}(\sigma)$. Then $\mathrm{o}(\boldsymbol{\tau}^\frown\sigma) = \mathrm{o}(\mathrm{h}_\sigma(\boldsymbol{\tau}^\frown\zeta))\cdot\sigma$.

2.2.2.1: $\eta_{r_0+1} < \zeta$. Then the i.h. applies to $\boldsymbol{\tau}^\frown\zeta$.

2.2.2.2: $\eta_{r_0+1} = \zeta$. Then we argue as before, since

$$\mathrm{h}_\gamma(\boldsymbol{\alpha}^\frown\beta) = \mathrm{h}_\gamma(\boldsymbol{\tau}^\frown\eta_{r_0+1}) \le_{\text{lex}} \mathrm{h}_\sigma(\boldsymbol{\tau}^\frown\eta_{r_0+1}),$$

so that the i.h. (if necessary) applies to $\mathrm{h}_\sigma(\boldsymbol{\tau}^\frown\eta_{r_0+1})$.

2.2.2.3: $\eta_{r_0+1} > \zeta$. Then ζ is an element of $\mathrm{ts}^{\eta_{r_0}}(\eta_{r_0+1})$, and by a monotonicity argument as in 2.2.1 we obtain the claim as a consequence of

$$\mathrm{o}(\mathrm{h}_\sigma(\boldsymbol{\tau}^\frown\zeta)) < \mathrm{o}(\mathrm{h}_\gamma(\boldsymbol{\alpha}^\frown\beta)) < \mathrm{o}(\boldsymbol{\tau}^\frown\sigma).$$

This concludes the proof for $k = 1$, and for $k > 1$ the claim now follows easily. □

4. Enumerating Relativized Connectivity Components

Recall Definition 4.4 of [5]. We are now going to characterize the functions κ and ν by giving an alternative definition which is considerably less intertwined. The first step is to define the restrictions of $\kappa^{\boldsymbol{\alpha}}$ and $\nu^{\boldsymbol{\alpha}}$ to additive principal indices. Recall part 3 of Lemma 2.3.

Definition 4.1: Let $\boldsymbol{\alpha} \in \mathrm{RS}$ where $\boldsymbol{\alpha} = (\alpha_1, \ldots, \alpha_n)$, $n \geq 0$, $\alpha_0 := 1$. We define $\kappa^{\boldsymbol{\alpha}}_\beta$ and $\nu^{\boldsymbol{\alpha}}_\beta$ for additive principal β as follows, writing κ_β instead of $\kappa^{()}_\beta$.

Case 1: $n = 0$. For $\beta < 1^\infty$ define

$$\kappa_\beta := \mathrm{o}((\beta)).$$

Case 2: $n > 0$. For $\beta \leq \mu_{\alpha_n}$, i.e. $\boldsymbol{\alpha}^\frown\beta \in \mathrm{TS}$, define

$$\nu^{\boldsymbol{\alpha}}_\beta := \mathrm{o}(\boldsymbol{\alpha}^\frown\beta).$$

$\kappa^{\boldsymbol{\alpha}}_\beta$ for $\beta \leq \lambda_{\alpha_n}$ is defined by cases. If $\beta \leq \alpha_n$ let $i \in \{0, \ldots, n-1\}$ be maximal such that $\alpha_i < \beta$. If $\beta > \alpha_n$ let $\beta =_{\mathrm{MNF}} \beta_1 \cdot \ldots \cdot \beta_k$ and set $\beta' := (1/\beta_1) \cdot \beta$.

$$\kappa^{\boldsymbol{\alpha}}_\beta := \begin{cases} \kappa^{\boldsymbol{\alpha}\restriction_i}_\beta & \text{if } \beta \leq \alpha_n \\ \mathrm{o}(\boldsymbol{\alpha}) \cdot \beta' & \text{if } \beta_1 = \alpha_n \ \&\ k > 1 \\ \mathrm{o}(\boldsymbol{\alpha}^\frown\beta) & \text{if } \beta_1 > \alpha_n. \end{cases}$$

Remark. Note that in the case $n > 0$ we have the following inequalities between $\kappa^{\boldsymbol{\alpha}}_\beta$ and $\nu^{\boldsymbol{\alpha}}_\beta$, which are consequences of the monotonicity of o proved in Theorems 3.19 and 3.20.

(1) If $\beta \leq \alpha_n$ then $\kappa^{\boldsymbol{\alpha}}_\beta \leq \kappa^{\boldsymbol{\alpha}\restriction_{n-1}}_{\alpha_n} = \mathrm{o}(\boldsymbol{\alpha})$. Later we will define $\nu^{\boldsymbol{\alpha}}_0 := \mathrm{o}(\boldsymbol{\alpha})$.
(2) If $\beta_1 = \alpha_n$ and $k > 1$ then $\kappa^{\boldsymbol{\alpha}}_\beta = \mathrm{o}(\boldsymbol{\alpha}) \cdot \beta' = \nu^{\boldsymbol{\alpha}}_{\beta'}$, which is less than $\nu^{\boldsymbol{\alpha}}_\beta$ if $\beta \leq \mu_{\alpha_n}$.
(3) Otherwise we have $\kappa^{\boldsymbol{\alpha}}_\beta = \nu^{\boldsymbol{\alpha}}_\beta$.

Corollary 4.2:

(1) κ and ν are strictly increasing with respect to their $<_{\mathrm{lex}}$-ordered arguments $\boldsymbol{\alpha}^\frown\beta \in \mathrm{TS}$.
(2) Each branch $\kappa^{\boldsymbol{\alpha}}$ (where $\boldsymbol{\alpha} \in \mathrm{RS}$) and $\nu^{\boldsymbol{\alpha}}$ (where $\boldsymbol{\alpha} \in \mathrm{RS} - \{()\}$) is continuous at arguments $\beta \in \mathbb{L}$.

Proof: This is a consequence of Corollary 3.21. □

We now prepare for the conservative extension of κ and ν to their entire domain as well as the definition of dp which is in accordance with Definition 4.4 of [5].

Definition 4.3: Let $\boldsymbol{\tau} \in \mathrm{RS}$, $\boldsymbol{\tau} = (\tau_1, \ldots, \tau_n)$, $n \geq 0$, $\tau_0 := 1$. The term system $\mathrm{T}^{\boldsymbol{\tau}}$ is obtained from T^{τ_n} by successive substitution of parameters in (τ_i, τ_{i+1}) by their T^{τ_i}-representations, for $i = n-1, \ldots, 1$. The parameters τ_i are represented by the terms $\vartheta^{\tau_i}(0)$. The length $\mathrm{l}^{\boldsymbol{\tau}}(\alpha)$ of a $\mathrm{T}^{\boldsymbol{\tau}}$-term α is defined inductively by

(1) $\mathrm{l}^{\boldsymbol{\tau}}(0) := 0$,
(2) $\mathrm{l}^{\boldsymbol{\tau}}(\beta) := \mathrm{l}^{\boldsymbol{\tau}}(\gamma) + \mathrm{l}^{\boldsymbol{\tau}}(\delta)$ if $\beta =_{\mathrm{NF}} \gamma + \delta$, and
(3) $\mathrm{l}^{\boldsymbol{\tau}}(\vartheta(\eta)) := \begin{cases} 1 & \text{if} \quad \eta = 0 \\ \mathrm{l}^{\boldsymbol{\tau}}(\eta) + 4 & \text{if} \quad \eta > 0 \end{cases}$

where $\vartheta \in \{\vartheta^{\tau_i} \mid 0 \leq i \leq n\} \cup \{\vartheta_{i+1} \mid i \in \mathbb{N}\}$.

Remark. Recall Equation (2.1) as well as Lemma 2.13 and Definitions 2.14 and 2.18 of [9].

(1) For $\beta = \vartheta^{\tau_n}(\Delta + \eta) \in \mathbb{E}$ such that $\beta \leq \mu_{\tau_n}$ we have

$$\mathrm{l}^{\boldsymbol{\tau}}(\Delta) = \mathrm{l}^{\boldsymbol{\tau}^\frown\beta}(\iota_{\tau_n,\beta}(\Delta)) < \mathrm{l}^{\boldsymbol{\tau}}(\beta). \tag{4.1}$$

(2) For $\beta \in \mathrm{T}^{\boldsymbol{\tau}} \cap \mathbb{P}^{>1} \cap \Omega_1$ let $\tau \in \{\tau_0, \ldots, \tau_n\}$ be maximal such that $\tau < \beta$. Clearly,

$$\mathrm{l}^{\boldsymbol{\tau}}(\bar{\beta}) < \mathrm{l}^{\boldsymbol{\tau}}(\beta), \tag{4.2}$$

cf. Subsection 2.3, and

$$\mathrm{l}^{\boldsymbol{\tau}}(\zeta^{\tau}_{\beta}) < \mathrm{l}^{\boldsymbol{\tau}}(\beta), \tag{4.3}$$

In case of $\beta \notin \mathbb{E}$ we have

$$\mathrm{l}^{\boldsymbol{\tau}}(\log(\beta)), \mathrm{l}^{\boldsymbol{\tau}}(\log((1/\tau)\cdot\beta)) < \mathrm{l}^{\boldsymbol{\tau}}(\beta), \tag{4.4}$$

and for $\beta \in \mathbb{E}$ we have

$$\mathrm{l}^{\boldsymbol{\tau}^\frown\beta}(\lambda^{\tau}_{\beta}) < \mathrm{l}^{\boldsymbol{\tau}}(\beta). \tag{4.5}$$

Finally, the definition of the enumeration functions of relativized connectivity components can be completed. This is easily seen to be a sound, elementary recursive definition.

Definition 4.4: (cf. 4.4 of [5]) Let $\boldsymbol{\alpha} \in \mathrm{RS}$ where $\boldsymbol{\alpha} = (\alpha_1, \ldots, \alpha_n)$, $n \geq 0$, and set $\alpha_0 := 1$. We define the functions

$$\kappa^{\boldsymbol{\alpha}}, \mathrm{dp}_{\boldsymbol{\alpha}} : \mathrm{dom}(\kappa^{\boldsymbol{\alpha}}) \to 1^{\infty},$$

where $\mathrm{dom}(\kappa^{\boldsymbol{\alpha}}) := 1^\infty$ if $n = 0$ and $\mathrm{dom}(\kappa^{\boldsymbol{\alpha}}) := [0, \lambda_{\alpha_n}]$ if $n > 0$, simultaneously by recursion on $\mathrm{l}^{\boldsymbol{\alpha}}(\beta)$, extending Definition 4.1. The clauses extending the definition of $\kappa^{\boldsymbol{\alpha}}$ are as follows.

(1) $\kappa^{\boldsymbol{\alpha}}_0 := 0$, $\kappa^{\boldsymbol{\alpha}}_1 := 1$,
(2) $\kappa^{\boldsymbol{\alpha}}_\beta := \kappa^{\boldsymbol{\alpha}}_\gamma + \mathrm{dp}_{\boldsymbol{\alpha}}(\gamma) + \kappa^{\boldsymbol{\alpha}}_\delta$ for $\beta =_{\mathrm{NF}} \gamma + \delta$.

$\mathrm{dp}_{\boldsymbol{\alpha}}$ is defined as follows, using ν as already defined on TS.

(1) $\mathrm{dp}_{\boldsymbol{\alpha}}(0) := 0$, $\mathrm{dp}_{\boldsymbol{\alpha}}(1) := 0$, and $\mathrm{dp}_{\boldsymbol{\alpha}}(\alpha_n) := 0$ in case of $n > 0$,
(2) $\mathrm{dp}_{\boldsymbol{\alpha}}(\beta) := \mathrm{dp}_{\boldsymbol{\alpha}}(\delta)$ if $\beta =_{\mathrm{NF}} \gamma + \delta$,
(3) $\mathrm{dp}_{\boldsymbol{\alpha}}(\beta) := \mathrm{dp}_{\boldsymbol{\alpha}\restriction_{n-1}}(\beta)$ if $n > 0$ for $\beta \in \mathbb{P} \cap (1, \alpha_n)$,
(4) for $\beta \in \mathbb{P}^{>\alpha_n} - \mathbb{E}$ let $\gamma := (1/\alpha_n) \cdot \beta$ and $\log(\gamma) =_{\mathrm{ANF}} \gamma_1 + \dots + \gamma_m$ and set

$$\mathrm{dp}_{\boldsymbol{\alpha}}(\beta) := \kappa^{\boldsymbol{\alpha}}_{\gamma_1} + \mathrm{dp}_{\boldsymbol{\alpha}}(\gamma_1) + \dots + \kappa^{\boldsymbol{\alpha}}_{\gamma_m} + \mathrm{dp}_{\boldsymbol{\alpha}}(\gamma_m),$$

(5) for $\beta \in \mathbb{E}^{>\alpha_n}$ let $\boldsymbol{\gamma} := (\alpha_1, \dots, \alpha_n, \beta)$, and set

$$\mathrm{dp}_{\boldsymbol{\alpha}}(\beta) := \nu^{\boldsymbol{\gamma}}_{\mu^{\alpha_n}_\beta} + \kappa^{\boldsymbol{\gamma}}_{\lambda^{\alpha_n}_\beta} + \mathrm{dp}_{\boldsymbol{\gamma}}(\lambda^{\alpha_n}_\beta).$$

Definition 4.5: (cf. 4.4 of [5]) Let $\boldsymbol{\alpha} \in \mathrm{RS}$ where $\boldsymbol{\alpha} = (\alpha_1, \dots, \alpha_n)$, $n > 0$, and set $\alpha_0 := 1$. We define

$$\nu^{\boldsymbol{\alpha}} : \mathrm{dom}(\nu^{\boldsymbol{\alpha}}) \to 1^\infty$$

where $\mathrm{dom}(\nu^{\boldsymbol{\alpha}}) := [0, \mu_{\alpha_n}]$, extending Definition 4.1 and setting $\alpha := \mathrm{o}(\boldsymbol{\alpha})$, by

(1) $\nu^{\boldsymbol{\alpha}}_0 := \alpha$,
(2) $\nu^{\boldsymbol{\alpha}}_\beta := \nu^{\boldsymbol{\alpha}}_\gamma + \kappa^{\boldsymbol{\alpha}}_{\varrho^{\alpha_n}_\gamma} + \mathrm{dp}_{\boldsymbol{\alpha}}(\varrho^{\alpha_n}_\gamma) + \check{\chi}^{\alpha_n}(\gamma) \cdot \alpha$ if $\beta = \gamma + 1$,
(3) $\nu^{\boldsymbol{\alpha}}_\beta := \nu^{\boldsymbol{\alpha}}_\gamma + \kappa^{\boldsymbol{\alpha}}_{\varrho^{\alpha_n}_\gamma} + \mathrm{dp}_{\boldsymbol{\alpha}}(\varrho^{\alpha_n}_\gamma) + \nu^{\boldsymbol{\alpha}}_\delta$ if $\beta =_{\mathrm{NF}} \gamma + \delta \in \mathrm{Lim}$.

In the sequel we want to establish the results of Lemma 4.5 of [5] for the new definitions within a weak theory, avoiding the long transfinite induction used in the corresponding proof in [5]. Then the agreement of the definitions of κ, dp and ν in [5] and here can be shown in a weak theory as well. This includes also Lemma 4.7 of [5] and extends to the relativization of tracking sequences to contexts as lined out in Definition 4.13 through Lemma 4.17 of [5].

Lemma 4.6: *Let* $\boldsymbol{\alpha} = (\alpha_1, \dots, \alpha_n) \in \mathrm{RS}$ *and set* $\alpha_0 := 1$.

(1) Let $\gamma \in \mathrm{dom}(\kappa^{\boldsymbol{\alpha}}) \cap \mathbb{P}$. *If* $\gamma =_{\mathrm{MNF}} \gamma_1 \cdot \ldots \cdot \gamma_k \geq \alpha_n$, *setting* $\gamma' := (1/\gamma_1) \cdot \gamma$, *we have*

$$(\kappa^{\boldsymbol{\alpha}}_\gamma + \mathrm{dp}_{\boldsymbol{\alpha}}(\gamma)) \cdot \omega = \begin{cases} \mathrm{o}(\boldsymbol{\alpha}) \cdot \gamma' \cdot \omega \text{ if } \gamma_1 = \alpha_n \\ \mathrm{o}(\boldsymbol{\alpha}^\frown\gamma \cdot \omega) \text{ otherwise.} \end{cases}$$

If $\gamma < \alpha_n$ *we have* $(\kappa^{\boldsymbol{\alpha}}_\gamma + \mathrm{dp}_{\boldsymbol{\alpha}}(\gamma)) \cdot \omega < \mathrm{o}(\boldsymbol{\alpha})$.

(2) For $\gamma \in \mathrm{dom}(\kappa^{\boldsymbol{\alpha}}) - (\mathbb{E} \cup \{0\})$ *we have*

$$\mathrm{dp}_{\boldsymbol{\alpha}}(\gamma) < \kappa^{\boldsymbol{\alpha}}_\gamma.$$

(3) For $\gamma \in \mathbb{E}^{>\alpha_n}$ *such that* $\mu_\gamma < \gamma$ *we have*

$$\mathrm{dp}_{\boldsymbol{\alpha}}(\gamma) < \mathrm{o}(\boldsymbol{\alpha}^\frown\gamma) \cdot \mu_\gamma \cdot \omega.$$

(4) For $\gamma \in \mathrm{dom}(\kappa^{\boldsymbol{\alpha}}) \cap \mathbb{E}^{>\alpha_n}$ *we have*

$$\kappa^{\boldsymbol{\alpha}}_\gamma \cdot \omega \leq \mathrm{dp}_{\boldsymbol{\alpha}}(\gamma) \quad \text{and} \quad \mathrm{dp}_{\boldsymbol{\alpha}}(\gamma) \cdot \omega =_{\mathrm{NF}} \mathrm{o}(\mathrm{h}_\omega(\boldsymbol{\alpha}^\frown\gamma)) \cdot \omega.$$

(5) Let $\gamma \in \mathrm{dom}(\nu^{\boldsymbol{\alpha}}) \cap \mathbb{P}$, $\gamma =_{\mathrm{MNF}} \gamma_1 \cdot \ldots \cdot \gamma_k$. *We have*

$$(\nu^{\boldsymbol{\alpha}}_\gamma + \kappa^{\boldsymbol{\alpha}}_{\varrho^{\alpha_n}_\gamma} + \mathrm{dp}_{\boldsymbol{\alpha}}(\varrho^{\alpha_n}_\gamma)) \cdot \omega = \begin{cases} \mathrm{o}(\boldsymbol{\alpha}) \cdot \gamma \cdot \omega & \text{if } \gamma_1 \leq \alpha_n \\ \mathrm{o}(\mathrm{h}_\omega(\boldsymbol{\alpha}^\frown\gamma)) \cdot \omega & \text{if } \gamma \in \mathbb{E}^{>\alpha_n} \\ \mathrm{o}(\boldsymbol{\alpha}^\frown\gamma) \cdot \omega & \text{otherwise.} \end{cases}$$

Proof: The lemma is shown by simultaneous induction on $\mathrm{l}^{\boldsymbol{\alpha}}(\gamma)$ over all parts.

Ad 1. The claim is immediate if $\gamma = \alpha_n$, and if $\gamma_1 = \alpha_n$ and $k > 1$, we have $\kappa^{\boldsymbol{\alpha}}_\gamma = \mathrm{o}(\boldsymbol{\alpha}) \cdot \gamma'$ and by part 2 the claim follows. Now assume that $\gamma_1 > \alpha_n$, whence $\kappa^{\boldsymbol{\alpha}}_\gamma = \mathrm{o}(\boldsymbol{\alpha}^\frown\gamma)$. The case $\gamma \notin \mathbb{E}$ is handled again by part 2. If $\gamma \in \mathbb{E}$, we apply part 4 to see that

$$(\kappa^{\boldsymbol{\alpha}}_\gamma + \mathrm{dp}_{\boldsymbol{\alpha}}(\gamma)) \cdot \omega = \mathrm{dp}_{\boldsymbol{\alpha}}(\gamma) \cdot \omega = \mathrm{o}(\mathrm{h}_\omega(\boldsymbol{\alpha}^\frown\gamma)) \cdot \omega,$$

and since $\mu_\gamma \geq \omega$, the latter is equal to $\mathrm{o}(\boldsymbol{\alpha}^\frown\gamma \cdot \omega)$.

Now consider the situation where $\gamma < \alpha_n$. Let $i \in [0, \ldots, n-1]$ be maximal such that $\alpha_i < \gamma$. The same argument as above yields the corresponding claim for $\boldsymbol{\alpha}_{\restriction_i}$ instead of $\boldsymbol{\alpha}$, and by the monotonicity of o we see that the resulting ordinal is strictly below $\mathrm{o}(\boldsymbol{\alpha}_{\restriction_{i+1}}) \leq \mathrm{o}(\boldsymbol{\alpha})$.

Note that in the case $\gamma \cdot \omega \in \mathrm{dom}(\kappa^{\boldsymbol{\alpha}})$ we have $(\kappa^{\boldsymbol{\alpha}}_\gamma + \mathrm{dp}_{\boldsymbol{\alpha}}(\gamma)) \cdot \omega = \kappa^{\boldsymbol{\alpha}}_{\gamma \cdot \omega}$ as a direct consequence of the definitions.

Ad 2. We may assume that $\gamma > \alpha_n$ (otherwise replace n by the suitable $i < n$ and $\boldsymbol{\alpha}$ by $\boldsymbol{\alpha}_{\restriction_i}$). Set $\gamma' := \log((1/\alpha_n) \cdot \gamma) =_{\mathrm{ANF}} \gamma_1 + \cdots + \gamma_m$, so that $\gamma = \alpha_n \cdot \omega^{\gamma_1 + \cdots + \gamma_m}$ and $\mathrm{dp}_{\boldsymbol{\alpha}}(\gamma) = \sum_{i=1}^m (\kappa^{\boldsymbol{\alpha}}_{\gamma_i} + \mathrm{dp}_{\boldsymbol{\alpha}}(\gamma_i))$. According to the definition, $\kappa^{\boldsymbol{\alpha}}_{\gamma}$ is either $\mathrm{o}(\boldsymbol{\alpha})\cdot\omega^{\gamma_1+\cdots+\gamma_m}$ if $\gamma_1\le\alpha_n$, or $\mathrm{o}(\boldsymbol{\alpha}^\frown\gamma)$ if $\gamma_1>\alpha_n$. In the case $\gamma_i\cdot\omega<\gamma$ for $i=1,\dots,m$ an application of part 1 of the i.h. to the γ_i yields the claim, thanks to the monotonicity of o. Otherwise we must have $\gamma=\gamma_1\cdot\omega$ where $\gamma_1\in\mathbb{E}^{>\alpha_n}$, and applying part 1 of the i.h. to γ_1 yields

$$\mathrm{dp}_{\boldsymbol{\alpha}}(\gamma)=\kappa^{\boldsymbol{\alpha}}_{\gamma_1}+\mathrm{dp}_{\boldsymbol{\alpha}}(\gamma_1)+1<(\kappa^{\boldsymbol{\alpha}}_{\gamma_1}+\mathrm{dp}_{\boldsymbol{\alpha}}(\gamma_1))\cdot\omega=\mathrm{o}(\boldsymbol{\alpha}^\frown\gamma)=\kappa^{\boldsymbol{\alpha}}_{\gamma}.$$

Ad 3. Let $\lambda_\gamma=_{\mathrm{ANF}}\lambda_1+\cdots+\lambda_r$, $\lambda\in\{\lambda_1,\dots,\lambda_r\}$, and note that $\lambda\le\gamma\cdot\mu_\gamma$. Setting $\lambda':=(1/\mathrm{mf}(\lambda))\cdot\lambda$, we have $\lambda'\le\mu_\gamma$ and applying part 1 of the i.h. to λ we obtain $(\kappa^{\boldsymbol{\alpha}}_{\lambda}+\mathrm{dp}_{\boldsymbol{\alpha}}(\lambda))\cdot\omega\le\mathrm{o}(\boldsymbol{\alpha}^\frown\gamma)\cdot\mu_\gamma\cdot\omega$. Since $\nu^{\boldsymbol{\alpha}^\frown\gamma}_{\mu_\gamma}=\mathrm{o}(\boldsymbol{\alpha}^\frown(\gamma,\mu_\gamma))=\mathrm{o}(\boldsymbol{\alpha}^\frown\gamma)\cdot\mu_\gamma$, we obtain the claim.

Ad 4. The inequality is seen by a quick inspection of the respective definitions. We have $\kappa^{\boldsymbol{\alpha}}_{\gamma}=\mathrm{o}(\boldsymbol{\alpha}^\frown\gamma)$, and since $\mu_\gamma\ge\omega$ we obtain

$$\mathrm{o}(\boldsymbol{\alpha}^\frown\gamma)\cdot\omega=\mathrm{o}(\boldsymbol{\alpha}^\frown(\gamma,\omega))\le\mathrm{o}(\boldsymbol{\alpha}^\frown(\gamma,\mu_\gamma))\le\mathrm{dp}_{\boldsymbol{\alpha}}(\gamma).$$

In order to verify the claimed equation, note that

$$\mathrm{h}_\omega(\boldsymbol{\alpha}^\frown\gamma)=\boldsymbol{\alpha}^\frown\mathrm{sk}_\omega(\gamma),$$

where $\mathrm{sk}_\omega(\gamma)=(\delta_1,\dots,\delta_{l+1})$ consists of a maximal strictly increasing chain $\boldsymbol{\delta}:=(\delta_1,\dots,\delta_l)=(\gamma,\mu_\gamma,\mu_{\mu_\gamma},\dots)$ of $\mathbb{E}$-numbers and $\delta_{l+1}=\mu_{\delta_l}\notin\mathbb{E}^{>\delta_l}$. We have $\lambda_{\delta_i}=\varrho_{\mu_{\delta_i}}+\zeta_{\delta_i}$ where $\zeta_{\delta_i}<\delta_i$ and, for $i<l$, $\varrho_{\mu_{\delta_i}}=\mu_{\delta_i}=\delta_{i+1}\in\mathbb{E}^{>\delta_i}$. Applying the i.h. to (the additive decompositions) of these terms λ_{δ_i} we obtain

$$\mathrm{dp}_{\boldsymbol{\alpha}}(\gamma)\cdot\omega=\mathrm{dp}_{\boldsymbol{\alpha}^\frown\boldsymbol{\delta}_{\restriction l-1}}(\delta_l)\cdot\omega=(\nu^{\boldsymbol{\alpha}^\frown\boldsymbol{\delta}}_{\mu_{\delta_l}}+\kappa^{\boldsymbol{\alpha}^\frown\boldsymbol{\delta}}_{\lambda_{\delta_l}}+\mathrm{dp}_{\boldsymbol{\alpha}^\frown\boldsymbol{\delta}}(\lambda_{\delta_l}))\cdot\omega,$$

and consider the additive decomposition of the term λ_{δ_l}. The components below δ_l from ζ_{δ_l} are easily handled using the inequality of part 1 of the i.h., while for $\mu_{\delta_l}=_{\mathrm{MNF}}\mu_1\cdot\ldots\cdot\mu_j$ we have $\varrho_{\mu_{\delta_l}}\le\delta_l\cdot\log(\mu_{\delta_l})=\delta_l\cdot(\log(\mu_1)+\cdots+\log(\mu_j))$ (where the only possible difference is δ_l) and consider the summands separately. Let $\mu\in\{\mu_1,\dots,\mu_j\}$.

Case 1: $\varrho_{\mu_{\delta_l}}\in\mathbb{E}^{>\delta_l}$. Let $\log(\mu_{\delta_l})=\lambda+k$ where $\lambda\in\mathrm{Lim}\cup\{0\}$ and $k<\omega$. We must have $\chi^{\delta_l}(\lambda)=1$, since otherwise $\varrho_{\mu_{\delta_l}}=\mu_{\delta_l}\in\mathbb{E}^{>\delta_l}$, which would contradict the maximality of the length of $\boldsymbol{\delta}$. It follows that $k=1$, hence $\mu_{\delta_l}=\lambda\cdot\omega$, $\lambda=\varrho_{\mu_{\delta_l}}\in\mathbb{E}^{>\delta_l}$, and applying the i.h. to λ yields

$$(\kappa^{\boldsymbol{\alpha}^\frown\boldsymbol{\delta}}_{\lambda}+\mathrm{dp}_{\boldsymbol{\alpha}^\frown\boldsymbol{\delta}}(\lambda))\cdot\omega=\mathrm{o}(\mathrm{h}_\omega(\boldsymbol{\alpha}^\frown\boldsymbol{\delta}^\frown\lambda))\cdot\omega=\mathrm{o}(\boldsymbol{\alpha}^\frown\boldsymbol{\delta}^\frown\lambda\cdot\omega)=\nu^{\boldsymbol{\alpha}^\frown\boldsymbol{\delta}}_{\mu_{\delta_l}},$$

whence $\mathrm{dp}_{\boldsymbol{\alpha}}(\gamma)\cdot\omega=\mathrm{o}(\mathrm{h}_\omega(\boldsymbol{\alpha}^\frown\gamma))\cdot\omega$ as claimed.

Case 2: Otherwise.

Subcase 2.1: $\mu < \delta_l$. Then applying part 2 of the i.h. to $\delta_l \cdot \log(\mu)$ we see that

$$\mathrm{dp}_{\boldsymbol{\alpha}^\frown\boldsymbol{\delta}}(\delta_l \cdot \log(\mu)) < \kappa^{\boldsymbol{\alpha}^\frown\boldsymbol{\delta}}_{\delta_l\cdot\log(\mu)} = \mathrm{o}(\boldsymbol{\alpha}^\frown\boldsymbol{\delta}) \cdot \log(\mu).$$

Subcase 2.2: $\mu = \delta_l$. We calculate $\kappa^{\boldsymbol{\alpha}^\frown\boldsymbol{\delta}}_{\delta_l^2} = \mathrm{o}(\boldsymbol{\alpha}^\frown\boldsymbol{\delta}) \cdot \delta_l$ and $\mathrm{dp}_{\boldsymbol{\alpha}^\frown\boldsymbol{\delta}}(\delta_l^2) = \mathrm{o}(\boldsymbol{\alpha}^\frown\boldsymbol{\delta})$.

Subcase 2.3: $\mu > \delta_l$.

2.3.1: $\mu \notin \mathbb{E}^{>\delta_l}$. Then $\delta_l < \delta_l \cdot \log(\mu) \notin \mathbb{E}^{>\delta_l}$, hence by the i.h., applied to $\delta_l \cdot \log(\mu)$, which is a summand of λ_{δ_l},

$$\mathrm{dp}_{\boldsymbol{\alpha}^\frown\boldsymbol{\delta}}(\delta_l \cdot \log(\mu)) < \kappa^{\boldsymbol{\alpha}^\frown\boldsymbol{\delta}}_{\delta_l\cdot\log(\mu)} \le \nu^{\boldsymbol{\alpha}^\frown\boldsymbol{\delta}}_{\mu} \le \nu^{\boldsymbol{\alpha}^\frown\boldsymbol{\delta}}_{\mu_{\delta_l}}.$$

2.3.2: $\mu \in \mathbb{E}^{>\delta_l}$. Then we have $\delta_l \cdot \log(\mu) = \mu$, $\mu \cdot \omega \le \mu_{\delta_l}$, and applying the i.h. to μ

$$\mathrm{dp}_{\boldsymbol{\alpha}^\frown\boldsymbol{\delta}}(\mu) \cdot \omega = \mathrm{o}(\mathrm{h}_\omega(\boldsymbol{\alpha}^\frown\boldsymbol{\delta}^\frown\mu)) \cdot \omega = \nu^{\boldsymbol{\alpha}^\frown\boldsymbol{\delta}}_{\mu\cdot\omega} \le \nu^{\boldsymbol{\alpha}^\frown\boldsymbol{\delta}}_{\mu_{\delta_l}}.$$

Ad 5. We have $\varrho_\gamma \le \alpha_n \cdot (\log(\gamma_1) + \cdots + \log(\gamma_k))$.

Case 1: $\gamma \in \mathbb{E}^{>\alpha_n}$. Here we have $\varrho_\gamma = \gamma$ and $\nu^{\boldsymbol{\alpha}}_\gamma = \kappa^{\boldsymbol{\alpha}}_\gamma < \mathrm{dp}_{\boldsymbol{\alpha}}(\gamma) = \nu^{\boldsymbol{\alpha}^\frown\gamma}_{\mu_\gamma} + \kappa^{\boldsymbol{\alpha}^\frown\gamma}_{\lambda_\gamma} + \mathrm{dp}_{\boldsymbol{\alpha}^\frown\gamma}(\lambda_\gamma)$, and by part 4 we have $\mathrm{dp}_{\boldsymbol{\alpha}}(\gamma) \cdot \omega = \mathrm{o}(\mathrm{h}_\omega(\boldsymbol{\alpha}^\frown\gamma)) \cdot \omega$.

Case 2: $\gamma_1 \le \alpha_n$. Here we argue similarly as in the proof of part 4, case 2. However, the access to the i.h. is different. Clearly, $\kappa^{\boldsymbol{\alpha}}_{\alpha_n\cdot\log(\gamma_i)} = \mathrm{o}(\boldsymbol{\alpha}) \cdot \log(\gamma_i)$. For given i, let $\log(\log(\gamma_i)) =_{\mathrm{ANF}} \xi_1 + \cdots + \xi_s$. In the case $\gamma_i = \alpha_n$ we have $\mathrm{dp}_{\boldsymbol{\alpha}}(\alpha_n^2) = \mathrm{o}(\boldsymbol{\alpha})$. Now assume that $\gamma_i < \alpha_n$. An application of part 1 of the i.h. to ξ_j yields $(\kappa^{\boldsymbol{\alpha}}_{\xi_j} + \mathrm{dp}_{\boldsymbol{\alpha}}(\xi_j)) \cdot \omega < \mathrm{o}(\boldsymbol{\alpha})$ for $j = 1, \ldots, s$. Therefore

$$(\kappa^{\boldsymbol{\alpha}}_{\alpha_n\cdot\log(\gamma_i)} + \mathrm{dp}_{\boldsymbol{\alpha}}(\alpha_n \cdot \log(\gamma_i))) \cdot \omega = \kappa^{\boldsymbol{\alpha}}_{\alpha_n\cdot\log(\gamma_i)} \cdot \omega = \mathrm{o}(\boldsymbol{\alpha}) \cdot \log(\gamma_i) \cdot \omega.$$

These considerations show that we obtain

$$(\nu^{\boldsymbol{\alpha}}_\gamma + \kappa^{\boldsymbol{\alpha}}_{\varrho^{\alpha_n}_\gamma} + \mathrm{dp}_{\boldsymbol{\alpha}}(\varrho^{\alpha_n}_\gamma)) \cdot \omega = \nu^{\boldsymbol{\alpha}}_\gamma \cdot \omega = \mathrm{o}(\boldsymbol{\alpha}) \cdot \gamma \cdot \omega.$$

Case 3: Otherwise.

Subcase 3.1: $\varrho_\gamma \in \mathbb{E}^{>\alpha_n}$. Since $\gamma \notin \mathbb{E}^{>\alpha_n}$ we have $\log(\gamma) = \lambda + 1$ where $\lambda \in \mathbb{E}^{>\alpha_n}$ and $\chi^{\alpha_n}(\lambda) = 1$. Hence $\gamma = \lambda \cdot \omega$ and $\varrho_\gamma = \lambda$, for which part 4 of the i.h. yields

$$(\kappa^{\boldsymbol{\alpha}}_\lambda + \mathrm{dp}_{\boldsymbol{\alpha}}(\lambda)) \cdot \omega = \mathrm{dp}_{\boldsymbol{\alpha}}(\lambda) \cdot \omega = \mathrm{o}(\mathrm{h}_\omega(\boldsymbol{\alpha}^\frown\lambda)) \cdot \omega = \mathrm{o}(\boldsymbol{\alpha}^\frown\gamma),$$

implying the claim.

Subcase 3.2: $\varrho_\gamma \notin \mathbb{E}^{>\alpha_n}$. Here we extend the argumentation from case 2, where the situation $\gamma_i \le \alpha_n$ has been resolved. In the case $\gamma_i \in \mathbb{E}^{>\alpha_n}$ we have $\gamma_i \cdot \omega \le \gamma$ and apply part 4 of the i.h. to γ_i to obtain

$$(\kappa^{\boldsymbol{\alpha}}_{\alpha_n \cdot \log(\gamma_i)} + \mathrm{dp}_{\boldsymbol{\alpha}}(\alpha_n \cdot \log(\gamma_i))) \cdot \omega = \mathrm{dp}_{\boldsymbol{\alpha}}(\gamma_i) \cdot \omega = \mathrm{o}(\boldsymbol{\alpha}^\frown \gamma_i \cdot \omega) \le \mathrm{o}(\boldsymbol{\alpha}^\frown \gamma).$$

We are left with the cases where $\gamma_i \in \mathbb{M}^{>\alpha_n} - \mathbb{E}$. Writing $\log(\log(\gamma_i)) =_{\mathrm{ANF}} \xi_1 + \cdots + \xi_s$, which resides in $(\alpha_n, \log(\gamma_i))$, we have $\mathrm{dp}_{\boldsymbol{\alpha}}(\alpha_n \cdot \log(\gamma_i)) = \sum_{j=1}^{s}(\kappa^{\boldsymbol{\alpha}}_{\xi_j} + \mathrm{dp}_{\boldsymbol{\alpha}}(\xi_j))$, where for each j part 1 of the i.h. applied to ξ_j together with the monotonicity of $\kappa^{\boldsymbol{\alpha}}$ on additive principal arguments yield

$$(\kappa^{\boldsymbol{\alpha}}_{\xi_j} + \mathrm{dp}_{\boldsymbol{\alpha}}(\xi_j)) \cdot \omega = \kappa^{\boldsymbol{\alpha}}_{\xi_j \cdot \omega} \le \kappa^{\boldsymbol{\alpha}}_{\alpha_n \cdot \log(\gamma_i)} < \nu^{\boldsymbol{\alpha}}_{\gamma},$$

and we conclude as in case 2. □

Corollary 4.7: *Let $\boldsymbol{\alpha} \in \mathrm{RS}$. We have*

(1) $\kappa^{\boldsymbol{\alpha}}_{\gamma \cdot \omega} = (\kappa^{\boldsymbol{\alpha}}_{\gamma} + \mathrm{dp}_{\boldsymbol{\alpha}}(\gamma)) \cdot \omega$ *for* $\gamma \in \mathbb{P}$ *such that* $\gamma \cdot \omega \in \mathrm{dom}(\kappa^{\boldsymbol{\alpha}})$.
(2) $\nu^{\boldsymbol{\alpha}}_{\gamma \cdot \omega} = (\nu^{\boldsymbol{\alpha}}_{\gamma} + \kappa^{\boldsymbol{\alpha}}_{\varrho^{\alpha_n}_{\gamma}} + \mathrm{dp}_{\boldsymbol{\alpha}}(\varrho^{\alpha_n}_{\gamma})) \cdot \omega$ *for* $\gamma \in \mathbb{P}$ *such that* $\gamma \cdot \omega \in \mathrm{dom}(\nu^{\boldsymbol{\alpha}})$.

$\kappa^{\boldsymbol{\alpha}}$ and for $\boldsymbol{\alpha} \ne ()$ also $\nu^{\boldsymbol{\alpha}}$ are strictly monotonically increasing and continuous.

Proof: Parts 1 and 2 follow from Definitions 3.14 and 4.1 using parts 1 and 5 of Lemma 4.6, respectively. In order to see general monotonicity and continuity we can build upon Corollary 4.2. The missing argument is as follows. For any $\beta \in \mathbb{P}$ in the respective domain and any $\gamma =_{\mathrm{ANF}} \gamma_1 + \cdots + \gamma_m < \beta$ we have $\kappa^{\boldsymbol{\alpha}}_{\gamma} < \kappa^{\boldsymbol{\alpha}}_{\beta}$ using part 1, and $\nu^{\boldsymbol{\alpha}}_{\gamma} < \nu^{\boldsymbol{\alpha}}_{\beta}$ using part 2, since $\gamma_i \cdot \omega \le \beta$ for $i = 1, \ldots, m$. □

Theorem 4.8: *Let $\boldsymbol{\alpha} = (\alpha_1, \ldots, \alpha_n) \in \mathrm{RS}$, $n \ge 0$, and set $\alpha_0 := 1$. For $\beta \in \mathbb{P}$ let $\delta := (1/\bar{\beta}) \cdot \beta$, so that $\beta =_{\mathrm{NF}} \bar{\beta} \cdot \delta$ if $\beta \notin \mathbb{M}$.*

(1) For all $\beta \in \mathrm{dom}(\kappa^{\boldsymbol{\alpha}}) \cap \mathbb{P}^{>\alpha_n}$ we have

$$\kappa^{\boldsymbol{\alpha}}_{\beta} = \kappa^{\boldsymbol{\alpha}}_{\bar{\beta}+1} \cdot \delta.$$

(2) For all $\beta \in \mathrm{dom}(\nu^{\boldsymbol{\alpha}}) \cap \mathbb{P}^{>\alpha_n}$ (where $n > 0$) we have

$$\nu^{\boldsymbol{\alpha}}_{\beta} = \nu^{\boldsymbol{\alpha}}_{\bar{\beta}+1} \cdot \delta.$$

Hence, the definitions of κ, ν, and dp *given in* [5] *and here fully agree.*

Proof: We rely on the monotonicity of o. Note that Corollary 4.7 has already shown the theorem for β of the form $\gamma \cdot \omega$, i.e. successors of additive principal numbers. Let $\beta =_{\mathrm{MNF}} \beta_1 \cdot \ldots \cdot \beta_k \in \mathbb{P}^{>\alpha_n}$.

Case 1: $k = 1$. Then $\delta = \beta$, and setting $n_0 := n_0(\boldsymbol{\alpha}^\frown\beta)$ and $\gamma := \gamma(\boldsymbol{\alpha}^\frown\beta)$ according to Definition 3.14, we have

$$\kappa^{\boldsymbol{\alpha}}_{\beta} = \nu^{\boldsymbol{\alpha}}_{\beta} = \mathrm{o}(\boldsymbol{\alpha}^\frown\beta) = \begin{cases} \beta & \text{if } n_0 = 0 \\ \mathrm{o}(\boldsymbol{\alpha}') \cdot \beta & \text{if } n_0 > 0, \end{cases}$$

where $\boldsymbol{\alpha}' := \mathrm{h}_\beta(\boldsymbol{\alpha}_{\restriction n_0-1}{}^\frown\gamma)$, and parts 1 and 5 of Lemma 4.6 yield

$$\kappa^{\boldsymbol{\alpha}}_{\bar{\beta}+1} \cdot \beta = \nu^{\boldsymbol{\alpha}}_{\bar{\beta}+1} \cdot \beta = \begin{cases} \mathrm{o}(\boldsymbol{\alpha}) \cdot \beta & \text{if } \bar{\beta} = \alpha_n \\ \mathrm{o}(\mathrm{h}_\omega(\boldsymbol{\alpha}^\frown\bar{\beta})) \cdot \beta & \text{if } \bar{\beta} > \alpha_n. \end{cases}$$

If $n_0 = 0$ the claim is immediate since $\mathrm{o}(\boldsymbol{\alpha}) \cdot \beta = \beta$ if $\bar{\beta} = \alpha_n$ and $1 < \mathrm{o}(\mathrm{h}_\omega(\boldsymbol{\alpha}^\frown\bar{\beta})) < \mathrm{o}(\boldsymbol{\alpha}^\frown\beta) = \beta$ if $\bar{\beta} > \alpha_n$. Now assume that $n_0 > 0$.

Subcase 1.1: $\bar{\beta} = \alpha_n$. This implies $n_0 \le n$ and therefore $\boldsymbol{\alpha}' <_{\mathrm{lex}} \boldsymbol{\alpha} <_{\mathrm{lex}} \boldsymbol{\alpha}^\frown\beta$. By the monotonicity of o we obtain

$$\mathrm{o}(\boldsymbol{\alpha}') < \mathrm{o}(\boldsymbol{\alpha}^\frown\beta) = \mathrm{o}(\boldsymbol{\alpha}') \cdot \beta,$$

which implies the claim since $\beta \in \mathbb{M}$.

Subcase 1.2: $\bar{\beta} > \alpha_n$. This implies $\bar{\beta} \in \mathbb{E}^{>\alpha_n}$ and $\kappa^{\boldsymbol{\alpha}}_{\bar{\beta}+1} \cdot \beta = \nu^{\boldsymbol{\alpha}}_{\bar{\beta}+1} \cdot \beta = \mathrm{o}(\mathrm{h}_\omega(\boldsymbol{\alpha}^\frown\bar{\beta})) \cdot \beta$, where $\mathrm{h}_\omega(\boldsymbol{\alpha}^\frown\bar{\beta}) <_{\mathrm{lex}} \boldsymbol{\alpha}^\frown\beta$.

1.2.1: $n_0 \le n$. Then we obtain $\boldsymbol{\alpha}' <_{\mathrm{lex}} \boldsymbol{\alpha} <_{\mathrm{lex}} \mathrm{h}_\omega(\boldsymbol{\alpha}^\frown\bar{\beta}) <_{\mathrm{lex}} \boldsymbol{\alpha}^\frown\beta$ and hence

$$\mathrm{o}(\boldsymbol{\alpha}') < \mathrm{o}(\mathrm{h}_\omega(\boldsymbol{\alpha}^\frown\bar{\beta})) < \mathrm{o}(\boldsymbol{\alpha}^\frown\beta) = \mathrm{o}(\boldsymbol{\alpha}') \cdot \beta,$$

which implies the claim.

1.2.2: $n_0 = n+1$. Then we have $\gamma \le \bar{\beta}$, $\boldsymbol{\alpha}' = \mathrm{h}_\beta(\boldsymbol{\alpha}^\frown\gamma)$, and using Corollary 3.13 it follows that

$$\kappa^{\boldsymbol{\alpha}}_{\beta} = \nu^{\boldsymbol{\alpha}}_{\beta} = \mathrm{o}(\boldsymbol{\alpha}^\frown\beta) = \mathrm{o}(\boldsymbol{\alpha}') \cdot \beta > \mathrm{o}(\mathrm{h}_\omega(\boldsymbol{\alpha}^\frown\bar{\beta})) \ge \mathrm{o}(\mathrm{h}_\beta(\boldsymbol{\alpha}^\frown\gamma)),$$

which again implies the claim.

Case 2: $k > 1$. Then we have $\bar{\beta} =_{\mathrm{MNF}} \beta_1 \cdot \ldots \cdot \beta_{k-1} \ge \alpha_n$, $\delta = \beta_k$, and set $\beta' := (1/\beta_1) \cdot \beta$ and $\bar{\beta}' := (1/\beta_1) \cdot \bar{\beta}$.

Subcase 2.1: $\beta_1 = \alpha_n$. Using Lemma 4.6 we obtain

$$\kappa^{\boldsymbol{\alpha}}_{\bar{\beta}+1} \cdot \delta = \mathrm{o}(\boldsymbol{\alpha}) \cdot \beta' = \kappa^{\boldsymbol{\alpha}}_{\beta}$$

and

$$\nu^{\boldsymbol{\alpha}}_{\bar{\beta}+1} \cdot \delta = \mathrm{o}(\boldsymbol{\alpha}) \cdot \beta = \nu^{\boldsymbol{\alpha}}_{\beta}.$$

Subcase 2.2: $\beta_1 > \alpha_n$. Then we have $\kappa^{\boldsymbol{\alpha}}_{\beta} = \nu^{\boldsymbol{\alpha}}_{\beta}$.

2.2.1: $\bar{\beta} \notin \mathbb{E}^{>\alpha_n}$. Then by the involved definitions

$$\kappa^{\boldsymbol{\alpha}}_{\bar{\beta}+1} \cdot \delta = \nu^{\boldsymbol{\alpha}}_{\bar{\beta}+1} \cdot \delta = \mathrm{o}(\boldsymbol{\alpha}^\frown\bar{\beta}) \cdot \delta = \nu^{\boldsymbol{\alpha}}_{\beta} = \kappa^{\boldsymbol{\alpha}}_{\beta}.$$

2.2.2: $\bar{\beta} \in \mathbb{E}^{>\alpha_n}$. This implies $k = 2$, $\delta = \beta_2$, and we see that

$$\kappa^{\boldsymbol{\alpha}}_{\bar{\beta}+1} \cdot \delta = \nu^{\boldsymbol{\alpha}}_{\bar{\beta}+1} \cdot \delta = \mathrm{o}(\mathrm{h}_\omega(\boldsymbol{\alpha}^\frown\bar{\beta})) \cdot \delta.$$

In the case $\delta > \mu_{\beta_1}$ we have $\mathrm{h}_\delta(\boldsymbol{\alpha}^\frown\bar{\beta}) = \boldsymbol{\alpha}^\frown\bar{\beta}$ and hence obtain uniformly

$$\kappa^{\boldsymbol{\alpha}}_{\beta} = \nu^{\boldsymbol{\alpha}}_{\beta} = \mathrm{o}(\mathrm{h}_\delta(\boldsymbol{\alpha}^\frown\bar{\beta})) \cdot \delta.$$

By Corollary 3.12 we have $\mathrm{h}_\delta(\boldsymbol{\alpha}^\frown\bar{\beta}) \le_{\mathrm{lex}} \mathrm{h}_\omega(\boldsymbol{\alpha}^\frown\bar{\beta})$, hence

$$\mathrm{o}(\mathrm{h}_\delta(\boldsymbol{\alpha}^\frown\bar{\beta})) \le \mathrm{o}(\mathrm{h}_\omega(\boldsymbol{\alpha}^\frown\bar{\beta})) < \mathrm{o}(\boldsymbol{\alpha}^\frown\beta) = \mathrm{o}(\mathrm{h}_\delta(\boldsymbol{\alpha}^\frown\bar{\beta})) \cdot \delta,$$

which implies the claim since $\delta \in \mathbb{M}$. □

Lemma 4.9: *Let* $\boldsymbol{\alpha} = (\alpha_1, \ldots, \alpha_n) \in \mathrm{RS}$, $n > 0$.

(1) For all β *such that* $\boldsymbol{\alpha}^\frown\beta \in \mathrm{TS}$ *we have*

$$\mathrm{o}(\boldsymbol{\alpha}^\frown\beta) < \mathrm{o}(\boldsymbol{\alpha}) \cdot \widehat{\alpha_n}.$$

(2) For all γ *such that* $\boldsymbol{\alpha}^\frown\gamma \in \mathrm{RS}$ *we have*

$$\mathrm{o}(\mathrm{h}_\omega(\boldsymbol{\alpha}^\frown\gamma)) < \mathrm{o}(\boldsymbol{\alpha}^\frown\gamma) \cdot \widehat{\gamma}.$$

Proof: We prove the lemma by simultaneous induction on $\mathrm{lSeq}(\boldsymbol{\alpha}^\frown\beta)$ and $\mathrm{lSeq}(\mathrm{h}_\omega(\boldsymbol{\alpha}^\frown\gamma))$, respectively.

Ad 1. Let $\beta =_{\mathrm{MNF}} \beta_1 \cdot \ldots \cdot \beta_k$ and $\beta' := (1/\beta_1) \cdot \beta$. Note that $\beta \le \mu_{\alpha_n} < \widehat{\alpha}$.

Case 1: $\beta_1 \le \alpha_n$. Immediate, since $\mathrm{o}(\boldsymbol{\alpha}^\frown\beta) = \mathrm{o}(\boldsymbol{\alpha}) \cdot \beta$.

Case 2: $k > 1$ where $\beta_1 \in \mathbb{E}^{>\alpha_n}$ and $\beta_2 \le \mu_{\beta_1}$. Then we have $\mathrm{o}(\boldsymbol{\alpha}^\frown\beta) = \mathrm{o}(\mathrm{h}_{\beta_2}(\boldsymbol{\alpha}^\frown\beta_1)) \cdot \beta'$ and apply the i.h. to $\boldsymbol{\alpha}^\frown\beta_1$, which clearly satisfies $\mathrm{lSeq}(\boldsymbol{\alpha}^\frown\beta_1) <_{\mathrm{lex}} \mathrm{lSeq}(\boldsymbol{\alpha}^\frown\beta)$. By Corollaries 3.13 and 3.21 we have $\mathrm{o}(\mathrm{h}_{\beta_2}(\boldsymbol{\alpha}^\frown\beta_1)) \le \mathrm{o}(\mathrm{h}_\omega(\boldsymbol{\alpha}^\frown\beta_1))$, and by the i.h., parts 1 (for $\boldsymbol{\alpha}^\frown\beta_1$) and 2 (for $\mathrm{h}_\omega(\boldsymbol{\alpha}^\frown\beta_1)$) we obtain

$$\mathrm{o}(\mathrm{h}_\omega(\boldsymbol{\alpha}^\frown\beta_1)) < \mathrm{o}(\boldsymbol{\alpha}^\frown\beta_1) \cdot \widehat{\beta_1} < \mathrm{o}(\boldsymbol{\alpha}) \cdot \widehat{\alpha_n},$$

where we have used that $\widehat{\beta_1} \le \widehat{\alpha_n}$ according to Lemma 3.17 of [5]. This implies the desired inequality.

Case 3: Otherwise. Let $n_0 := n_0(\boldsymbol{\alpha}^\frown\beta)$ and $\gamma := \gamma(\boldsymbol{\alpha}^\frown\beta)$ according to

Definition 3.14.

Subcase 3.1: $n_0 = 0$. Immediate.

Subcase 3.2: $n_0 > 0$. By definition we have $\mathrm{o}(\boldsymbol{\alpha}^\frown\beta) = \mathrm{o}(\mathrm{h}_{\beta_1}(\boldsymbol{\alpha}_{\restriction_{n_0-1}}{}^\frown\gamma))\cdot\beta$.

3.2.1: $n_0 \leq n$. The monotonicity of o then yields $\mathrm{o}(\boldsymbol{\alpha}^\frown\beta) \leq \mathrm{o}(\boldsymbol{\alpha})\cdot\beta < \mathrm{o}(\boldsymbol{\alpha})\cdot\widehat{\alpha_n}$.

3.2.2: $n_0 = n+1$. Then γ is the immediate predecessor of β in $\mathrm{ts}^{\alpha_n}(\beta)$. We apply the i.h. for parts 1 (to $\boldsymbol{\alpha}^\frown\gamma$) and 2 (to $\mathrm{h}_\omega(\boldsymbol{\alpha}^\frown\gamma)$) and argue as in case 2 to see that $\mathrm{o}(\boldsymbol{\alpha}^\frown\beta) = \mathrm{o}(\mathrm{h}_{\beta_1}(\boldsymbol{\alpha}^\frown\gamma))\cdot\beta < \mathrm{o}(\boldsymbol{\alpha})\cdot\widehat{\alpha_n}$.

Ad 2. We have $\mathrm{h}_\omega(\boldsymbol{\alpha}^\frown\gamma) = \boldsymbol{\alpha}^\frown\mathrm{sk}_\omega(\gamma)$, and setting $\mathrm{sk}_\omega(\gamma) =: (\delta_1,\ldots,\delta_{l+1})$ we obtain a strictly increasing sequence of $\mathbb{E}$-numbers $\boldsymbol{\delta} := (\delta_1,\ldots,\delta_l) = (\gamma,\mu_\gamma,\mu_{\mu_\gamma},\ldots)$ such that $\delta_{l+1} = \mu_{\delta_l} \notin \mathbb{E}^{>\delta_l}$. For $i = 1,\ldots,l$ we have

$$\mathrm{h}_\omega(\boldsymbol{\alpha}^\frown\boldsymbol{\delta}_{\restriction_i}) = \boldsymbol{\alpha}^\frown\boldsymbol{\delta}^\frown\delta_{l+1},$$

$$\mathrm{lSeq}(\boldsymbol{\alpha}^\frown\boldsymbol{\delta}_{\restriction_i}) \leq_{\mathrm{lex}} \mathrm{lSeq}(\mathrm{h}_\omega(\boldsymbol{\alpha}^\frown\gamma)),$$

and the i.h., part 1 (up to $\mathrm{h}_\omega(\boldsymbol{\alpha}^\frown\gamma)$), yields

$$\mathrm{o}(\boldsymbol{\alpha}^\frown\boldsymbol{\delta}_{\restriction_{i+1}}) < \mathrm{o}(\boldsymbol{\alpha}^\frown\boldsymbol{\delta}_{\restriction_i})\cdot\widehat{\delta_i}.$$

Appealing to Lemma 3.17 of [5] we obtain

$$\widehat{\delta_l} \leq \widehat{\delta_{l-1}} \leq \cdots \leq \widehat{\delta_1} = \widehat{\gamma},$$

and finally conclude that $\mathrm{o}(\mathrm{h}_\omega(\boldsymbol{\alpha}^\frown\gamma)) < \mathrm{o}(\boldsymbol{\alpha}^\frown\gamma)\cdot\widehat{\gamma}$. □

Corollary 4.10: *For all* $\boldsymbol{\alpha}^\frown\gamma \in \mathrm{RS}$ *the ordinal* $\mathrm{o}(\boldsymbol{\alpha}^\frown\gamma)\cdot\widehat{\gamma}$ *is a strict upper bound of*

$$\mathrm{Im}(\kappa^{\boldsymbol{\alpha}^\frown\gamma}),\ \mathrm{Im}(\nu^{\boldsymbol{\alpha}^\frown\gamma}),\ \mathrm{dp}_{\boldsymbol{\alpha}}(\gamma),\ \textit{and}\ \nu^{\boldsymbol{\alpha}^\frown\gamma}_{\mu_\gamma} + \kappa^{\boldsymbol{\alpha}^\frown\gamma}_{\lambda_\gamma} + \mathrm{dp}_{\boldsymbol{\alpha}^\frown\gamma}(\lambda_\gamma).$$

Proof: This directly follows from Lemmas 4.6 and 4.9. □

5. Revisiting Tracking Chains

5.1. *Preliminary remarks*

Our preparations in the previous sections are almost sufficient to demonstrate that the characterization of $\mathcal{C}_2$ provided in Section 7 of [5] is elementary recursive. We first provide a brief argumentation based on the previous sections showing that the structure $\mathcal{C}_2$ is elementary recursive. In the following subsections we will elaborate on the characterization of $\le_1$ and $\le_2$ within $\mathcal{C}_2$, further illuminating the structure.

In Section 5 of [5] the termination of the process of maximal extension (see Definition 5.2 of [5]) is seen when applying the 1^τ-measure from the second step on, as clause 2.3.1 of Definition 5.2 in [5] can only be applied at the beginning of the process of maximal extension. Lemma 5.4, part a), of [5] is not needed in full generality, the 1^τ-measure suffices, cf. Lemma 5.5 of [5]. The proof of Lemma 5.12 of [5], parts a), b), and c), actually proceeds by induction on the number of 1-step extensions, an "induction on $\mathrm{cs}'(\boldsymbol{\alpha})$ along $<_{\mathrm{lex}}$" is not needed.

In Section 6 of [5] the proof of Lemma 6.2 actually proceeds by induction on the length of the additive decomposition of α. Definition 6.1 of [5], which assigns to each $\alpha < 1^\infty$ its unique tracking chain $\mathrm{tc}(\alpha)$, involves the evaluation function o (in the guise of $\bar{\cdot}$, cf. Definition 5.9 of [5]), which we have shown to be elementary recursive.

5.2. *Pre-closed and spanning sets of tracking chains*

For the formal definition of tracking chains recall Definition 5.1 of [5]. We will rely on its detailed terminology, including the notation $\boldsymbol{\alpha} \subseteq \boldsymbol{\beta}$ when $\boldsymbol{\alpha}$ is an initial chain of $\boldsymbol{\beta}$.

The following definitions of pre-closed and spanning sets of tracking chains provide a generalization of the notion of maximal extension, denoted by me, cf. Definition 5.2 of [5].

Definition 5.1: (Pre-closedness) Let $M \subseteq_{\mathrm{fin}} \mathrm{TC}$. M is *pre-closed* if and only if M

(1) is *closed under initial chains:* if $\boldsymbol{\alpha} \in M$ and $(i,j) \in \mathrm{dom}(\boldsymbol{\alpha})$ then $\boldsymbol{\alpha}_{\restriction(i,j)} \in M$,

(2) is *ν-index closed:* if $\boldsymbol{\alpha} \in M$, $m_n > 1$, $\alpha_{n,m_n} =_{\mathrm{ANF}} \xi_1 + \cdots + \xi_k$ then $\boldsymbol{\alpha}[\mu_{\tau'}], \boldsymbol{\alpha}[\xi_1 + \cdots + \xi_l] \in M$ for $1 \le l \le k$,

(3) *unfolds minor $\le_2$-components:* if $\boldsymbol{\alpha} \in M$, $m_n > 1$, and $\tau < \mu_{\tau'}$ then:

  3.1. $\boldsymbol{\alpha}_{\restriction n-1}{}^\frown(\alpha_{n,1}, \ldots, \alpha_{n,m_n}, \mu_\tau) \in M$ in the case $\tau \in \mathbb{E}^{>\tau'}$, and
  3.2. otherwise $\boldsymbol{\alpha}^\frown(\varrho^{\tau'}_\tau) \in M$, provided that $\varrho^{\tau'}_\tau > 0$,

(4) is *κ-index closed:* if $\boldsymbol{\alpha} \in M$, $m_n = 1$, and $\alpha_{n,1} =_{\mathrm{ANF}} \xi_1 + \cdots + \xi_k$, then:

  4.1. if $m_{n-1} > 1$ and $\xi_1 = \tau_{n-1,m_{n-1}} \in \mathbb{E}^{>\tau_{n-1,m_{n-1}-1}}$ then $\boldsymbol{\alpha}_{\restriction n-2}{}^\frown(\alpha_{n-1,1}, \ldots, \alpha_{n-1,m_{n-1}}, \mu_{\xi_1}) \in M$, else $\boldsymbol{\alpha}_{\restriction n-1}{}^\frown(\xi_1) \in M$, and

4.2. $\boldsymbol{\alpha}_{\restriction_{n-1}}{}^\frown(\xi_1 + \cdots + \xi_l) \in M$ for $l = 2, \ldots, k$,

(5) *maximizes* me-μ-*chains:* if $\boldsymbol{\alpha} \in M$, $m_n \geq 1$, and $\tau \in \mathbb{E}^{>\tau'}$, then:

5.1. if $m_n = 1$ then $\boldsymbol{\alpha}_{\restriction_{n-1}}{}^\frown(\alpha_{n,1}, \mu_\tau) \in M$, and

5.2. if $m_n > 1$ and $\tau = \mu_{\tau'} = \lambda_{\tau'}$ then $\boldsymbol{\alpha}_{\restriction_{n-1}}{}^\frown(\alpha_{n,1}, \ldots, \alpha_{n,m_n}, \mu_\tau) \in M$.

Remark. Pre-closure of some $M \subseteq_{\text{fin}}$ TC is obtained by closing under clauses 1–5 in this order once, hence finite: in clause 5 note that μ-chains are finite since the ht-measure of terms strictly decreases with each application of μ. Note further that intermediate indices are of the form $\lambda_{\tau'}$, whence we have a decreasing l-measure according to inequality 4.5 in the remark following Definition 4.3.

Definition 5.2: (Spanning sets of tracking chains) $M \subseteq_{\text{fin}}$ TC is *spanning* if and only if it is pre-closed and closed under

(6) *unfolding of* $\leq_1$-*components:* for $\boldsymbol{\alpha} \in M$, if $m_n = 1$ and $\tau \notin \mathbb{E}^{\geq\tau'}$ (i.e. $\tau = \tau_{n,m_n} \notin \mathbb{E}_1$, $\tau' = \tau_n^\star$), let

$$\log((1/\tau') \cdot \tau) =_{\text{ANF}} \xi_1 + \cdots + \xi_k,$$

if otherwise $m_n > 1$ and $\tau = \mu_{\tau'}$ such that $\tau < \lambda_{\tau'}$ in the case $\tau \in \mathbb{E}^{>\tau'}$, let

$$\lambda_{\tau'} =_{\text{ANF}} \xi_1 + \cdots + \xi_k.$$

Set $\xi := \xi_1 + \cdots + \xi_k$, unless $\boldsymbol{\alpha}^\frown(\xi_1 + \cdots + \xi_k) \notin$ TC, in which case we set $\xi := \xi_1 + \cdots + \xi_{k-1}$. Then the closure of $\{\boldsymbol{\alpha}^\frown(\xi)\}$ under clauses 4 and 5 is contained in M, provided that $\xi \neq 0$.

Remark. Closure of some $M \subseteq_{\text{fin}}$ TC under clauses 1–6 is a finite process since pre-closure is finite and since the κ-indices added in clause 6 strictly decrease in l-measure.

Semantically, the above notion of spanning sets of tracking chains and closure under clauses 1–6 leaves some redundancy in the form that certain κ-indices could be omitted. This will be adressed elsewhere, since the current formulation is advantageous for technical reasons.

Definition 5.3: (Relativization) Let $\boldsymbol{\alpha} \in \text{TC} \cup \{()\}$ and $M \subseteq_{\text{fin}}$ TC be a set of tracking chains that properly extend $\boldsymbol{\alpha}$. M is *pre-closed above* $\boldsymbol{\alpha}$ if and only if it is pre-closed with the modification that clauses 1–5 only apply when the respective resulting tracking chains $\boldsymbol{\beta}$ properly extend $\boldsymbol{\alpha}$. M is *weakly spanning above* $\boldsymbol{\alpha}$ if and only if M is pre-closed above $\boldsymbol{\alpha}$ and closed under clause 6.

Lemma 5.4: *If M is spanning (weakly spanning above some $\boldsymbol{\alpha}$), then it is closed under* me *(closed under* me *for proper extensions of $\boldsymbol{\alpha}$).*

Proof: This follows directly from the definitions involved. □

5.3. *Characterizing $\leq_1$ and $\leq_2$ in $\mathcal{C}_2$*

The purpose of this section is to provide a detailed picture of the restriction of $\mathcal{R}_2$ to 1^∞ on the basis of the results of [5] and to conclude with the extraction of an elementary recursive arithmetical characterization of this structure given in terms of tracking chains which we will refer to as $\mathcal{C}_2$.

We begin with a few observations that follow from the results in Section 7 of [5] and explain the concept of tracking chains. The evaluations of all initial chains of some tracking chain $\boldsymbol{\alpha} \in \mathrm{TC}$ form a $<_1$-chain. Evaluations of initial chains $\boldsymbol{\alpha}_{\restriction_{i,j}}$ where $(i,j) \in \mathrm{dom}(\boldsymbol{\alpha})$ and $j = 2, \ldots, m_i$ with fixed index i form $<_2$-chains. Recall that indices $\alpha_{i,j}$ are κ-indices for $j = 1$ and ν-indices otherwise, cf. Definitions 5.1 and 5.9 of [5].

According to Theorem 7.9 of [5], an ordinal $\alpha < 1^\infty$ is $\leq_1$-minimal if and only if its tracking chain consists of a single κ-index, i.e. if its tracking chain $\boldsymbol{\alpha}$ satisfies $(n, m_n) = (1,1)$. Clearly, the least $\leq_1$-predecessor of any ordinal $\alpha < 1^\infty$ with tracking chain $\boldsymbol{\alpha}$ is $\mathrm{o}(\boldsymbol{\alpha}_{\restriction_{1,1}}) = \kappa_{\alpha_{1,1}}$. According to Corollary 7.11 of [5] the ordinal 1^∞ is $\leq_1$-minimal. An ordinal $\alpha > 0$ has a non-trivial $\leq_1$-reach if and only if $\tau_{n,1} > \tau_n^\star$, hence in particular when $m_n > 1$, cf. condition 2 in Definition 5.1 of [5].

We now turn to a characterization of the *greatest immediate* $\leq_1$-*successor*, $\mathrm{gs}(\alpha)$, of an ordinal $\alpha < 1^\infty$ with tracking chain $\boldsymbol{\alpha}$. Recall the notations ρ_i and $\boldsymbol{\alpha}[\xi]$ from Definition 5.1 of [5]. The largest α-$\leq_1$-minimal ordinal is the root of the λth α-$\leq_1$-component for $\lambda := \rho_n \dot{-} 1$. Therefore, if α has a non-trivial $\leq_1$-reach, its greatest immediate $\leq_1$-successor $\mathrm{gs}(\alpha)$ has the tracking chain $\boldsymbol{\alpha}^\frown(\lambda)$, *unless* we either have $\tau_{n,m_n} < \mu_{\tau_n'} \;\&\; \chi^{\tau_n'}(\tau_{n,m_n}) = 0$, where $\mathrm{tc}(\mathrm{gs}(\alpha)) = \boldsymbol{\alpha}[\alpha_{n,m_n} + 1]$, or $\boldsymbol{\alpha}^\frown(\lambda)$ is in conflict with condition 5 of Definition 5.1 of [5], in which case we have $\mathrm{tc}(\mathrm{gs}(\alpha)) = \boldsymbol{\alpha}_{\restriction_{n-1}}{}^\frown(\alpha_{n,1}, \ldots, \alpha_{n,m_n}, 1)$. In case α does not have any $<_1$-successor, we set $\mathrm{gs}(\alpha) := \alpha$.

α is $\leq_2$-minimal if and only if for its tracking chain $\boldsymbol{\alpha}$ we have $m_n \leq 2$ and $\tau_n^\star = 1$, and α has a non-trivial $\leq_2$-reach if and only if $m_n > 1$ and $\tau_{n,m_n} > 1$. Note that any $\alpha \in \mathrm{Ord}$ with a non-trivial $\leq_2$-reach is the proper supremum of its $<_1$-predecessors, hence 1^∞ does not possess any $<_2$-successor. Iterated closure under the relativized notation system T^τ for

$\tau = 1^\infty, (1^\infty)^\infty, \ldots$ results in the infinite $<_2$-chain through Ord. Its $<_1$-root is 1^∞, the root of the "master main line" of $\mathcal{R}_2$, outside the core of $\mathcal{R}_2$, i.e. 1^∞.

According to part (a) of Theorem 7.9 of [5] α has a greatest $<_1$-predecessor if and only if it is not $\le_1$-minimal and has a trivial $\le_2$-reach (i.e. does not have any $<_2$-successor). This is the case if and only if either $m_n = 1$ and $n > 1$, where we have $\mathrm{pred}_1(\alpha) = \mathrm{o}_{n-1,m_{n-1}}(\boldsymbol{\alpha})$, or $m_n > 1$ and $\tau_{n,m_n} = 1$. In this latter case α_{n,m_n} is of a form $\xi + 1$ for some $\xi \ge 0$, and using again the notation from Definition 5.1 of [5] we have $\mathrm{pred}_1(\alpha) = \mathrm{o}(\boldsymbol{\alpha}[\xi])$ if $\chi^{\tau_{n,m_n-1}}(\xi) = 0$, whereas $\mathrm{pred}_1(\alpha) = \mathrm{o}(\mathrm{me}(\boldsymbol{\alpha}[\xi]))$ in the case $\chi^{\tau_{n,m_n-1}}(\xi) = 1$.

Recall Definition 7.12 of [5], defining for $\boldsymbol{\alpha} \in \mathrm{TC}$ the notation $\boldsymbol{\alpha}^\star$ and the index pair $\mathrm{gbo}(\boldsymbol{\alpha}) =: (n_0, m_0)$, which according to Corollary 7.13 of [5] enables us to express the $\le_1$-reach $\mathrm{lh}(\alpha)$ of $\alpha := \mathrm{o}(\boldsymbol{\alpha})$, cf. Definition 7.7 of [5], by

$$\mathrm{lh}(\alpha) = \mathrm{o}(\mathrm{me}(\boldsymbol{\beta}^\star)), \tag{5.1}$$

where $\boldsymbol{\beta} := \boldsymbol{\alpha}_{\restriction n_0, m_0}$, which in the case $m_0 = 1$ is equal to $\mathrm{o}(\mathrm{me}(\boldsymbol{\beta})) = \mathrm{o}_{n_0,1}(\boldsymbol{\alpha}) + \mathrm{dp}_{\bar{\tau}_{n_0,0}}(\tau_{n_0,1})$ and in the case $m_0 > 1$ equal to $\mathrm{o}(\mathrm{me}(\boldsymbol{\beta}[\mu_{\tau_{n_0,m_0-1}}]))$. Note that if $\mathrm{cml}(\boldsymbol{\alpha}^\star)$ does not exist we have

$$\mathrm{lh}(\alpha) = \mathrm{o}(\mathrm{me}(\boldsymbol{\alpha}^\star)),$$

and the tracking chain $\boldsymbol{\beta}$ of any ordinal β such that $\mathrm{o}(\boldsymbol{\alpha}^\star) \le_1 \beta$ is then an extension of $\boldsymbol{\alpha}$, $\boldsymbol{\alpha} \subseteq \boldsymbol{\beta}$, as will follow from Lemma 5.11.

The relation $\le_1$ can be characterized by

$$\alpha \le_1 \beta \quad \Leftrightarrow \quad \alpha \le \beta \le \mathrm{lh}(\alpha), \tag{5.2}$$

showing that $\le_1$ is a forest contained in $\le$, which *respects* the ordering $\le$, i.e. if $\alpha \le \beta \le \gamma$ and $\alpha \le_1 \gamma$ then $\alpha \le_1 \beta$. On the basis of Lemma 5.4, Equation (5.1) has the following

Corollary 5.5: *Let $M \subseteq_{\mathrm{fin}} \mathrm{TC}$ be spanning (weakly spanning above some $\boldsymbol{\alpha} \in \mathrm{TC}$) and $\boldsymbol{\beta} \in M$, $\beta := \mathrm{o}(\boldsymbol{\beta})$. Then*

$$\mathrm{tc}(\mathrm{lh}(\beta)) \in M,$$

provided that $\mathrm{o}(\boldsymbol{\beta}_{\restriction \mathrm{gbo}(\boldsymbol{\beta})})$ is a proper extension of $\boldsymbol{\alpha}$ in the case that M is weakly spanning above $\boldsymbol{\alpha}$.

We now recall how to retrieve the greatest $<_2$-predecessor of an ordinal below 1^∞, if it exists, and the iteration of this procedure to obtain the maximum chain of $<_2$-predecessors. Recall Definition 5.3 and Lemma 5.10 of [5]. Using the following proposition we can prove two other useful characterizations of the relationship $\alpha \le_2 \beta$.

Proposition 5.6: *Let $\alpha < 1^\infty$ with $\mathrm{tc}(\alpha) =: \boldsymbol{\alpha}$. We define a sequence $\boldsymbol{\sigma} \in \mathrm{RS}$ as follows.*

(1) If $m_n \le 2$ and $\tau_n^\star = 1$, whence α is $\le_2$-minimal according to Theorem 7.9 of [5], set $\boldsymbol{\sigma} := ()$. Otherwise,
(2) if $m_n > 2$, whence $\mathrm{pred}_2(\alpha) = \mathrm{o}_{n,m_n-1}(\boldsymbol{\alpha})$ with base τ_{n,m_n-2} according to Theorem 7.9 of [5], we set $\boldsymbol{\sigma} := \mathrm{cs}(\boldsymbol{\alpha}_{\restriction_{n,m_n-2}})$,
(3) and if $m_n \le 2$ and $\tau_n^\star > 1$, whence $\mathrm{pred}_2(\alpha) = \mathrm{o}_{i,j+1}(\boldsymbol{\alpha})$ with base $\tau_{i,j}$ where $(i,j) := n^\star$, again according to Theorem 7.9 of [5], we set $\boldsymbol{\sigma} := \mathrm{cs}(\boldsymbol{\alpha}_{\restriction_{i,j}})$.

Each σ_i is then of a form $\tau_{k,l}$ where $1 \le l < m_k$, $1 \le k \le n$. The corresponding $<_2$-predecessor of α is $\mathrm{o}_{k,l+1}(\boldsymbol{\alpha}) =: \beta_i$. We obtain sequences $\boldsymbol{\sigma} = (\sigma_1, \ldots, \sigma_r)$ and $\boldsymbol{\beta} = (\beta_1, \ldots, \beta_r)$ with $\beta_1 <_2 \cdots <_2 \beta_r <_2 \alpha$, where $r = 0$ if α is $\le_2$-minimal, so that $\mathrm{Pred}_2(\alpha) = \{\beta_1, \ldots, \beta_r\}$ and hence $\beta <_2 \alpha$ if and only if $\beta \in \mathrm{Pred}_2(\alpha)$, showing that $\le_2$ is a forest contained in $\le_1$.

Lemma 5.7: *Let $\alpha, \beta < 1^\infty$ with tracking chains $\mathrm{tc}(\alpha) = \boldsymbol{\alpha} = (\boldsymbol{\alpha}_1, \ldots, \boldsymbol{\alpha}_n)$, $\boldsymbol{\alpha}_i = (\alpha_{i,1}, \ldots, \alpha_{i,m_i})$, $1 \le i \le n$, and $\mathrm{tc}(\beta) = \boldsymbol{\beta} = (\boldsymbol{\beta}_1, \ldots, \boldsymbol{\beta}_l)$, $\boldsymbol{\beta}_i = (\beta_{i,1}, \ldots, \beta_{i,k_i})$, $1 \le i \le l$. Assume further that $\boldsymbol{\alpha} \subseteq \boldsymbol{\beta}$ with associated chain $\boldsymbol{\tau}$ and that $m_n > 1$. Set $\tau := \tau_{n,m_n-1}$. The following are equivalent:*

(1) $\alpha \le_2 \beta$
(2) $\tau \le \tau_{j,1}$ for $j = n+1, \ldots, l$
(3) $\tilde{\tau} \mid \beta$.

Proof: Note that $\boldsymbol{\alpha} \subseteq \boldsymbol{\beta}$ is a necessary condition for $\alpha \le_2 \beta$, cf. Proposition 5.6.

1 ⇒ 2: Iterating the operation $(\cdot)'$ (which for $\tau_{j,1}$ is $\tau_{j,1}' = \tau_j^\star$) then reaches τ. This implies $\tau_{j,1} \ge \tau$ for $j = n+1, \ldots, l$.

2 ⇒ 1: Iterating the procedure to find the greatest $\le_2$-predecessor, cf. Proposition 5.6, from β downward satisfies $\tau_j^\star \ge \tau$ at each $j \in (n, l]$, where we therefore have $(n, m_n - 1) \le_{\mathrm{lex}} j^\star$.

2 ⇒ 3: Note that according to Lemma 5.10(c) of [5] we have

$$\mathrm{ts}(\tilde{\tau}) = \mathrm{cs}(\boldsymbol{\alpha}_{\restriction_{n,m_n-1}}).$$

We argue by induction along $<_{\text{lex}}$ on (l, k_l). The case $k_l > 1$ is trivial since β is then a multiple of $\mathrm{o}(\boldsymbol{\beta}_{\restriction_{l,k_l-1}})$. Assume now that $k_l = 1$, so that $\tau_{l,1} \geq \tau$. Let $(u,v) := l^\star$, so $(n, m_n - 1) \leq_{\text{lex}} (u,v)$ and $\tau_l^\star \geq \tau$. By Lemma 5.10(b) of [5] we have

$$\tilde{\tau}_{l,1} = \kappa^{\tilde{\tau}_l^\star}_{\tau_{l,1}}.$$

If $(n, m_n - 1) = (u,v)$ we obtain $\tilde{\tau} \mid \kappa^{\tilde{\tau}}_{\tau_{l,1}}$ since $\tau_{l,1} \geq \tau$. By the i.h. we have $\tilde{\tau} \mid \mathrm{o}(\boldsymbol{\beta}_{\restriction_{u,v}})$, so $\tilde{\tau} \mid \tilde{\tau}_{u,v}$, and since $\tau_{l,1} \geq \tau_l^\star$ we also have $\tilde{\tau} \mid \tilde{\tau}_{l,1}$, which implies that $\tilde{\tau} \mid \beta$, cf. Definition 4.1.

3 ⇒ 2: Assume there exists a maximal $j \in \{n+1, \ldots, l\}$ such that $\tau_{j,1} < \tau$. Then we have $\tau_j^\star \leq \tau_{j,1} < \tau$, so $(u,v) := j^\star <_{\text{lex}} (n, m_n - 1)$. Let $\mathrm{ts}(\tilde{\tau}) =: (\sigma_1, \ldots, \sigma_s)$ and recall Theorem 3.19.

Case 1: $\tau_{j,1} \notin \mathbb{E}^{>\tau_j^\star}$. Then it follows that $(j,1) = (l, k_l)$ since j was chosen maximally. In the case $\tau_j^\star = 1$ we have $\tau_{j,1} < \sigma_1$ and hence $\tilde{\tau}_{j,1} = \kappa_{\tau_{j,1}} < \tilde{\tau}$. Otherwise, i.e. $\tau_j^\star = \tau_{u,v} > 1$, we obtain $\mathrm{ts}(\tilde{\tau}_{u,v}) = \mathrm{cs}(\boldsymbol{\alpha}_{\restriction_{u,v}}) <_{\text{lex}} \mathrm{ts}(\tilde{\tau})$, so

$$\tilde{\tau}_{j,1} = \kappa^{\mathrm{ts}(\tilde{\tau}_{u,v})}_{\tau_{j,1}} < \tilde{\tau}.$$

Case 2: $\tau_{j,1} \in \mathbb{E}^{>\tau_j^\star}$. Then $\mathrm{ts}(\tilde{\tau}_{j,1}) = \mathrm{cs}(\boldsymbol{\beta}_{\restriction_{j,1}}) = \mathrm{cs}(\boldsymbol{\beta}_{\restriction_{u,v}})^\frown \tau_{j,1} <_{\text{lex}} \mathrm{ts}(\tilde{\tau})$, hence $\tilde{\tau}_{j,1} < \tilde{\tau}$. In the case $\tau_{l,k_l} \in \mathbb{E}^{>\tau'_{l,k_l}}$ we see that $\mathrm{ts}(\tilde{\tau}_{j,1}) \subseteq \mathrm{ts}(\tilde{\tau}_{l,k_l})$, hence $\tilde{\tau}_{l,k_l} < \tilde{\tau}$. The other cases are easier, cf. case 1. □

Applying the mappings tc, cf. Section 6 of [5], and o, which we verified in the present article to be elementary recursive, cf. Subsection 5.1, we are now able to formulate the arithmetical characterization of $\mathcal{C}_2$.

Corollary 5.8: *The structure $\mathcal{C}_2$ is characterized elementary recursively by*

(1) $(1^\infty, \leq)$ *is the standard ordering of the classical notation system* $1^\infty = \mathrm{T}^1 \cap \Omega_1$, *cf.* [7],
(2) $\alpha \leq_1 \beta$ *if and only if* $\alpha \leq \beta \leq \mathrm{lh}(\alpha)$ *where* lh *is given by Equation (5.1), and*
(3) $\alpha \leq_2 \beta$ *if and only if condition 2 of Lemma 5.7 holds.* □

Corollary 5.9: *Let* $M \subseteq_{\text{fin}} \mathrm{TC}$ *be spanning (weakly spanning above some* $\boldsymbol{\alpha} \in \mathrm{TC}$*). Then* M *is closed under* lh_2.

Proof: This follows from Lemma 5.4 using Lemma 5.7, cf. Corollaries 5.6 and 7.13 of [5]. □

Recall Definition 5.13 of [5] which characterizes the standard linear ordering $\le$ on 1^∞ by an ordering $\le_{\mathrm{TC}}$ on the corresponding tracking chains. We can formulate a characterization of the relation $\le_1$ (below 1^∞) in terms of the corresponding tracking chains as well. This follows from an inspection of the ordering $\le_{\mathrm{TC}}$ in combination with the above statements. Let $\alpha, \beta < 1^\infty$ with tracking chains $\mathrm{tc}(\alpha) = \boldsymbol{\alpha} = (\boldsymbol{\alpha}_1, \ldots, \boldsymbol{\alpha}_n)$, $\boldsymbol{\alpha}_i = (\alpha_{i,1}, \ldots, \alpha_{i,m_i})$, $1 \le i \le n$, and $\mathrm{tc}(\beta) = \boldsymbol{\beta} = (\boldsymbol{\beta}_1, \ldots, \boldsymbol{\beta}_l)$, $\boldsymbol{\beta}_i = (\beta_{i,1}, \ldots, \beta_{i,k_i})$, $1 \le i \le l$. We have $\alpha \le_1 \beta$ if and only if either $\boldsymbol{\alpha} \subseteq \boldsymbol{\beta}$ or there exists $(i,j) \in \mathrm{dom}(\boldsymbol{\alpha}) \cap \mathrm{dom}(\boldsymbol{\beta})$, $j < \min\{m_i, k_i\}$, such that $\boldsymbol{\alpha}_{\restriction_{i,j}} = \boldsymbol{\beta}_{\restriction_{i,j}}$ and $\alpha_{i,j+1} < \beta_{i,j+1}$, and we either have $\chi^{\tau_{i,j}}(\alpha_{i,j+1}) = 0 \;\&\; (i, j+1) = (n, m_n)$ or $\chi^{\tau_{i,j}}(\alpha_{i,j+1}) = 1 \;\&\; \alpha \le_1 \mathrm{o}(\mathrm{me}(\boldsymbol{\alpha}_{\restriction_{i,j+1}})) <_1 \beta$. Iterating this argument and recalling Lemma 5.5 of [5] we obtain the following

Proposition 5.10: *Let α and β with tracking chains $\boldsymbol{\alpha}$ and $\boldsymbol{\beta}$, respectively, as above. We have $\alpha \le_1 \beta$ if and only if either $\boldsymbol{\alpha} \subseteq \boldsymbol{\beta}$ or there exists the $<_{\mathrm{lex}}$-increasing chain of index pairs $(i_1, j_1+1), \ldots, (i_s, j_s+1) \in \mathrm{dom}(\boldsymbol{\alpha})$ of maximal length $s \ge 1$ where $j_r \ge 1$ for $r = 1, \ldots, s$, such that $(i_1, j_1+1) \in \mathrm{dom}(\boldsymbol{\beta})$, $\boldsymbol{\alpha}_{\restriction_{i_1,j_1}} = \boldsymbol{\beta}_{\restriction_{i_1,j_1}}$, $\alpha_{i_1,j_1+1} < \beta_{i_1,j_1+1}$,*

$$\boldsymbol{\alpha}_{\restriction_{i_s,j_s+1}} \subseteq \boldsymbol{\alpha} \subseteq \mathrm{me}(\boldsymbol{\alpha}_{\restriction_{i_s,j_s+1}}),$$

and $\chi^{\tau_{i_r,j_r}}(\alpha_{i_r,j_r+1}) = 1$ at least whenever $(i_r, j_r+1) \ne (n, m_n)$. Setting $\alpha_r := \mathrm{o}(\boldsymbol{\alpha}_{\restriction_{i_r,j_r+1}})$ for $r = 1, \ldots, s$ as well as $\alpha_r^+ := \mathrm{o}(\mathrm{me}(\boldsymbol{\alpha}_{\restriction_{i_r,j_r+1}}))$ for r such that $\chi^{\tau_{i_r,j_r}}(\alpha_{i_r,j_r+1}) = 1$ and $\alpha_s^+ := \alpha$ if $\chi^{\tau_{i_s,j_s}}(\alpha_{i_s,j_s+1}) = 0$ we have

$$\alpha_1 <_2 \cdots <_2 \alpha_s \le_2 \alpha \le_1 \alpha_s^+ <_1 \cdots <_1 \alpha_1^+ <_1 \mathrm{o}(\boldsymbol{\beta}_{\restriction_{i_1,j_1+1}}) \le_1 \beta.$$

For $\beta = \mathrm{lh}(\alpha)$ the cases $\boldsymbol{\alpha} \subseteq \boldsymbol{\beta}$ and $s = 1$ with $(i_1, j_1+1) = (n, m_n)$ correspond to the situation $\mathrm{gbo}(\boldsymbol{\alpha}) = (n, m_n)$, while otherwise we have $\mathrm{gbo}(\boldsymbol{\alpha}) = (i_1, j_1+1)$.

Remark. Note that the above index pairs characterize the relevant submaximal ν-indices in the initial chains of $\boldsymbol{\alpha}$ with respect to $\boldsymbol{\beta}$ and leave the intermediate steps of maximal (me-) extension along the iteration. Using Lemma 5.5 of [5] we observe that the sequence $\tau_{i_1,j_1}, \ldots, \tau_{i_s,j_s}$ of bases in the above proposition satisfies

$$\tau_{i_1,j_1} < \cdots < \tau_{i_s,j_s} \quad \text{and} \quad \tau_{i_s,j_s} < \tau_{i,1} \quad \text{for every } i \in (i_s, n], \tag{5.3}$$

so that in the case where $\alpha <_1 \beta$ and $\boldsymbol{\alpha} \not\subseteq \boldsymbol{\beta}$ we have $\alpha <_1 \mathrm{gs}(\alpha) \le_1 \beta$ with $\tau_{i_s,j_s} \mid \rho_n \dot{-} 1$.

Lemma 5.11: *The relation $\subseteq$ of initial chain on* TC *respects the ordering* $\le_{\mathrm{TC}}$ *and hence also the characterization of* $\le_1$ *on* TC.

Proof: Suppose tracking chains $\boldsymbol{\alpha}, \boldsymbol{\beta}, \boldsymbol{\gamma} \in \mathrm{TC}$ satisfy $\boldsymbol{\alpha} \le_{\mathrm{TC}} \boldsymbol{\beta} \le_{\mathrm{TC}} \boldsymbol{\gamma}$ and $\boldsymbol{\alpha} \subseteq \boldsymbol{\gamma}$. The case where any two chains are equal is trivial. We may therefore assume that $\mathrm{o}(\boldsymbol{\alpha}) < \mathrm{o}(\boldsymbol{\beta}) < \mathrm{o}(\boldsymbol{\gamma})$ as $(\mathrm{TC}, \le_{\mathrm{TC}})$ and $(1^\infty, <)$ are order-isomorphic. Since $\boldsymbol{\alpha} \subseteq \boldsymbol{\gamma}$ we have $\mathrm{o}(\boldsymbol{\alpha}) <_1 \mathrm{o}(\boldsymbol{\gamma})$, hence $\mathrm{o}(\boldsymbol{\alpha}) <_1 \mathrm{o}(\boldsymbol{\beta})$ as $\le_1$ respects $\le$. Assume toward contradiction that $\boldsymbol{\alpha} \not\subseteq \boldsymbol{\beta}$, whence by Proposition 5.10 there exists $(i, j+1) \in \mathrm{dom}(\boldsymbol{\alpha}) \cap \mathrm{dom}(\boldsymbol{\beta})$ such that $\boldsymbol{\alpha}_{\restriction_{i,j}} = \boldsymbol{\beta}_{\restriction_{i,j}}$ and $\alpha_{i,j+1} < \beta_{i,j+1}$. But then for any $\boldsymbol{\delta} \in \mathrm{TC}$ such that $\boldsymbol{\alpha} \subseteq \boldsymbol{\delta}$ we have

$$\mathrm{o}(\boldsymbol{\delta}) < \mathrm{o}(\boldsymbol{\alpha}^+) \le \mathrm{o}(\boldsymbol{\beta}) < \mathrm{o}(\boldsymbol{\gamma}),$$

where the tracking chain $\boldsymbol{\alpha}^+ := \boldsymbol{\alpha}_{\restriction_{i,j+1}}[\alpha_{i,j+1} + 1]$ is the result of changing the terminal index $\alpha_{i,j+1}$ of $\boldsymbol{\alpha}_{\restriction_{i,j+1}}$ to $\alpha_{i,j+1} + 1$. □

We may now return to the issue of closure under lh, which has a convenient sufficient condition on the basis of the following

Definition 5.12: A tracking chain $\boldsymbol{\alpha} \in \mathrm{TC}$ is called *convex* if and only if every ν-index in $\boldsymbol{\alpha}$ is maximal, i.e. given by the corresponding μ-operator.

Corollary 5.13: *Let $\boldsymbol{\alpha} \in \mathrm{TC}$ be convex and $M \subseteq_{\mathrm{fin}} \mathrm{TC}$ be weakly spanning above $\boldsymbol{\alpha}$. Then M is closed under* lh.

Proof: This is a consequence of Proposition 5.10, Corollary 5.5, Lemma 5.4, and Equation (5.1). □

While it is easy to observe that in $\mathcal{R}_2$ the relation $\le_1$ is a forest that respects $\le$ and the relation $\le_2$ is a forest contained in $\le_1$ which respects $\le_1$, we can now conclude that this also holds for the arithmetical formulations of $\le_1$ and $\le_2$ in $\mathcal{C}_2$, without referring to the results in Section 7 of [5].

Corollary 5.14: *Consider the arithmetical characterizations of $\le_1$ and $\le_2$ on 1^∞. The relation $\le_2$ respects $\le_1$, i.e. whenever $\alpha \le_1 \beta \le_1 \gamma < 1^\infty$ and $\alpha \le_2 \gamma$, then $\alpha \le_2 \beta$.*

Proof: In the case $\boldsymbol{\beta} \subseteq \boldsymbol{\gamma}$ this directly follows from Lemma 5.7, while otherwise we additionally employ Proposition 5.10 and Equation (5.3). □

6. Conclusion

In the present article, the arithmetical characterization of the structure $\mathcal{C}_2$, which was established in Theorem 7.9 and Corollary 7.13 of [5], has been analyzed and shown to be elementary recursive. We have seen that finite isomorphism types of $\mathcal{C}_2$ (in its arithmetical formulation) are contained in the class of "respecting forests", cf. [2], over the language $(\leq, \leq_1, \leq_2)$. In a subsequent article [10] we will establish the converse by providing an effective assignment of isominimal realizations in $\mathcal{C}_2$ to arbitrary respecting forests. We will provide an algorithm to find pattern notations for the ordinals in $\mathcal{C}_2$, and will conclude that the union of isominimal realizations of respecting forests is indeed the core of $\mathcal{R}_2$, i.e. the structure $\mathcal{C}_2$ in its semantical formulation based on Σ_i-elementary substructures, $i = 1, 2$. As a corollary we will see that the well-quasi orderedness of respecting forests with respect to coverings, which was shown by Carlson in [3], implies (in a weak theory) transfinite induction up to the proof-theoretic ordinal 1^∞ of $\mathrm{KP}\ell_0$.

We expect that the approaches taken here and in our treatment of the structure $\mathcal{R}_1^+$, see [8] and [9], will naturally extend to an analysis of the structure $\mathcal{R}_2^+$ and possibly to structures of patterns of higher order. A subject of ongoing work is to verify that the core of $\mathcal{R}_2^+$ matches the proof-theoretic strength of a limit of KPI-models.

Acknowledgements

I would like to express my gratitude to Professor Ulf Skoglund for encouragement and support of my research. I would like to thank Dr. Steven D. Aird for editing the manuscript.

References

1. T. J. Carlson: *Ordinal Arithmetic and Σ_1-Elementarity.* Archive for Mathematical Logic 38 (1999) 449-460.
2. T. J. Carlson: *Elementary Patterns of Resemblance.* Annals of Pure and Applied Logic 108 (2001) 19-77.
3. T. J. Carlson: *Generalizing Kruskal's theorem to pairs of cohabitating trees.* Archive for Mathematical Logic 55 (2016) 37–48.
4. T. J. Carlson and G. Wilken: *Normal Forms for Elementary Patterns.* The Journal of Symbolic Logic 77 (2012) 174-194.
5. T. J. Carlson and G. Wilken: *Tracking Chains of Σ_2-Elementarity.* Annals of Pure and Applied Logic 163 (2012) 23-67.
6. W. Pohlers: *Proof Theory. The First Step into Impredicativity.* Springer, Berlin (2009).

7. G. Wilken: *Ordinal Arithmetic based on Skolem Hulling.* Annals of Pure and Applied Logic 145 (2007) 130-161.
8. G. Wilken: Σ_1-*Elementarity and Skolem Hull Operators.* Annals of Pure and Applied Logic 145 (2007) 162-175.
9. G. Wilken: *Assignment of Ordinals to Elementary Patterns of Resemblance.* The Journal of Symbolic Logic 72 (2007) 704-720.
10. G. Wilken: *Pure patterns of order 2.* Submitted.

GROUPS WITH ORDERINGS OF ARBITRARY ALGORITHMIC COMPLEXITY

Jennifer Chubb[a]

Department of Mathematics & Statistics
University of San Francisco
San Francisco, CA 94117, USA
jcchubb@usfca.edu

Mieczyslaw K. Dabkowski

Department of Mathematical Sciences
University of Texas at Dallas
Richardson, TX, USA
mdab@utdallas.edu

Valentina Harizanov[b]

Department of Mathematics
George Washington University
Washington, DC, USA
harizanv@gwu.edu

We give general sufficient conditions that a computable group admits bi-orderings of arbitrary computability-theoretic complexity in a strong sense. We apply this result to show that structures from a large class of computable, finitely presented, residually nilpotent groups admit bi-orderings in every *truth-table degree*, a refinement of the Turing degrees. This class includes a wide variety of important groups such as finitely generated free groups, surface groups and certain nilpotent groups.

[a]Chubb was partially supported by the University of San Francisco Faculty Development Fund and by a travel grant from the Association for Women in Mathematics.

[b]Harizanov was partially supported by the NSF grant 1202328, by the CCFF award and CCAS Dean's research chair award of the George Washington University. Harizanov thanks the organizers of the research workshop Sets and Computations at the IMS of the National University of Singapore for inviting her and partially supporting her participation.

1. Introduction

Ordered algebraic structures are ubiquitous in mathematics and have been studied since Dedekind, Hölder, and Hilbert, and it is important to understand their constructive properties. Here, we consider linear orderings of the elements of a given group that respect the algebraic structure and study the algorithmic complexity of the admitted orderings.

A linear ordering of the elements of a group, $\prec$, is a *left ordering* if for any group elements a, b, and c, we have

$$a \prec b \implies ca \prec cb.$$

In other words, the ordering is invariant under the group acting on the left. If $\prec$ is simultaneously left-invariant and right-invariant, it is a *bi-ordering*. In the discussion that follows, statements about *orderings* apply to both left orderings and bi-orderings.

The classical study of collections of orderings of various types admitted by a given algebraic structure has a long history, and although much is understood for certain classes of structures, there are basic questions that remain unanswered. (See, for example, Fuchs [20], Kopytov and Medvedev [30], and Mura and Rhemtulla [36] for introduction and survey of early results.)

The corresponding computability-theoretic properties of the orderings of algebraic structures have also been investigated. In 1979, Metakides and Nerode [35] showed that families of orderings of computable fields are in exact correspondence to the collection of effectively closed subsets of the Cantor space, that is, the Π^0_1 subsets of 2^ω. A subset of 2^ω is a Π^0_1 *class* (also called *effectively closed*) if it is the collection of infinite paths through a computable subtree of $2^{<\omega}$. By a *subtree*, we mean a subset of $2^{<\omega}$ that is closed under initial segments. A subtree of $2^{<\omega}$ is computable if its set of nodes is computable.

Theorem 1.1: (Metakides and Nerode [35], Theorems 7.1 and 7.3.)

(1) The class of orderings of a recursively presented field is a Π^0_1 class.
(2) Let C be a non-empty Π^0_1 class of sets in 2^ω. Then there is a recursively presented formally real field F, and a homeomorphism from C onto the order space of F, which is Turing degree (in fact, many-one degree) preserving.

A large body of research in computability theory exists surrounding the properties of Π^0_1 classes and their members (see, for example, [7]), and since these classes characterize the collections of orderings of computable fields, this problem is well-studied. The situation for groups, however, is more complicated.

Although the collection of left orderings or bi-orderings of a computable group always forms a Π^0_1 class, not every Π^0_1 class can be so realized. Solomon explained in [45, 46] that no such correspondence between collections of orderings and Π^0_1 classes can exist for the collections of orderings of computable groups as a consequence of the existence of a Π^0_1 class containing pairwise Turing-incomparable elements. He went on to show that the degrees of a computably bounded Π^0_1 class cannot, in general, be realized by the degrees of the orderings of a computable torsion-free abelian group. These negative results motivate continued study of computability-theoretic properties of orderings admitted by a given computable group, particularly a non-abelian group.

Progress has been made for certain classes of groups. For example, computable, torsion-free, abelian groups of finite rank greater than 1 admit orderings in every Turing degree, and those with infinite rank admit orderings in every Turing degree above that of the halting set, $\mathbf{0}'$ (see [45, 46]). In 1986, Downey and Kurtz [12] constructed a computable abelian group of infinite rank admitting no computable ordering. The collection of Turing degrees realized by the orderings of this (or any such) group cannot, however, exactly coincide with the cone of degrees above $\mathbf{0}'$ (this is a consequence of the Jockusch-Soare low basis theorem [27]).

For computable, infinite-rank, torsion-free, abelian groups, it is always possible to find an ordering in every Turing degree capable of computing the dependence algorithm for the computable divisible closure of the group. It is natural to wonder if the Turing degree of the requisite low ordering of the group presented by Downey and Kurtz sits in the cone of degrees above the corresponding dependence algorithm and is thus naturally realized, in a certain sense. The construction in [12] may however be modified (see [10]) to ensure that the degree of the dependence algorithm of the computable divisible closure of the group produced is of complete degree (that of the halting set), and so the low orderings of this group do not arise as a consequence of the existence of a low degree dependence algorithm.

More recently, in [28], Kach, Lange, and Solomon showed that there are computable abelian torsion-free groups with computable orderings but not orderings of every Turing degree, demonstrating that the spectrum of Turing degrees achieved by admitted orderings need not be upwards closed.

There are some results of similar flavor for non-abelian groups. Solomon [45, 46] showed that computable, torsion-free, properly n-step nilpotent groups admit bi-orderings of all Turing degrees at or above $\mathbf{0}^{(n)}$. In [23], it was shown that it is possible to construct a computable copy of the free group with countably infinitely many generators admitting no computable left orderings.

In this article, we consider a large class of residually nilpotent groups that includes finitely generated free groups as well as surface groups, and show that they admit bi-orderings of every truth-table degree (see Definition 2.2). We also show that the members of a certain class of nilpotent groups similarly admit bi-orderings in every truth-table degree. In Section 2, we review basic definitions and facts about orderings on groups, and about strong reducibilities and strong degrees in computability theory. In Section 3, we establish general conditions that suffice for a computable group to admit bi-orderings in every truth-table degree. In Section 4, we apply this result to certain classes of groups of importance in algebra and low-dimensional topology. In an Appendix we provide a worked-out example illustrating the technical result in Lemma 5.2. See [44] and [39] for background in general computability theory and notation.

2. Basic Definitions and Facts

2.1. *Orderings of groups*

A (partial) left ordering $\prec$ of a group G is a binary, irreflexive, antisymmetric, transitive relation on the elements of G, which is left-invariant with respect to the group operation. That is, for every $a, b \in G$ and all $c \in G$, we have

$$a \prec b \implies c \cdot a \prec c \cdot b,$$

where $\cdot$ denotes the group operation. We will usually omit explicit notation for a group operation and write ab for $a \cdot b$. A (partial) right-ordering is defined similarly, and a (partial) bi-ordering is one that is simultaneously left- and right-invariant. When the word "partial" is not used, we mean a total ordering.

An equivalent definition for a left ordering, a right ordering, and a bi-ordering may be given in terms of the *positive cone* of the ordering. For a left ordering, right ordering, or a bi-ordering $\prec$ on G, the corresponding positive cone $P_\prec$ is the set of all elements of G that are greater than the identity element e_G in this ordering, together with the identity itself. That is,

$$g \in P_{\prec} \iff (g \succ e_G \vee g = e_G).$$

Note that specifying a positive cone is sufficient to specify an ordering: To determine whether $a \prec b$ holds, we need only check whether $a^{-1}b \in P_{\prec}$ for a left ordering or a bi-ordering, or whether $ba^{-1} \in P_{\prec}$ for a right-ordering. It is easy to see that $\prec$ and $P_{\prec}$ have the same m-degree when G is a computable group. (The cone $P_{\prec}$ is 1-reducible to $\prec$, but the reverse reducibility can only be an m-reduction in general.)

A subset P of G is the positive cone of a bi-ordering of G if the following conditions hold of P:

(1) $P \cdot P \subseteq P$, i.e., P is a sub-semigroup of G;
(2) $P \cap P^{-1} = \{e_G\}$, where P^{-1} is the set of inverses of elements of P; such a semigroup is said to be *pure*;
(3) $P \cup P^{-1} = G$; and
(4) for each $g \in G$, $gPg^{-1} \subseteq P$, i.e., P is a *normal* sub-semigroup of G.

We will often identify an ordering with its positive cone.

2.2. *Strong reducibilities*

There are many notions of strong reducibilities (strong in the sense that they imply Turing reducibility). Here we are concerned with truth-table reducibility, which is most conveniently defined in terms of another strong reduction notion: weak truth-table (or bounded Turing) reducibility. Informally, a set A is weak truth-table reducible to another set B if there is a procedure for computing A using oracle B that is *predictable* in that there is a computable function specifying the number of bits of the oracle B that will be needed in the computation of A from B for a given input. The set A is truth-table reducible to B if there is a procedure for computing A from oracle B that, in addition to being predictable, is robust: If another oracle is used in the execution of the algorithm in place of B, the procedure will still halt, though not necessarily with correct output regarding membership in A. The formal definitions are as follows for sets of natural numbers A and B. We use φ_e to denote the function computed by the eth program (Turing machine) in some systematic enumeration of these, and φ_e^A for the function computed by the eth oracle program using oracle A. For any set A, we use $A \upharpoonright n$ to denote the first n bits of A.

Definition 2.1: A set A is *weak truth-table reducible* to set B if there is a computable function h and an index e so that $A(x) = \varphi_e^{B\upharpoonright h(x)}(x)$. We write $A \leq_{wtt} B$ when this is the case, and $A \equiv_{wtt} B$ when both $A \leq_{wtt} B$ and $B \leq_{wtt} A$.

The stronger notion, *tt*-reducibility, is a refinement of *wtt*-reducibility.

Definition 2.2: A set A is *truth-table reducible* to B if $A \leq_{wtt} B$ via a function φ_e^B and computable function h, having the additional property that for any string $\sigma \in 2^{<\omega}$ of length $h(x)$, $\varphi_e^\sigma(x)\downarrow$. We write $A \leq_{tt} B$ when this is the case, and $A \equiv_{tt} B$ when both $A \leq_{tt} B$ and $B \leq_{tt} A$.

The relations $\leq_{wtt}$ and $\leq_{tt}$ are pre-orderings of the power set of the natural numbers, and the induced equivalence classes are called *wtt*-degrees and *tt*-degrees, respectively. We will write $deg(A)$ for the Turing degree of the set A, and $deg_{tt}(A)$ for its *tt*-degree:

$$deg_{tt}(A) = \{B \in 2^\omega \mid A \equiv_{tt} B\}.$$

3. Bi-Orderings of Arbitrary tt-Degree

In [11], the authors investigated when a computable group admits left orderings of every Turing degree in an upper cone, where the base of the cone computes a particular family of finite subsets of the group. In particular, they presented general conditions that suffice to ensure a computable group admits left orderings of every Turing degree.

Theorem 3.1: [11] *Let G be a computable group. Let $\mathcal{P}$ be a computable family of finite subsets of $G - \{e\}$ satisfying the following conditions for every $p \in \mathcal{P}$:*

(1) $e \notin \mathrm{sgr}(p)$, where $\mathrm{sgr}(p)$ is the sub-semigroup generated by p;
(2) $(\exists r_0, r_1 \in \mathcal{P})(\exists g \in G)[r_0, r_1 \supset p \wedge g \in r_0 \wedge g^{-1} \in r_1]$; and
(3) $(\forall g \in G - \{e\})(\exists r \in \mathcal{P})[r \supseteq p \wedge (g \in r \vee g^{-1} \in r)]$.

Then G admits a left ordering of every Turing degree.

In [10], Chubb showed that when this family is computably enumerable (c.e.), the same conditions suffice to ensure that G admits a left ordering in every *tt*-degree. Here, we show that stronger conditions provide for bi-orderings in every *tt*-degree, and in the next section give some applications of our general result to specific natural families of orderable groups.

Notation 3.2: For an arbitrary subset C of elements of a group G, let $S(C)$ be the normal sub-semigroup of G generated by the elements in C.

For an ordering $\prec$ of group G, we denote by $P_{\prec}^{+}$ the set of strictly positive elements, i.e., $P_{\prec}^{+} =_{def} P_{\prec} - \{e_G\}$. We call this the strictly positive cone of $\prec$ in G.

Theorem 3.3: *Let G be a computable group, and $\mathcal{P}$ a c.e. family of finite subsets of $G - \{e\}$ satisfying the following conditions for every $p \in \mathcal{P}$:*

(1) $e \notin S(p)$;
(2) $(\exists r_0, r_1 \in \mathcal{P})(\exists g \in G)[r_0, r_1 \supset p \wedge g \in r_0 \wedge g^{-1} \in r_1]$; and
(3) $(\forall g \in G - \{e\})(\exists r \in \mathcal{P})[r \supseteq p \wedge (g \in r \vee g^{-1} \in r)]$.

Then G admits a bi-ordering in every tt-degree.

Proof: We construct a map $\mathcal{T} : 2^{<\omega} \to \mathcal{P}$ so that

$$\sigma \sqsubset \tau \implies \mathcal{T}(\sigma) \subseteq \mathcal{T}(\tau),$$

where $\sigma \sqsubset \tau$ denotes that σ is an initial segment of τ. For any $X \in 2^{\omega}$ we will have that $P_X^{+} = \bigcup_n \mathcal{T}(X \restriction n)$ is a pure, normal, sub-semigroup of G that contains exactly one of g and g^{-1} for each non-identity element $g \in G$. In other words, P_X^{+} is the strictly positive cone of a bi-ordering of G. We will see that $P_X^{+} \equiv_{tt} X$, and so $P_X = P_X^{+} \cup \{e\}$ is a bi-ordering of G of the same (arbitrary) *tt*-degree as X.

Let $G - \{e\} = \{g_0, g_1, \ldots\}$ and $\mathcal{P} = \{p_0, p_1, \ldots\}$ be computable enumerations of $G - \{e\}$ and $\mathcal{P}$, respectively.

Construction

Stage 0. Set $\mathcal{T}(\langle\,\rangle) = p_0$.

Stage $s+1$. At the beginning of this stage, we have $\mathcal{T}$ defined on $2^{\leq s}$. For each σ of length s, find the first r_0 and r_1 in $\mathcal{P}$, and the first $g \in G$ witnessing the satisfaction of condition (2) for $p = \mathcal{T}(\sigma)$. We are free to arrange that $g \in r_1$ and $g^{-1} \in r_0$. Next, find the first g_{j_0} and g_{j_1} appearing in the enumeration of $G - \{e\}$ so that neither of these elements nor their inverses are in r_0 and r_1, respectively. Let r_0' (r_1', respectively) be the first element of $\mathcal{P}$ extending r_0 (r_1, respectively) containing g_{j_0} or $g_{j_0}^{-1}$ (g_{j_1} or $g_{j_1}^{-1}$, respectively). Such r_i' exist by condition (3).

Set $\mathcal{T}(\sigma^\frown i) = r_i'$ for $i = 0, 1$.

This completes the construction of $\mathcal{T}$.

Now, let $\mathbf{x}$ be an arbitrary tt-degree, and X a set with $deg_{tt}(X) = \mathbf{x}$. Define $P_X^+ = \bigcup_s \mathcal{T}(X \upharpoonright s)$ and $P_X = P_X^+ \cup \{e\}$. Our first task is to show that P_X is the positive cone of a bi-ordering of G by verifying that the four conditions given above hold. Second, we check that P_X has the required computability-theoretic properties.

Observe that for any non-identity $g \in G$, there is a stage s so that either $g \in \mathcal{T}(X \upharpoonright s)$ or $g^{-1} \in \mathcal{T}(X \upharpoonright s)$; in fact, if $g = g_i$, then $s = i+1$. So, we have $P_X \cup P_X^{-1} = G$. Next, $P_X \cdot P_X \subseteq P_X$ since if for some non-trivial $a, b \in P_X$ we have $ab \notin P_X$, then necessarily $(ab)^{-1} \in P_X$. All three elements, a, b, and $(ab)^{-1}$, must be contained in $p = \mathcal{T}(X \upharpoonright s)$ for some s, and this element $p \in \mathcal{P}$ fails to satisfy condition (1) of the theorem, contrary to assumption. Continuing, $P_X \cap P_X^{-1} = \{e\}$ since e is in both sets, and a non-trivial element in the intersection results again in a violation of condition (1) by the image of some finite initial segment of X under $\mathcal{T}$. Finally, we verify that P_X is normal in G. Assuming otherwise, there must be $g \in G$ and $a \in P_X$ so that $gag^{-1} \notin P_X$. In this case we would have $(gag^{-1})^{-1} = ga^{-1}g^{-1} \in P_X$. Let s be a stage so that both a and $ga^{-1}g^{-1}$ are elements of $p = \mathcal{T}(X \upharpoonright s)$. Then in $S(p)$ we have both of these elements, as well as the conjugate of $ga^{-1}g^{-1}$ by g^{-1} since $S(p)$ is normal. However, this conjugate is a^{-1}, and we see that this again results in a contradiction to condition (1) of the theorem for $p \in \mathcal{P}$.

To see that $P_X \leq_{tt} X$, we observe the following. To determine if $g_i \in P_X$, we use the fact that for each σ of length $i+1$, either g_i or g_i^{-1} is in $\mathcal{T}(\sigma)$. So we have

$$g_i \in P_X \iff g_i \in \mathcal{T}(X \upharpoonright (i+1)).$$

Note that this is a tt-reduction since the initial segment of length $i+1$ of any set X is in the domain of $\mathcal{T}$.

For the reverse reduction, $X \leq_{tt} P_X$, we check whether $i \in X$ via the following algorithm.

Construct $\mathcal{T}$ until the domain includes all nodes of length $i+1$. If $P_X \supseteq \mathcal{T}(\sigma)$ for some σ of length $i+1$, then $i \in X$ if and only if the last bit of σ is 1. If P_X does not contain the image of any σ of length $i+1$, output 0.

If P_X is, in fact, the partial ordering defined by the construction above, this algorithm yields the correct answer to the membership question on X. If it is not, the algorithm halts, but may not yield a correct answer. Thus, the algorithm gives a valid tt-reduction. $\square$

4. Orderings of $\mathbb{Z}^k$

Let $\mathbb{Z}^2$ be the group of pairs of integers under coordinate-wise addition. In [41], Sikora characterized the upper cones of orderings of $\mathbb{Z}^2$ as half-planes (see Figure 1, and in [46], Solomon showed that there are orderings of this group in every Turing degree.

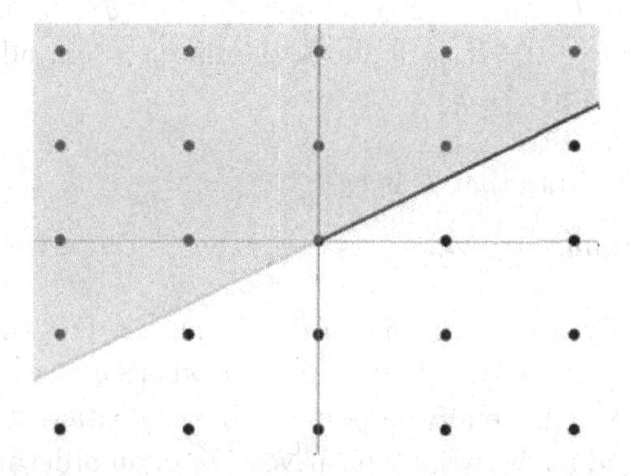

Fig. 1. The lattice points in the shaded area and on the black portion of the line separating the half-planes constitute the positive cone of an ordering of $\mathbb{Z}^2$.

In fact, an ordering of $\mathbb{Z}^2$ will have the same Turing degree as the real that is the slope of the line through the origin, which separates the positive and the negative cone (see Lemma 8.6 of [46]). Here, we construct a family of the sort described in Theorem 3.3 to obtain the following result.

Theorem 4.1: *The group $\mathbb{Z}^2$ admits orderings in every tt-degree.*

Proof: Let $\{g_0, g_1, \ldots\}$ be a computable enumeration of $\mathbb{Z}^2 - \{(0,0)\}$ where $g_i = (x_i, y_i)$. We will construct a family of finite subsets of $\mathbb{Z}^2$ having the properties described in Theorem 3.3. We do this by finite approximation.

Let $q > 0$. For the purposes of this construction, we say a point $g_s = (x_s, y_s)$ is *above the line* $y = qx$ if $y_s > qx_s$, or $y_s = qx_s$ and $x_s > 0$. Otherwise, it is *below the line.* In what follows, q will always be a rational number and equal to y_i/x_i for some given x_i and y_i in the positive integers. Note that determining whether a given point of $\mathbb{Z}^2$ is above or below such a line is computable.

Construction

Stage 0. Let $\mathcal{P}_0 = \emptyset$.
Stage $s+1$. Define $\mathcal{P}_{s+1}$ as the elements of $\mathcal{P}_s$ together with the following finite sets:

- All sets $\{g_i\}$ where $0 < i \le s$ and g_i has $x_i > 0$ and $y_i \ge 0$;
- For each $p \in \mathcal{P}_s$ and each g_t with $t \le s$, let $q = \min\{y/x \mid (x,y) \in p \wedge x > 0 \wedge y \ge 0\}$. If g_t is above the line $y = qx$, add $p \cup \{g_t\}$, and otherwise, add $p \cup \{-g_t\}$.

Let $\mathcal{P} = \bigcup_{i \in \omega} \mathcal{P}_i$. Note that $\mathcal{P}$ is c.e.

End of construction.

Now, we verify the conditions of Theorem 3.3. For any $p \in \mathcal{P}$, all of the elements of p are above the line $y = qx$ where $q = \min\{y/x \mid (x,y) \in p \wedge x > 0 \wedge y \ge 0\}$. Since all the points in p are contained in a half-plane, they are contained in the strictly positive cone of an ordering of $\mathbb{Z}^2$, so $S(p)$ does not contain the identity $(0,0)$, and hence $\mathcal{P}$ satisfies condition (1).

Next, let $g = g_{s_1}$ be a non-identity element of G and $p \in \mathcal{P}_{s_2}$. At stage $t = \max(s_1, s_2) + 1$, either $p \cup \{g\}$ or $p \cup \{-g\}$ will be added to $\mathcal{P}_t$. So, $\mathcal{P}$ satisfies condition (3).

Finally, we will verify condition (2). Let $p \in \mathcal{P}_s$, and $g_j = (x_j, y_j) \in p$ so that $q_{max} = y_j/x_j = \min\{y/x \mid (x,y) \in p \wedge x, y > 0\}$. If p contains any points in the third quadrant, let $q_{min} = \max\{(y/x \mid (x,y) \in p \wedge x, y < 0\}$, otherwise, let $q_{min} = 0$. Note that $q_{min} \lneq q_{max}$. Find the first $g_k = (x_k, y_k)$ in the enumeration so that g_k is in the first quadrant of the plane and $q_{min} \lneq y_k/x_k \lneq q_{max}$, and let $g_{k'} = -g_k$. Let $t = max\{j, k, k'\}$, and let n be the cardinality of p. Then by the end of stage $t+1$, $p \cup \{-g_k\}$ will be added to $\mathcal{P}$, and by the end of stage $t+n+1$, $p \cup \{g_k\}$ will have be added to $\mathcal{P}$. □

As a corollary, we obtain the following results.

Corollary 4.2: *Every computable group isomorphic to $\mathbb{Z}^k$ for $k > 1$ admits orderings in all tt-degrees. Every computable group isomorphic to $\mathbb{Z}^\omega$, which has a computable basis admits orderings in all tt-degrees.*

Proof: Consider $\mathbb{Z}^k$. Let $\mathbf{x}$ be a *tt*-degree, and $\prec$ an ordering of $\mathbb{Z}^2$ of *tt*-degree $\mathbf{x}$. For any non-identity element $g = (x_1, \ldots, x_k)$ of $\mathbb{Z}^k$, declare g to be greater than the identity in $\mathbb{Z}^k$ if and only if $(0,0) \prec (x_1, x_2)$ in $\mathbb{Z}^2$, or if $x_1 = x_2 = 0$ and the first non-zero entry of g is positive. Clearly, this ordering is *tt*-equivalent to $\prec$.

Similarly, we obtain the same result for a computable isomorphic copy of $\mathbb{Z}^\omega$ with a computable basis. □

We will use Corollary 4.2 in the proof of Theorem 5.5 at the end of the next section.

5. Orderings of Finitely Presented Residually Nilpotent Groups

We now apply Theorem 3.3 to a large class of finitely presented, residually nilpotent groups, which are not nilpotent. Recall that for a group G, the *lower central series* of G is the descending sequence of subgroups of G, indexed by ordinals, $\{\gamma_\alpha(G)\}_{\alpha \geq 1}$, defined as follows:

$$\gamma_1(G) = G,$$
$$\gamma_{\alpha+1}(G) = [G, \gamma_\alpha(G)],$$
$$\gamma_\beta(G) = \bigcap_{\alpha<\beta} \gamma_\alpha(G) \text{ when } \beta \text{ is a limit ordinal.}$$

Here, as usual, for subgroups A and B of G, the commutator $[A, B]$ is the subgroup

$$[A, B] = \langle a^{-1}b^{-1}ab \mid a \in A, b \in B\rangle.$$

A group G is *residually nilpotent* if every non-identity element has a non-identity homomorphic image in some nilpotent group. For any group G, the subgroup $\gamma_\omega(G)$ is the smallest subgroup of G for which $G/\gamma_\omega(G)$ is residually nilpotent. Thus, G is residually nilpotent if and only if $\gamma_\omega(G)$ is the trivial group. It is well-known that finitely generated nilpotent groups are supersolvable. Furthermore, as it was shown in [26], supersolvable groups are residually finite. It follows that finitely generated nilpotent groups are finitely presented and have decidable word problem. It then follows that finitely presented residually nilpotent groups have decidable word problem, hence they have computable isomorphic copies.

Let $F_n = \langle x_1, x_2, \ldots, x_n \mid \ \rangle$ denote the free group with basis $\{x_1, x_2, \ldots, x_n\}$.

Theorem 5.1: *Let G be a computable, finitely presented, torsion-free group. Let*

$$G = \gamma_1(G) \geq \gamma_2(G) \geq \cdots$$

be the lower central series of G. Assume that $\gamma_\omega(G)$ is trivial, and for every $i \in \{1, 2, \ldots\}$, $\gamma_i(G)/\gamma_{i+1}(G)$ is non-trivial and torsion-free (i.e., G is residually nilpotent but not nilpotent). Then there are bi-orderings on G in all tt-degrees.

Before we start our proof of Theorem 5.1, we note that there are many examples of groups (which we will discuss later) to which Theorem 5.1 applies (for example, all fundamental groups of closed and oriented surfaces of genus $g \geq 2$).

Proof: Let a group

$$G = \langle x_1, \ldots, x_n \mid r_1, \ldots, r_k \rangle$$

be computable, torsion-free, and non-nilpotent with lower central series

$$G = \gamma_1(G) \geq \gamma_2(G) \geq \cdots \geq \gamma_\omega(G) = \{e\}.$$

The quotient groups of the lower central series of G are finite-rank (always true for finitely generated groups), and abelian (always true). In addition, assume that the quotient groups of the lower central series are torsion-free.

For $i \geq 1$, let $\gamma_i(G)/\gamma_{i+1}(G)$ have rank k_i, and write $\{h_{i,1}, \ldots, h_{i,k_i}\}$ for the generators of $\gamma_i(G)/\gamma_{i+1}(G)$. Note that if we choose some of the generators to be positive and some to be negative, this choice will induce a lexicographical bi-ordering of $\gamma_i(G)/\gamma_{i+1}(G)$ since this quotient group is abelian. Furthermore, we can make such a choice for each quotient group of the lower central series so that we have a family of orderings, $\{\prec_i\}_{i\geq 1}$. Together these induce a *standard* bi-ordering $\prec$ on G if we declare $g \prec e$ if and only if there is an i so that $g\gamma_{i+1}(G) \prec_i \gamma_{i+1}(G)$ (see [37], [42], [20]).

Our result will be established by an application of Theorem 3.3, so we will construct a family $\mathcal{P}$ of finite subsets of G satisfying the conditions stated in that theorem. The family will consist of two kinds of finite subsets of G: basic sets, which we will call *seeds* (see definition below), and extensions of these sets, which we will call *sprouts*.

We will describe a scheme for uniformly effectively selecting representatives of the generating elements (cosets) of each of the quotient groups. The seeds will contain representatives of the generating cosets (or inverses of these) from one of the quotient groups, selected in a systematic way. Sprouts will be finite extensions of seeds, each extendible to a standard bi-ordering of G.

The lower central series provides a complete filtration of $G - \{e\}$ by uniformly computably enumerable sets. As a consequence of the fact that the groups constituting the lower central series of a free group F_n are uniformly computable as sets, we have that the elements of the lower central series of G are uniformly c.e. They are each the range of a computable function – the canonical homomorphism from F_n to G – with computable domain. The homomorphism is computable since the group G is computable (see [11]).

For $i \geq 1$, let $\{a_{i,1}, \ldots, a_{i,k_i}\}$ be the set of representatives of the elements of a generating basis for $\gamma_i(G)/\gamma_{i+1}(G)$. That is, for every i, j, we have $a_{i,j} \in h_{i,j}$. We show in Corollary 5.3 below that a choice of representatives of all of these sets, $\{a_{i,j} : i \geq 1 \wedge 1 \leq j \leq k_i\}$, can be made uniformly computably in i, and that the rank of each quotient group can be computed. Each $g \in \gamma_i(G) - \gamma_{i+1}(G)$ is a member of a unique non-trivial coset $C_g = g\gamma_{i+1}(G)$. Since the quotient group is abelian, note that

$$C_g = \left(\prod_{j=1}^{k_i} (a_{i,j}^{\varepsilon_{i,j}})^{c_j} \right) \gamma_{i+1}(G)$$

for some choice of $\varepsilon_{i,j} \in \{-1, 1\}$ and $c_j \in \mathbb{Z}$. So, since $\gamma_{i+1}(G)$ is c.e., we can also computably enumerate C_g.

Construction

We finitely approximate a c.e. family of finite subsets, $\mathcal{P}$, satisfying the conditions in Theorem 3.3 as follows.

Stage 0. $\mathcal{P} = \emptyset$

Stage $s+1$. First, add new seeds to $\mathcal{P}$ by adding all sets of the form $\{a_{s,1}^{\varepsilon_{s,1}}, \ldots, a_{s,k_s}^{\varepsilon_{s,k_s}}\}$ where $\varepsilon_{s,j} \in \{-1, 1\}$. (There are only finitely many of these sets, in fact, 2^{k_s}.)

Next, add sprouts to $\mathcal{P}$.

(1) For each seed σ in $\mathcal{P}$, add the set obtained by forming words of length $\leq s$ from the members of σ that represent positive elements in the induced lexicographic ordering of the corresponding quotient group. More precisely, if $\sigma = \{a_{s,1}^{\varepsilon_{s,1}}, \ldots, a_{s,k_s}^{\varepsilon_{s,k_s}}\}$, add all words of length $\leq s$ of the form $\prod_{j=1}^{k_s} (a_{i,j}^{\varepsilon_{i,j}})^{c_j}$, where the first non-zero c_j is positive.

(2) Add all consistent unions of sets already contained in $\mathcal{P}$. (A union would be *inconsistent* if it contained elements g_1 and g_2 with $g_1 g_2 = e$. Again, there are only finitely many sets of this type.)

(3) For each p already in $\mathcal{P}$, add p' constructed as follows: Let p' be the subset generated by adding to p all the elements of $gC_{g,s}$ for $g \in p$,

where $C_{g,s}$ is the set of the first s elements enumerated from the unique non-trivial element of the quotient group of the lower central series to which g belongs. (The appropriate C_g can be identified by considering the history of the construction of p.)

This completes the construction of $\mathcal{P}$. □

We first recall the definition of Fox derivative and discuss its basic properties. Let $\mathbb{Z}[F_n]$ denote its integral group ring of the free group F_n. As shown in [18], there is a unique map $\frac{\partial}{\partial x_j} : F_n \to \mathbb{Z}[F_n]$, $j = 1, 2, \ldots, n$, called *derivation*, which satisfies the following conditions:

$$\frac{\partial}{\partial x_j}(x_i) = \begin{cases} 1 \text{ if } j = i \\ 0 \text{ if } j \neq i \end{cases} \text{ for } i = 1, 2, \ldots, n;$$

$$\frac{\partial}{\partial x_j}(uv) = \frac{\partial}{\partial x_j}(u) + u\frac{\partial}{\partial x_j}(v) \text{ for } u, v \in F_n.$$

The map $\frac{\partial}{\partial x_j}$ can be extended linearly to a map $\frac{\partial}{\partial x_j} : \mathbb{Z}[F_n] \to \mathbb{Z}[F_n]$, $j = 1, 2, \ldots, n$, called *Fox derivative.* This allows us to define the mth order derivative for $m > 1$:

$$\frac{\partial^m}{\partial x_{i_m} \partial x_{i_{m-1}} \cdots \partial x_{i_1}}(u) = \frac{\partial}{\partial x_{i_m}} \left(\frac{\partial^{m-1}}{\partial x_{i_{m-1}} \cdots \partial x_{i_1}}(u) \right) \text{ for } u \in F_n.$$

Assume that $X = \{x_1, x_2, \ldots, x_n\}$ is an ordered alphabet. Let X^* be the set of all words over X, and let X_k^* be the set of all words of length k. Define the set of Lyndon words of length k (where $k \geq 1$) to be the set W_k of words $w \in X_k^*$ for which w is lexicographically smaller (denoted by $\prec$) than any of its cyclic rearrangements[c]. Let

$$W = \bigcup_{k \geq 1} W_k$$

denote the set of all Lyndon words over alphabet X. As shown in [9], the size of the set W_k is given by

$$|W_k| = \frac{1}{k} \sum_{d|k} \mu\left(\frac{k}{d}\right) n^d,$$

[c]It is worth mentioning that the set W_k can also be defined as the set of words $w \in X_k^*$ such that w is strictly lexicographically smaller than any of its proper right factors.

where μ is the Möbius function on $\mathbb{Z}_+$ defined by:

$$\mu(m) = \begin{cases} 1 & \text{if } m = 1, \\ (-1)^i & \text{if } m = p_1 p_2 \cdots p_i \text{ where } p_1, \ldots, p_i \text{ are distinct primes,} \\ 0 & \text{if } m \text{ is divisible by a square.} \end{cases}$$

For example, let $X = \{x_1, x_2\}$ and $x_1 \prec x_2$. Then $w = x_1x_1x_2x_1x_2 \in W_4$ since $x_1x_1x_2x_1x_2 \prec x_1x_2x_1x_2x_1$, $x_1x_1x_2x_1x_2 \prec x_2x_1x_2x_1x_1$, $x_1x_1x_2x_1x_2 \prec x_1x_2x_1x_1x_2$ and $x_1x_1x_2x_1x_2 \prec x_2x_1x_1x_2x_1$. In this example we also have

$$W_1 = \{x_1, x_2\}, W_2 = \{x_1x_2\}, W_3 = \{x_1x_1x_2, x_1x_2x_2\}, \text{ and}$$
$$W_4 = \{x_1x_1x_1x_2, x_1x_1x_2x_2, x_1x_2x_2x_2\}.$$

In general, there is an efficient algorithm developed by Duval in [17] for finding elements of W_k for $k \geq 2$.

Define $\psi : \mathbb{Z}[F_n] \to \mathbb{Z}$ by $\psi\left(\sum_{u \in F_n} a_u u\right) = \sum_{u \in F_n} a_u$. Define $D_u^0 : F_n \to \mathbb{Z}$ as follows for $u = x_{i_1}x_{i_2}\cdots x_{i_m} \in X^*$ and $v \in F_n$:

$$D_u^0(v) = \psi\left(\frac{\partial^m}{\partial x_{i_1}\partial x_{i_2}\cdots\partial x_{i_m}}(v)\right).$$

For a Lyndon word $w \in W$, a factorization $w = uv$ is called the *standard factorization* of w iff v is a Lyndon word of the maximal length $|v| \geq 1$ (see, for instance, [9]). With each $w \in W$ we associate the *standard commutator* defined recursively as follows:

$$[w] = w \text{ if } w \in X, \text{ and}$$
$$[w] = [[u], [v]] \text{ if } w = uv \text{ is the standard factorization of } w.$$

Let $C_k = \{[w] \in F_n \mid w \in W_k\}$ be the standard commutators of weight $k \geq 1$. For instance, in the example above, we have

$$C_1 = \{x_1, x_2\}, C_2 = \{[x_1, x_2]\}, C_3 = \{[x_1, [x_1, x_2]], [[x_1, x_2], x_2]\}, \text{ and}$$
$$C_4 = \{[x_1, [x_1, [x_1, x_2]]], [[x_1, [x_1, x_2]], x_2], [[[x_1, x_2], x_2], x_2]\}.$$

Let

$$L_m(G) = \gamma_m(G)/\gamma_{m+1}(G).$$

It was shown in [9] that W_k has the following important properties.

(1) If $w \in W_k$ then: (a) For every $e \in X_k^*$, if $e \prec w$, then $D_e^0([w]) = 0$; (b) $D_w^0([w]) = 1$ (see Lemma 3.4 in [9]).

(2) For all $u \in F_n$, $u \in \gamma_k(F_n)$ iff $D^0_w(u) = 0$ for all $w \in \bigcup_{i=1}^{k-1} W_i$ (see Corollary 3.6 in [9]).
(3) The set $\{c\gamma_k(F_n) \mid c \in C_k\}$ is a basis of the free abelian group $L_k(F_n)$, $k \geq 1$ (see Theorem 3.5 in [9]).
(4) The maps $D^0_w : F_n \to \mathbb{Z}$ for $w \in W_k$ form a basis for the additive group $\mathrm{Hom}(L_k(F_n), \mathbb{Z})$ of homomorphisms of the multiplicative group $L_k(F_n)$ into the additive group $\mathbb{Z}$ (see Theorem 3.5 in [9]).

The following lemma is a direct consequence of an algorithm given in [9] for computing the presentation of $L_m(G)$. We modify this algorithm to make it uniform in $m \geq 1$.

Lemma 5.2: *Let $G = \langle x_1, x_2, \ldots, x_n \mid r_1, r_2, \ldots, r_k \rangle$ be a computable finitely presented group. A presentation of the quotient groups of the lower central series of G, $\{L_m(G)\}_{m<\omega}$, can be computed uniformly in m.*

Proof: Let $R = \{r_1, r_2, \ldots, r_k\}$. By $\langle\langle R \rangle\rangle$ we denote the normal closure of R in F_n. For $m \geq 1$, let $W_m = \{w_1, w_2, ..., w_s\}$ be the set of all Lyndon words of length m over the alphabet $X = \{x_1, x_2, \ldots, x_n\}$, and $C_m = \{b_1, b_2, \ldots, b_s\}$ be the set of the corresponding standard commutators. If $S_m = \{r_{m_1}, r_{m_2}, \ldots, r_{m_l}\} \subseteq F_n$ satisfies the condition

$$\langle\langle R \rangle\rangle \cap \gamma_m(F_n) = \langle\langle S_m \rangle\rangle \leq F_n,$$

then, according to [9], the group $L_m(G)$ has the following presentation

$$L_m(G) = \{\bar{b}_1, \bar{b}_2, \ldots, \bar{b}_s \mid \bar{r}_{m_1}, \bar{r}_{m_2}, \ldots, \bar{r}_{m_l}\},$$

where $\bar{b}_j = b_j\gamma_{m+1}(F_n)$, $1 \leq j \leq s$, and

$$\bar{r}_{m_j} = \prod_{k=1}^{s} b_k^{D^0_{w_k}(r_{m_j})}\gamma_{m+1}(F_n),\ 1 \leq j \leq l.$$

Hence, the presentation matrix for $L_m(G)$ is given by

$$M_m(G) = \begin{pmatrix} D^0_{w_1}(r_{m_1}) & D^0_{w_2}(r_{m_1}) & \cdots & D^0_{w_s}(r_{m_1}) \\ D^0_{w_1}(r_{m_2}) & D^0_{w_2}(r_{m_2}) & \cdots & D^0_{w_s}(r_{m_2}) \\ \vdots & \vdots & \cdots & \vdots \\ D^0_{w_1}(r_{m_l}) & D^0_{w_2}(r_{m_l}) & \cdots & D^0_{w_s}(r_{m_l}) \end{pmatrix}.$$

The set S_m, $m \geq 1$, defined above is constructed inductively on m as follows. For $m = 1$, one naturally sets $S_1 = \{r_1, r_2, \ldots, r_k\}$. Assume by induction that for $m \geq 1$, the set $S_m = \{r_{m_1}, r_{m_2}, \ldots, r_{m_l}\}$ has been determined. Let $v_1, v_2, \ldots, v_l$ be the rows of $M_m(G)$. The row space of the matrix $M_m(G)$ is generated over $\mathbb{Z}$ by linearly independent vectors $v_{i_1}, v_{i_2}, \ldots, v_{i_{p_m}}$, where $1 \leq i_1 < i_2 < \cdots < i_{p_m} \leq l$, and, to make our procedure uniform in m, we require that $v_{i_1}, v_{i_2}, \ldots, v_{i_{p_m}}$ be the first linearly independent rows chosen from $\{v_1, v_2, \ldots, v_l\}$, while $u_{k_1}, u_{k_2}, \ldots, u_{k_{q_m}}$, $1 \leq k_1 < k_2 < \cdots < k_{q_m} \leq l$, are the remaining (linearly dependent) vectors from $\{v_1, v_2, \ldots, v_l\}$. Clearly, we have $p_m + q_m = l$, and define the corresponding relations: $s_j = r_{m_{i_j}}$, for $1 \leq j \leq p_m$, and $t_j = r_{m_{k_j}}$, for $1 \leq j \leq q_m$. For every $i \in \{1, 2, \ldots, q_m\}$, there are unique integers $\beta_{i,k} \in \mathbb{Z}$ such that $z_i = t_i \left(\prod_{k=1}^{p} s_k^{\beta_{i,k}} \right)^{-1} \in \gamma_{m+1}(F_n)$. These coefficients can be found by taking $(\beta_{i,1}, \beta_{i,2}, \ldots, \beta_{i,p_m}) \in \mathbb{Z}^{p_m}$ and then checking whether

$$D_w^0(z_i) = 0 \text{ for all } w \in \bigcup_{j=1}^{m} W_j \text{ (see Corollary 3.6 in [9]).}$$

Let

$$S_{m+1} = \{z_1, z_2, \ldots, z_{q_m}\} \cup \{[s_k, x_j] \mid k \in \{1, 2, ..., p_m\} \wedge j \in \{1, 2, \ldots, n\}\}.$$

It follows from [8] (see Lemma A5) that

$$\langle\langle R \rangle\rangle \cap \gamma_{m+1}(F_n) = \langle\langle S_{m+1} \rangle\rangle \leq F_n.$$

It follows that the presentation matrix for $L_m(G)$ is given by

$$M'_m(G) = \begin{pmatrix} D_{w_1}^0(s_1) & D_{w_2}^0(s_1) & \cdots & D_{w_s}^0(s_1) \\ D_{w_1}^0(s_2) & D_{w_2}^0(s_2) & \cdots & D_{w_s}^0(s_2) \\ \vdots & \vdots & \cdots & \vdots \\ D_{w_1}^0(s_{p_m}) & D_{w_2}^0(s_{p_m}) & \cdots & D_{w_s}^0(s_{p_m}) \end{pmatrix}.$$

This completes the proof. □

For a finitely generated abelian group A there are nonnegative integers k, r, and if $k > 0$ integers $d_1, d_2, ..., d_k$ such that $1 \leq d_1 \mid d_2 \mid \cdots \mid d_k$ and

$$A \simeq \mathbb{Z}_{d_1} \oplus \mathbb{Z}_{d_2} \oplus \cdots \oplus \mathbb{Z}_{d_k} \oplus \mathbb{Z}^r.$$

The integers $d_1, d_2, \ldots, d_k$ are called the *elementary divisors* of A, and r is called the *rank* of A – these numbers are unique for A. For our purpose, we assume that $m \leq n$, and let $A = \langle x_1, x_2, \ldots, x_n \mid R_1, R_2, \ldots, R_m \rangle$, where $R_j = \sum_{i=1}^{n} m_{i,j} x_i$ for $m_{ij} \in \mathbb{Z}$ and $1 \leq j \leq m$, be a presentation of A.

Here, we outline a construction of an isomorphism $\varphi : A \to B$, where $B = \langle z_1, z_2, \ldots, z_k, y_1, y_2, \ldots, y_r \mid d_1 z_1, \ldots, d_k z_k \rangle$, and $1 \leq d_1 \mid d_2 \mid \cdots \mid d_k$, given the presentation matrix $M = (m_{i,j})_{1 \leq i \leq m,\ 1 \leq j \leq n}$ of A.

Recall that an integer $m \times n$ matrix $D = (d_{i,j})$ with rank $r \leq m$ is said to be in the *Smith normal form* if:

(i) D is diagonal,

(ii) $d_{ii} \in \mathbb{Z}_+$, where $1 \leq i \leq r$,

(iii) $d_{i,i} \mid d_{i+1,i+1}$, where $1 \leq i \leq r-1$, and

(iv) $d_{i,i} = 0$, where $r+1 \leq i \leq m$.

As shown in [43], for M, there are invertible integer matrices $U \in \mathrm{GL}(m, \mathbb{Z})$ and $V \in \mathrm{GL}(n, \mathbb{Z})$ such that $D = UMV$ is in the Smith normal form[d]. Let $A_n = \langle x_1, x_2, \ldots, x_n \mid \ \rangle$ denote the free abelian group of rank n and let $N = \mathrm{Span}_{\mathbb{Z}}(\{R_1, R_2, \ldots, R_m\})$. Clearly, $A = A_n/N$. Let $V = (v_{i,j})_{1 \leq i,j \leq n}$ and $V^{-1} = (\bar{v}_{i,j})_{1 \leq i,j \leq n}$. Define

$$z_i = \sum_{j=1}^{n} \bar{v}_{i,j} x_i \text{ for } i = 1, 2, \ldots, r, \text{ and}$$

$$y_{i-r} = \sum_{j=1}^{n} \bar{v}_{i,j} x_i \text{ for } i = r+1, r+2, \ldots, n.$$

Consider a free abelian group $B_n = \langle z_1, z_2, \ldots, z_r, y_1, y_2, \ldots, y_{n-r} \mid \ \rangle$ (with the basis $\{z_1, z_2, \ldots, z_r, y_1, y_2, \ldots, y_{n-r}\}$). Since $V \in \mathrm{GL}(n, \mathbb{Z})$, the map

$$\varphi : A_n \to B_n \text{ given by}$$
$$\varphi(x_i) = \sum_{j=1}^{r} v_{i,j} z_j + \sum_{j=r+1}^{n} v_{i,j} y_{j-r} \text{ for } i = 1, 2, \ldots, n,$$

is an isomorphism between free abelian groups of rank n. Let $\overline{R}_j = \varphi(R_j)$ for $j = 1, 2, \ldots, m$. Then

$$N' = \mathrm{Span}_{\mathbb{Z}}\left(\{\overline{R}_j \mid 1 \leq j \leq m\}\right) = \mathrm{Span}_{\mathbb{Z}}\left(\{d_{i,i} z_i \mid 1 \leq i \leq r\}\right),$$

and $\varphi(N) = N'$. Therefore, φ induces an isomorphism of the quotient groups $A = A_n/N$ and $B = B_n/N' = \langle z_1, z_2, \ldots, z_r, y_1, y_2, \ldots, y_{n-r} \mid d_1 z_1, \ldots, d_k z_r \rangle$, where $1 \leq d_1 \mid d_2 \mid \cdots \mid d_r$.

[d]There are many well-known algorithms for computing matrices D, U, and V (see, for instance, [29, 24]).

The following corollary is an easy consequence of Lemma 5.2 and above considerations.

Corollary 5.3: *Let G be as in the statement of Theorem 5.1. Then a basis of the quotient groups of the lower central series of G, $\{L_m(G)\}_{1\leq m<\omega}$, can be found uniformly computably in m.*

Lemma 5.4: *The family $\mathcal{P}$ of finite subsets satisfies the conditions of Theorem 3.3.*

Proof: Let $p \in \mathcal{P}$. To see that condition (1) holds, note that p can clearly be extended to the upper cone $P_<$ of some standard bi-ordering of G (all the quotient groups are orderable, and each of these orderings extends to some standard ordering of G). Since $P_< - \{e\}$ is a semigroup containing p, we must have that $e \notin S(p)$.

To see that condition (2) holds, let i be the largest natural number so that there is some $a_{i,j}^{\varepsilon_{i,j}} \in p$. At stage $i+1$, the seeds $\sigma_0 = \{a_{i+1,1}, \ldots, a_{i+1,k_{i+1}}\}$ and $\sigma_1 = \{a_{i+1,1}^{-1}, \ldots, a_{i+1,k_{i+1}}\}$ will be added to $\mathcal{P}$, and at stage $i+2$, both $r_0 = p \cup \sigma_0$ and $r_1 = p \cup \sigma_1$ will be added to $\mathcal{P}$. These both extend p, one contains $a_{i+1,1}$ and the other contains its inverse, $a_{i+1,1}^{-1}$.

Finally, as in condition (3), let g be a non-identity element of G, and assume that $g \in \gamma_i(G) - \gamma_{i+1}(G)$, and that (a finite set) p has been added to $\mathcal{P}$ at stage s. Then either p contains $\{a_{i,1}^{\varepsilon_{i,1}}, \ldots, a_{i,k_i}^{\varepsilon_{i,k_i}}\}$ for some choice of $\varepsilon_{i,j}$'s, if $s \geq i+2$, or an extension $p' \supset p$ contains such a set at stage $i+2$. Since these elements are representatives of generating sets of $\gamma_i(G)/\gamma_{i+1}(G)$, we can write

$$g = \left(\prod_{j=1}^{k_i} (a_{i,j}^{\varepsilon_{i,j}})^{c_j}\right) w$$

for some $w \in \gamma_{i+1}$ and integers c_j. If the first non-zero c_j is positive, then g will be enumerated into some p'' extending p' at some later stage as w appears in the enumeration of $\gamma_{i+1}(G)$. If the first non-zero c_j is negative, then g^{-1} is an element of

$$\left(\prod_{j=1}^{k_i} (a_{i,j}^{\varepsilon_{i,j}})^{-c_j}\right) \gamma_{i+1}(G)$$

(the quotient group is abelian), and so will be enumerated into some p'' extending p' at a later stage.

In conclusion, the conditions of Theorem 3.3 are satisfied, so it follows that G admits a bi-ordering in every *tt*-degree. □

Solomon [45, 46] showed that n-step nilpotent groups admit left orderings in every Turing degree above $\mathbf{0}^{(\mathbf{n})}$. We now consider finitely presented nilpotent groups satisfying an additional condition and show that they admit bi-orderings of every *tt*-degree. First, note that finitely presented nilpotent groups have a decidable word problem, so have computable isomorphic copies.

Theorem 5.5: *Let G be a computable, finitely presented, nilpotent group with lower central series*

$$\gamma_1(G) \geq \gamma_2(G) \geq \cdots \geq \gamma_n(G) = \{e\}$$

having the property that for each $i \in \{1, \ldots, n-1\}$, $\gamma_i(G)/\gamma_{i+1}(G)$ is torsion-free, and furthermore, for some $j < n$, $\gamma_j(G)/\gamma_{j+1}(G)$ is isomorphic to $\mathbb{Z}^k$ for $k \geq 2$. Then G admits bi-orderings of every tt-degree.

Proof: We can again make use of a set of canonical representatives of the cosets in the quotient groups of the lower central series of G, obtained from a basis as described above.

Let j be chosen so that the group $\gamma_j(G)/\gamma_{j+1}(G)$ (which is necessarily torsion-free and abelian of finite rank) has rank greater than 1. Then, by Corollary 4.2, this quotient group admits orderings of every *tt*-degree. Choose a (finite) family of orderings, $\{<_i\}_{i<n}$, on the quotient groups so that for all $i \neq j$, $<_i$ is computable, and $<_j$ has arbitrary *tt*-degree $\mathbf{x}$. Define the ordering $\prec$ on G by declaring

$$e_G \prec g \iff g \in \gamma_i(G) - \gamma_{i+1}(G) \text{ and } \gamma_{i+1} <_i g\gamma_{i+1}.$$

Then $\prec$ is a bi-ordering of G of *tt*-degree $\mathbf{x}$. To determine whether a non-identity element $g \in G$ is in the positive cone $P_\prec$, we need to find i so that $g \in \gamma_i(G) - \gamma_{i+1}(G)$ (this can be done computably), then find the canonical representative of its coset (also computable), and finally check to see if the representative is positive or negative according to $<_i$. In the case when $i \neq j$, this is a computable procedure. When $i = j$, we need only query the oracle $<_j$ about the representative of g in its coset. Note that this is a *tt*-reduction, since for any g we can compute the appropriate representative.

On the other hand, to *tt*-compute $<_j$ from $\prec$, given an element of the (computable) set of representatives of elements of $\gamma_j(G)/\gamma_{j+1}(G)$, we need only check whether it is positive according to $\prec$. □

Theorem 5.1 applies to many important classes of groups. One of them is the collection of fundamental groups $S_{g,0}$ of closed and oriented surfaces of genus $g \geq 2$. Such groups have presentations of the form

$$S_{g,0} = \langle x_1, y_1, \ldots, \ x_g, y_g \mid [x_1, y_1][x_2, y_2] \cdots [x_g, y_g] \rangle.$$

They are known to have a decidable word problem, are torsion-free and residually nilpotent (see [3, 19]). The quotient groups $\gamma_i(S_{g,0})/\gamma_{i+1}(S_{g,0})$, $i \geq 1$, are free abelian groups of finite rank[e]. In the Appendix, we give a detailed calculations using the method described in our proof of Lemma 5.2.

Other important examples of groups for which Theorem 5.1 applies can be found in the class of finitely generated one-relator parafree groups (see families of groups (5.1), (5.2), and (5.3) below). Recall that a group G is *parafree* if G is residually nilpotent and for some $k \geq 1$ and all n:

$$G/\gamma_n(G) \simeq F_k/\gamma_n(F_k).$$

It is well-known that a finitely presented parafree group G has a decidable word problem and, clearly, the quotients $\gamma_i(G)/\gamma_{i+1}(G)$ are free abelian groups of finite rank for all $i \geq 1$. In [4], Baumslag introduced an infinite family of finitely generated one-relator parafree but not free groups. These groups are given by the following presentations

$$G_{i,j} = \langle a, b, c \mid a = [c^i, a][c^j, b] \rangle, \text{ where } ij \neq 0. \tag{5.1}$$

Furthermore, let $\mathbb{F}_{n+2} = \langle s, t, x_1, \ldots, x_n \mid \quad \rangle$ be a free group of rank $n+2$, let $\mathbb{F}_{n+1} = \langle t, x_1, \ldots, x_n \mid \quad \rangle \leq \mathbb{F}_{n+2}$, and $w \in \mathbb{F}_{n+1}$. Assume that w involves x_1 and $w \in \mathbb{F}_{n+2}^{(k)}$, where $\mathbb{F}_{n+2}^{(k)}$ denotes kth term of the derived series of $\mathbb{F}_{n+2}$. Consider the family of groups given by the presentations

$$H_w = \langle s, t, x_1, \ldots, x_n \mid x_1 = w[s, t] \rangle.$$

The family of groups H_w was introduced in [5] by Baumslag. He proved that every H_w is parafree but not free. In particular, groups given by the following presentations

$$H_{i,j} = \langle a, s, t \mid a = [a^i, t^j][s, t] \rangle, \text{ where } i, j \geq 1, \tag{5.2}$$

[e] The quotients $\gamma_i(S_g)/\gamma_{i+1}(S_g)$, $i \geq 1$, are free abelian groups since the associated Lie ring of S_g is a free Lie ring by a result in [31].

are parafree but not free. Finally, in [6], Baumslag and Cleary introduced the family of parafree groups given by the following presentations

$$K_w = \langle t, x_1, \ldots, x_k \mid x_1^m w = t^n \rangle ,$$

where $m, n \in \mathbb{Z}_+$, $\gcd(m, n) = 1$, $x_1^m w$ is not a proper power in $\mathbb{F}_{k+1} = \langle t, x_1, \ldots, x_k \mid \quad \rangle$, and $w \in \mathbb{F}_{k+1}^{(1)}$. They showed that K_w is a free group iff $n = 1$ or $x_1^m w$ is a primitive element in $\mathbb{F}_{k+1}$. In particular, groups given by the presentations

$$K_{i,j} = \langle a, s, t \mid a^i [s, a] = t^j \rangle , \text{ where } i \geq 1,\ j \geq 2 \text{ and } \gcd(i, j) = 1, \tag{5.3}$$

are all parafree but not free. All of the above groups are one-relator groups for which $G/\gamma_n(G) \simeq F_2/\gamma_n(F_2)$, where G is $G_{i,j}, H_w, H_{i,j}, K_{i,j}$, respectively, and F_2 denotes the usual free group of rank 2.

Our last example of groups that satisfy assumptions of Theorem 5.1 are right-angled Artin groups (*RAAGs*). They are constructed as follows. Let Γ be a finite graph with vertices $V(\Gamma) = \{1, 2, \ldots, n\}$, where $n \geq 2$, and edges $E(\Gamma)$. Define a *right-angled Artin group* A_Γ by the following presentation

$$A_\Gamma = \langle x_1, \ldots, x_n \mid [x_i, x_j],\ (i, j) \in E(\Gamma) \rangle .$$

If Γ is not a complete graph on n vertices and $E(\Gamma) \neq \emptyset$, then A_Γ is neither free abelian of rank n nor a free group of rank n (if $E(\Gamma) = \emptyset$, then A_Γ is a free group, and if Γ is a complete graph on n vertices, then A_Γ is a free abelian group of rank n). As it was shown in [25, 47], all *RAAGs* are biautomatic (consequently have a decidable word problem), bi-orderable (consequently are torsion-free [15]), and are residually nilpotent [13]. We also note here that *RAAGs* play a very important role in geometric group theory. In particular, as it was shown in [14], A_Γ is the fundamental group of a 3-manifold M^3 iff every component of Γ is a tree[f] or a triangle. Furthermore, *RAAGs* also appear in the study of the fundamental groups of the configuration spaces of n distinct points on the graph formed by robots' tracks [1], [21].

We now consider the class of torsion-free nilpotent groups, which are very well understood. For example, it is well known that such groups are finitely presented and have decidable word, conjugacy, and isomorphism problems (a result due to Grunewald and Segal [40]). We start with the

[f]In the case when components of A_Γ are trees, $A_\Gamma \simeq \pi_1(M^3)$, where M^3 is a graph manifold [14].

most elementary example – a *free nilpotent group with n generators of class m*, which we denote by $F(n,m)$. Let W_{m+1} be the set of all Lyndon words of length $m \geq 1$ over the alphabet $X = \{x_1, x_2, \ldots, x_n\}$, and let $C_{m+1} = \{[w] \in F_n \mid w \in W_{m+1}\}$ be the corresponding set of all standard commutators. Consider the presentation of $F(n,m)$:

$$F(n,m) = F_n/\gamma_{m+1}(F_n) = \langle x_1, x_2, \ldots, x_n \mid C_{m+1}\rangle ,$$

Clearly, for $i = 1, 2, \ldots, m$, we have $\gamma_i(F(n,m))/\gamma_{i+1}(F(n,m)) \simeq \mathbb{Z}^{k_i}$, where $k_i = \frac{1}{i}\sum_{d|i} \mu\left(\frac{i}{d}\right) n^d$ and μ is the Möbius function. Since parafree groups are torsion-free and residually nilpotent, we see that the quotients of groups (5.1), (5.2), and (5.3) by the corresponding $(k+1)$st term of their lower central series are nilpotent of class k and torsion-free. In fact,

$$G_{i,j}/\gamma_{k+1}(G_{i,j}) \simeq H_{i,j}/\gamma_{k+1}(H_{i,j}) \simeq K_{i,j}/\gamma_{k+1}(K_{i,j}) \simeq F_2/\gamma_{k+1}(F_2).$$

Therefore, Theorem 5.5 applies to all groups mentioned above. Furthermore, as we mentioned before, for a surface groups $S_{g,0}$ of genus $g \geq 2$, we have $\bigcap_{j\geq 0} \gamma_j(S_{g,0}) = \{e\}$, and, for all $j \geq 1$, the quotient groups $\gamma_j(S_{g,0})/\gamma_{j+1}(S_{g,0})$ are free abelian of rank r_n, where

$$r_n = \frac{1}{n}\sum_{d|n} \mu(n/d) \left[\sum_{0\leq i\leq [d/2]} \frac{(-1)^i d}{d-i} \binom{d-i}{i} (2g)^{d-2i} \right].$$

Thus, $S_{g,0}/\gamma_{k+1}(S_{g,0})$ is torsion-free nilpotent of class $k \geq 1$. Hence we can apply Theorem 5.5 to all groups $S_{g,0}/\gamma_{k+1}(S_{g,0})$.

Finally, as we also mentioned before, right-angled Artin groups A_Γ are torsion-free and residually nilpotent so, in particular, we have that for all $k \geq 1$, the groups $A_\Gamma/\gamma_{k+1}(A_\Gamma)$ are torsion-free nilpotent and Theorem 5.5 can be applied.

Appendix

In this appendix we present an example with detailed computation for the purpose of illustrating Lemma 5.2. Let us consider the fundamental group of a surface of genus $g \geq 2$ with no boundary components

$$S_{g,0} = \langle x_1, y_1, \ldots, x_g, y_g \mid r_g\rangle , \text{ where } r_g = [x_1, y_1] \cdots [x_g, y_g] .$$

As it was shown independently by Baumslag [3] and Frederick [19], $S_{g,0}$ is residually nilpotent. Moreover, as a one-relator group, $S_{g,0}$ has a decidable word problem, and since r_g is cyclically reduced and not a proper power in $F_{2g} = \langle x_1, y_1, \ldots, x_g, y_g \mid \quad \rangle$, it follows that $S_{g,0}$ is also torsion-free. Furthermore, we see that $r_g \in \gamma_2(S_{g,0})$ and $r_g \notin \gamma_3(S_{g,0})$. It follows from [31] that the group $L_n(S_{g,0}) = \gamma_n(S_{g,0})/\gamma_{n+1}(S_{g,0})$, $n = 1, 2, \ldots$, is a torsion-free abelian group of rank m_n:

$$m_n = \frac{1}{n}\sum_{d|n} \mu(n/d)\left[\sum_{0\leq i\leq [d/2]} (-1)^i \frac{d}{d-i}\binom{d-i}{i}(2g)^{d-2i}\right],$$

where μ is the Möbius function. We now consider the simplest case when $g = 2$, i.e., $S_{2,0} = \langle x_1, y_1, x_2, y_2 \mid r\rangle$, where $r = [x_1, y_1][x_2, y_2]$. We have

$$m_n = \frac{1}{n}\sum_{d|n} \mu(n/d)\left[\sum_{0\leq j\leq [d/2]} (-1)^j \frac{d}{d-j}\binom{d-j}{j}(4)^{d-2j}\right] \quad (5.4)$$

for $n = 1, 2, 3, \ldots$, while for the free group $F_4 = \langle x_1, y_1, x_2, y_2 \mid \quad \rangle$ of rank 4 we have

$$k_n = \frac{1}{n}\sum_{d|n} \mu(\frac{n}{d})4^d. \quad (5.5)$$

From the identities (5.4) and (5.5) we obtain

k_1	k_2	k_3	k_4	m_1	m_2	m_3	m_4
4	6	20	60	4	5	16	52

Since $k_4 = 60$ and $m_4 = 52$ we show the computations only for $L_n(S_{2,0})$, $n = 1, 2, 3$. We will follow the notation used in Lemma 5.2.

For $L_1(S_{2,0})$, we have that $W_1 = \{x_1, y_1, x_2, y_2\}$ is the set of Lyndon words of length 1, where $x_1 \prec y_1 \prec x_2 \prec y_2$. We have the following basis $C_1 = \{a_1, a_2, a_3, a_4\}$, where $a_1 \in x_1\gamma_2(F_4)$, $a_2 \in y_1\gamma_2(F_4)$, $a_3 \in x_2\gamma_2(F_4)$, and $a_4 \in y_2\gamma_2(F_4)$, for $L_1(F_4) = \gamma_1(F_4)/\gamma_2(F_4) \simeq \mathbb{Z}^4$. Let $S_1 = \{r_{1,1}\}$, where $r_{1,1} = r$, and since

$$\begin{aligned}
D^0_{x_1}(r_{1,1}) &= \psi\left(-x_1^{-1} + x_1^{-1}y_1^{-1}\right) = 0,\\
D^0_{y_1}(r_{1,1}) &= \psi\left(-x_1^{-1}y_1^{-1} + x_1^{-1}y_1^{-1}x_1\right) = 0,\\
D^0_{x_2}(r_{1,1}) &= \psi\left(-[x_1, y_1]x_2^{-1} + [x_1, y_1]x_2^{-1}y_2^{-1}\right) = 0,\\
D^0_{y_2}(r_{1,1}) &= \psi\left(-[x_1, y_1]x_2^{-1}y_2^{-1} + [x_1, y_1]x_2^{-1}y_2^{-1}x_2\right) = 0,
\end{aligned}$$

we have

$\diagdown$	$D^0_{x_1}$	$D^0_{y_1}$	$D^0_{x_2}$	$D^0_{y_2}$
$r_{1,1}$	0	0	0	0

From Lemma 5.2 we obtain the following presentation matrix for $L_1(S_{2,0})$:

$$M_1(S_{2,0}) = (0\,0\,0\,0).$$

Therefore, if $r_{2,1} = r_{1,1}$, we have $S_2 = \{r_{2,1}\}$,

$$L_1(S_{2,0}) = \langle a_1, a_2, a_3, a_4 \mid \quad \rangle \simeq \mathbb{Z}^4$$

and it has the following basis $B_1 = \{a_1, a_2, a_3, a_4\}$.

For $L_2(S_{2,0})$, we have that $W_2 = \{x_1y_1, x_1x_2, x_1y_2, y_1x_2, y_1y_2, x_2y_2\}$ is the set of Lyndon words of length 2, where $x_1y_1 \prec x_1x_2 \prec x_1y_2 \prec y_1x_2 \prec y_1y_2 \prec x_2y_2$. We have the following basis $C_2 = \{a_1, a_2, \ldots, a_6\}$, where

$$\begin{aligned} &a_1 \in [x_1, y_1]\,\gamma_3(F_4),\ a_2 \in [x_1, x_2]\,\gamma_3(F_4),\ a_3 \in [x_1, y_2]\,\gamma_3(F_4),\\ &a_4 \in [y_1, x_2]\,\gamma_3(F_4),\ a_5 \in [y_1, y_2]\,\gamma_3(F_4),\ a_6 \in [x_2, y_2]\,\gamma_3(F_4), \end{aligned}$$

for $L_2(F_4) = \gamma_2(F_4)/\gamma_3(F_4) \simeq \mathbb{Z}^6$. Since $S_2 = \{r_2\}$, and

$$\begin{aligned} D^0_{x_1y_1}(r_{2,1}) &= \psi\left(D_{x_1}\left(D_{y_1}(r_{2,1})\right)\right)\\ &= \psi\left(D_{x_1}\left(-x_1^{-1}y_1^{-1} + x_1^{-1}y_1^{-1}x_1\right)\right)\\ &= \psi\left(x_1^{-1} - x_1^{-1} + x_1^{-1}y_1^{-1}\right) = 1,\\ D^0_{x_1x_2}(r_{2,1}) &= \psi\left(D_{x_1}\left(D_{x_2}(r_{2,1})\right)\right)\\ &= \psi\left(x_1^{-1} - x_1^{-1}y_1^{-1} - x_1^{-1} + x_1^{-1}y_1^{-1}\right) = 0,\\ D^0_{x_1y_2}(r_{2,1}) &= \psi\left(D_{x_1}\left(D_{y_2}(r_{2,1})\right)\right)\\ &= \psi\left(x_1^{-1} - x_1^{-1}y_1^{-1} - x_1^{-1} + x_1^{-1}y_1^{-1}\right) = 0,\\ D^0_{y_1x_2}(r_{2,1}) &= \psi\left(D_{y_1}\left(D_{x_2}(r_{2,1})\right)\right)\\ &= \psi\left(x_1^{-1}y_1^{-1} - x_1^{-1}y_1^{-1}x_1 - x_1^{-1}y_1^{-1} - x_1^{-1}y_1^{-1}x_1\right) = 0,\\ D^0_{y_1y_2}(r_{2,1}) &= \psi\left(D_{y_1}\left(D_{y_2}(r_{2,1})\right)\right)\\ &= \psi\left(x_1^{-1}y_1^{-1} - x_1^{-1}y_1^{-1}x_1 - x_1^{-1}y_1^{-1} - x_1^{-1}y_1^{-1}x_1\right) = 0,\\ D^0_{x_2y_2}(r_{2,1}) &= \psi\left(D_{x_2}\left(D_{y_2}(r_{2,1})\right)\right) = \psi\left([x_1, y_1]\,x_2^{-1}y_2^{-1}\right) = 1, \end{aligned}$$

we have

\	$D^0_{x_1y_1}$	$D^0_{x_1x_2}$	$D^0_{x_1y_2}$	$D^0_{y_1x_2}$	$D^0_{y_1y_2}$	$D^0_{x_2y_2}$
$r_{3,1}$	1	0	0	0	0	1

so, from Lemma 5.2, we obtain the following presentation matrix for $L_2(S_{2,0})$:

$$M_2(S_{2,0}) = (1\,0\,0\,0\,0\,1).$$

This matrix has the following Smith normal form

$$SM_2(S_{2,0}) = (1\,0\,0\,0\,0\,0).$$

By taking $z_1 = a_1 + a_6$, $z_2 = a_2$, $z_3 = a_3$, $z_4 = a_4$, $z_5 = a_5$, $z_6 = a_6$, we see that

$$\begin{aligned} L_2(S_{2,0}) &= \langle a_1, a_2, \ldots, a_6 \mid a_1 + a_6 = 0 \rangle \\ &= \langle z_2, z_3, \ldots, z_6 \mid \quad \rangle \simeq \mathbb{Z}^5, \end{aligned}$$

so $B_2 = \{z_2, z_3, \ldots, z_6\}$ is a basis for $L_2(S_{2,0})$, and if $r_{3,1} = [r_{2,1}, x_1]$, $r_{3,2} = [r_{2,1}, y_1]$, $r_{3,3} = [r_{2,1}, x_2]$, $r_{3,4} = [r_{2,1}, y_2]$, we have

$$S_3 = \{r_{3,1}, r_{3,2}, r_{3,3}, r_{3,4}\}.$$

For $L_3(S_{2,0})$, we have that $W_3 = \{x_1x_1y_1,\ x_1x_1x_2,\ x_1x_1y_2,\ x_1y_1y_1,\ x_1y_1x_2,\ x_1x_2y_1,\ x_1y_1y_2,\ x_1y_2y_1,\ x_1x_2x_2,\ x_1x_2y_2,\ x_1y_2x_2,\ x_1y_2y_2,\ y_1y_1x_2,\ y_1y_1y_2,\ y_1x_2x_2,\ y_1x_2y_2,\ y_1y_2x_2,\ y_1y_2y_2,\ x_2x_2y_2,\ x_2y_2y_2\}$ is the set of Lyndon words of length 3 ordered lexicographically. In order to compute all Fox derivatives, we need an additional property of the derivative, Property (3.3) on p. 90 in [18]. That is, if $X = \{x_1, x_2, \ldots, x_n\}$ is an ordered alphabet, X^* is the set of all words over the alphabet X, and X_m^* is the set of all words of length m, then for $a = a_1a_2\ldots a_m \in X_m^*$, $u \in \gamma_k(F_n)$, $v \in \gamma_l(F_n)$, where $k + l = m$, we have

$$D_a^0([u,v]) = D^0_{a_1a_2\ldots a_k}(u)\, D^0_{a_{k+1}a_{k+2}\ldots a_m}(v) - D^0_{a_1a_2\ldots a_l}(v)\, D^0_{a_{l+1}a_{l+2}\ldots a_m}(u).$$

Therefore, since $S_3 = \{r_{3,1}, r_{3,2}, r_{3,3}, r_{3,4}\}$ and $r_{2,1} \in \gamma_2(F_4)$, we have, for example,

$$\begin{aligned} D^0_{x_1x_1y_1}(r_{3,1}) &= D^0_{x_1x_1y_1}([r_{2,1}, x_1]) \\ &= D^0_{x_1x_1}(r_{2,1})\, D^0_{y_1}(x_1) - D^0_{x_1}(x_1)\, D^0_{x_1y_1}(r_{2,1}) \\ &= -D^0_{x_1}(x_1)\, D^0_{x_1y_1}(r_{2,1}) = -1, \end{aligned}$$

$$\begin{aligned} D^0_{x_1x_1x_2}(r_{3,1}) &= D^0_{x_1x_1x_2}([r_{2,1}, x_1]) \\ &= D^0_{x_1x_1}(r_{2,1})\, D^0_{x_2}(x_1) - D^0_{x_1}(x_1)\, D^0_{x_1x_2}(r_{2,1}) \\ &= -D^0_{x_1}(x_1)\, D^0_{x_1x_2}(r_{2,1}) = 0, \end{aligned}$$

$$\begin{aligned} D^0_{x_1x_1y_2}(r_{3,1}) &= D^0_{x_1x_1y_2}([r_{2,1}, x_1]) \\ &= D^0_{x_1x_1}(r_{2,1})\, D^0_{y_2}(x_1) - D^0_{x_1}(x_1)\, D^0_{x_1y_2}(r_{2,1}) \\ &= -D^0_{x_1}(x_1)\, D^0_{x_1y_2}(r_{2,1}) = 0, \end{aligned}$$

$$\begin{aligned} D^0_{x_1y_1y_1}(r_{3,1}) &= D^0_{x_1y_1y_1}([r_{2,1}, x_1]) \\ &= D^0_{x_1y_1}(r_{2,1})\, D^0_{y_1}(x_1) - D^0_{x_1}(x_1)\, D^0_{y_1y_1}(r_{2,1}) \\ &= -D^0_{x_1}(x_1)\, D^0_{y_1y_1}(r_{2,1}) = 0, \end{aligned}$$

$$\begin{aligned} D^0_{x_1y_1x_2}(r_{3,1}) &= D^0_{x_1y_1x_2}([r_{2,1}, x_1]) \\ &= D^0_{x_1y_1}(r_{2,1})\, D^0_{x_2}(x_1) - D^0_{x_1}(x_1)\, D^0_{y_1x_2}(r_{2,1}) \\ &= -D^0_{x_1}(x_1)\, D^0_{y_1x_2}(r_{2,1}) = 0, \end{aligned}$$

and, similarly, we find the remaining Fox derivatives. We obtain the following values for Fox derivatives:

\	$r_{3,1}$	$r_{3,2}$	$r_{3,3}$	$r_{3,4}$
$D^0_{x_1x_1y_1}$	−1	0	0	0
$D^0_{x_1x_1x_2}$	0	0	0	0
$D^0_{x_1x_1y_2}$	0	0	0	0
$D^0_{x_1y_1y_1}$	0	1	0	0
$D^0_{x_1y_1x_2}$	0	0	1	0
$D^0_{x_1x_2y_1}$	0	0	0	0
$D^0_{x_1y_1y_2}$	0	0	0	1
$D^0_{x_1y_2y_1}$	0	0	0	0
$D^0_{x_1x_2x_2}$	0	0	0	0
$D^0_{x_1x_2y_2}$	−1	0	0	0

\	$r_{3,1}$	$r_{3,2}$	$r_{3,3}$	$r_{3,4}$
$D^0_{x_1y_2x_2}$	1	0	0	0
$D^0_{x_1y_2y_2}$	0	0	0	0
$D^0_{y_1y_1x_2}$	0	0	0	0
$D^0_{y_1y_1y_2}$	0	0	0	0
$D^0_{y_1x_2x_2}$	0	0	0	0
$D^0_{y_1x_2y_2}$	0	−1	0	0
$D^0_{y_1y_2x_2}$	0	1	0	0
$D^0_{y_1y_2y_2}$	0	0	0	0
$D^0_{x_2x_2y_2}$	0	0	−1	0
$D^0_{x_2y_2y_2}$	0	0	0	1

It follows that the matrix of a presentation for the group $L_3(S_{2,0})$ is given by

$$M_3(S_{2,0}) = \begin{pmatrix} -1 & 0 & 0 & 0 & 0 & 0 & 0 & 0 & 0 & -1 & 1 & 0 & 0 & 0 & 0 & 0 & 0 & 0 & 0 & 0 \\ 0 & 0 & 0 & 1 & 0 & 0 & 0 & 0 & 0 & 0 & -1 & 1 & 0 & 0 & 0 & 0 & 0 & 0 & 0 & 0 \\ 0 & 0 & 0 & 0 & 1 & 0 & 0 & 0 & 0 & 0 & 0 & 0 & 0 & 0 & 0 & 0 & 0 & 0 & -1 & 0 \\ 0 & 0 & 0 & 0 & 0 & 0 & 1 & 0 & 0 & 0 & 0 & 0 & 0 & 0 & 0 & 0 & 0 & 0 & 0 & 1 \end{pmatrix}.$$

Let

a_1	$\in$	$[x_1, [x_1, y_1]]\,\gamma_4(F_4)$	a_6	$\in$	$[[x_1, x_2], y_1]\,\gamma_4(F_4)$
a_2	$\in$	$[x_1, [x_1, x_2]]\,\gamma_4(F_4)$	a_7	$\in$	$[x_1, [y_1, y_2]]\,\gamma_4(F_4)$
a_3	$\in$	$[x_1, [x_1, y_2]]\,\gamma_4(F_4)$	a_8	$\in$	$[[x_1, y_2], y_1]\,\gamma_4(F_4)$
a_4	$\in$	$[[x_1, y_1], y_1]\,\gamma_4(F_4)$	a_9	$\in$	$[[x_1, x_2], x_2]\,\gamma_4(F_4)$
a_5	$\in$	$[x_1, [y_1, x_2]]\,\gamma_4(F_4)$	a_{10}	$\in$	$[x_1, [x_2, y_2]]\,\gamma_4(F_4)$

a_{11}	$\in$	$[[x_1, y_2], x_2]\gamma_4(F_4)$	a_{16}	$\in$	$[y_1, [x_2, y_2]]\gamma_4(F_4)$
a_{12}	$\in$	$[[x_1, y_2], y_2]\gamma_4(F_4)$	a_{17}	$\in$	$[[y_1, y_2], x_2]\gamma_4(F_4)$
a_{13}	$\in$	$[y_1, [y_1, x_2]\gamma_4(F_4)]$	a_{18}	$\in$	$[[y_1, y_2], y_2]\gamma_4(F_4)$
a_{14}	$\in$	$[y_1, [y_1, y_2]]\gamma_4(F_4)$	a_{19}	$\in$	$[x_2, [x_2, y_2]]\gamma_4(F_4)$
a_{15}	$\in$	$[[y_1, x_2], x_2]\gamma_4(F_4)$	a_{20}	$\in$	$[[x_2, y_2], y_2]\gamma_4(F_4)$

We have the following basis $C_3 = \{a_i \mid i = 1, 2, \ldots, 20\}$ for $L_3(F_4) = \gamma_3(F_4)/\gamma_4(F_4) \simeq \mathbb{Z}^{20}$. From the presentation matrix we obtain the following presentation for the group $L_3(S_{2,0})$:

$$L_3(S_{2,0}) = \langle a_1, \ldots, a_{20} \mid -a_1 - a_{10} + a_{11} = 0,\ a_4 - a_{11} + a_{12} = 0,\\ a_5 - a_{19} = 0,\ a_7 + a_{20} = 0\rangle.$$

The Smith normal form for the matrix $M_3(S_{2,0})$ is

$$SM_3(S_{2,0}) = \begin{pmatrix} 1\,0\,0\,0\,0\,0\,0\,0\,0\,0\,0\,0\,0\,0\,0\,0\,0\,0\,0\,0 \\ 0\,1\,0\,0\,0\,0\,0\,0\,0\,0\,0\,0\,0\,0\,0\,0\,0\,0\,0\,0 \\ 0\,0\,1\,0\,0\,0\,0\,0\,0\,0\,0\,0\,0\,0\,0\,0\,0\,0\,0\,0 \\ 0\,0\,0\,1\,0\,0\,0\,0\,0\,0\,0\,0\,0\,0\,0\,0\,0\,0\,0\,0 \end{pmatrix} = M_3(S_{2,0})V,$$

where $V^{-1} = (v_{i,j})_{1\le i,j\le 20}$ and V^{-1} has the following nonzero entries $v_{1,1} = -1$, $v_{1,10} = -1$, $v_{1,11} = 1$, $v_{2,4} = 1$, $v_{2,11} = -1$, $v_{2,12} = 1$, $v_{3,5} = 1$, $v_{3,19} = -1$, $v_{4,7} = 1$, $v_{4,20} = 1$, $v_{5,3} = 1$, $v_{6,6} = 1$, $v_{7,2} = 1$, $v_{8,8} = \ldots = v_{20,20} = 1$. We let $z_1 = -a_1 - a_{10} + a_{11}$, $z_2 = a_4 - a_{11} + a_{12}$, $z_3 = a_5 - a_{19}$, $z_4 = a_7 + a_{20}$, $z_5 = a_3$, $z_6 = a_6$, $z_7 = a_2$, $z_8 = a_8$, $\ldots$, $z_{20} = a_{20}$. In the new generators, the presentation of $L_3(S_{2,0})$ is

$$\begin{aligned} L_3(S_{2,0}) &= \langle z_1, z_2, \ldots, z_{20} \mid z_1 = 0,\ z_2 = 0,\ z_3 = 0,\ z_4 = 0\rangle \\ &= \langle z_5, z_6, \ldots, z_{20} \mid \quad \rangle \simeq \mathbb{Z}^{16}, \end{aligned}$$

so $B_3 = \{z_5, z_6, \ldots, z_{20}\}$ is a basis for $L_3(S_{2,0})$. If

$$\begin{aligned} r_{4,1} &= [r_{3,1}, x_1],\ r_{4,2} = [r_{3,1}, y_1],\ r_{4,3} = [r_{3,1}, x_2], r_{4,4} = [r_{3,1}, y_2], \\ r_{4,5} &= [r_{3,2}, x_1],\ r_{4,6} = [r_{3,2}, y_1],\ r_{4,7} = [r_{3,2}, x_2], r_{4,8} = [r_{3,2}, y_2], \\ r_{4,9} &= [r_{3,3}, x_1],\ r_{4,10} = [r_{3,3}, y_1],\ r_{4,11} = [r_{3,3}, x_2], r_{4,12} = [r_{3,3}, y_2], \\ r_{4,13} &= [r_{3,4}, x_1],\ r_{4,14} = [r_{3,4}, y_1],\ r_{4,15} = [r_{3,4}, x_2], r_{4,16} = [r_{3,4}, y_2], \end{aligned}$$

we have

$$S_4 = \{r_{4,1}, r_{4,2}, \ldots, r_{4,16}\}.$$

We will end our computations at this place since the presentation matrix for $L_4(S_{2,0})$ is a 16×60 matrix.

References

1. A. Abrams and R. Ghrist, Finding topology in a factory: configuration spaces, *American Mathematical Monthly* 109 (2002), pp. 140–150.
2. J.W. Alexander, Ordered sets, complexes and the problem of compactification, *Proceedings of the National Academy of Science* 25 (1939), pp. 296–298.
3. G. Baumslag, On generalised free products, *Mathematische Zeitschrift* 78 (1962), pp. 423–438.
4. G. Baumslag, Some groups that are just about free, *Bulletin of the American Mathematical Society* 73 (1967), pp. 621–622.
5. G. Baumslag, Musings on Magnus, in: *The Mathematical Legacy of Wilhelm Magnus: Groups, Geometry and Special Functions* (Brooklyn, NY, 1992), Contemporary Mathematics 169, W. Abikoff, J.S. Birman, and K. Kuiken, editors, pp. 99–106, Providence, RI, American Mathematical Society, 1994.
6. G. Baumslag and S. Cleary, Parafree one-relator groups, *Journal of Group Theory* 9 (2006), pp. 191–201.
7. D. Cenzer and J.B. Remmel, *Effectively Closed Sets*, manuscript.
8. K.-T. Chen, Integration in free groups, *Annals of Mathematics* 54 (1951), pp. 147–162.
9. K.-T. Chen, R.H. Fox and R.C. Lyndon, Free Differential Calculus IV. The quotient groups of the lower central series, *Annals of Mathematics* 68 (1958), pp. 81–95.
10. J. Chubb, *Ordered Structures and Computability Theory*, PhD dissertation, George Washington University, 2009.
11. M.A. Dabkowska, M.K. Dabkowski, V.S. Harizanov, and A.A. Togha, Spaces of orders and their Turing degree spectra, *Annals of Pure and Applied Logic* 161 (2010), pp. 1134–1143.
12. R.G. Downey and S.A. Kurtz, Recursion theory and ordered groups, *Annals of Pure and Applied Logic* 32 (1986), pp. 137–151.
13. C.G. Droms, *Graph groups*, PhD dissertation, Syracuse University, 1983.
14. C. Droms, Graph groups, coherence, and three-manifolds, *Journal of Algebra* 106 (1987), pp. 484–489.
15. G. Duchamp, J-Y. Thibon, Simple orderings for free partially commutative groups, *International Journal of Algebra and Computation* 2 (1992), pp. 351–355.
16. J.-P. Duval, Factorizing words over an ordered alphabet, *Journal of Algorithms* 4 (1983), pp. 363–381.
17. J.-P. Duval, Génération d'une section des classes de conjugaison et arbre des mots de Lyndon de longueur bornée, *Theoretical Computer Science* 60 (1988), pp. 255–283 (in French).
18. R.H. Fox, Free differential calculus. I. Derivation in the free group ring, *Annals of Mathematics* 57 (1953), pp. 547–560.
19. K.N. Frederick, The Hopfian property for a class of fundamental groups, *Communications on Pure and Applied Mathematics* 16 (1963), pp. 1–8.
20. L. Fuchs, *Partially Ordered Algebraic Systems*, Pergamon Press, Oxford, 1963.

21. R. Ghrist, Configuration spaces and braid groups on graphs in robotics, in: *Knots, Braids, and Mapping Class Groups – Papers Dedicated to Joan S. Birman* (New York, 1998), AMS/IP Studies in Advanced Mathematics 24, J. Gilman, W.W. Menasco, and X-S. Lin, editors, pp. 29–40, American Mathematical Society, Providence, RI, 2001.
22. W.A. de Graaf and W. Nickel, Constructing faithful representations of finitely-generated torsion-free nilpotent groups, *Journal of Symbolic Computation* 33 (2002), pp. 31–41.
23. V. Harizanov, J.F. Knight, C. McCoy, V. Puzarenko, D.R. Solomon, and J. Wallbaum, Turing degrees and topology of orders on F_∞, preprint.
24. B. Hartley and T.O. Hawkes, *Rings, Modules and Linear Algebra*, Chapman and Hall, London, 1980.
25. S. Hermiller and J. Meier, Algorithms and geometry for graph products of groups, *Journal of Algebra* 171 (1995), pp. 230–257.
26. K.A. Hirsch, On infinite soluble groups. III, *Proceedings of the London Mathematical Society* 49 (1946), pp. 184–194.
27. C.G. Jockusch, Jr. and R. I. Soare, Π_1^0 classes and degrees of theories, *Transactions of the American Mathematical Society* 173 (1972), pp. 33–56.
28. A.M. Kach, K. Lange, and R. Solomon, Degrees of orders on torsion-free abelian groups, *Annals of Pure and Applied Logic*, 164 (2013), pp. 822–836.
29. R. Kannan and A. Bachem, Polynomial algorithms for computing the Smith and Hermite normal forms of an integer matrix, *SIAM Journal on Computing* 8 (1979), pp. 499–507.
30. V.M. Kopytov and N.Ya. Medvedev, *Right-Ordered Groups*, Siberian School of Algebra and Logic, Consultants Bureau, New York, 1996.
31. J.P. Labute, On the descending central series for groups with single defining relation, *Journal of Algebra* 14 (1970), pp. 16–23.
32. J. Łoś, On the existence of linear order in a group, *Bulletin de l'Académie Polonaise des Sciences Cl.* Ill (1954), pp. 21–23.
33. R.C. Lyndon and P.E. Schupp, *Combinatorial Group Theory*, Springer-Verlag, Berlin, 1977.
34. J. Malestein and A. Putman, On the self-intersections of curves deep in the lower central series of a surface group, *Geometriae Dedicata* 149 (2010), pp. 73–84.
35. G. Metakides and A. Nerode, Effective content of field theory, *Annals of Mathematical Logic* 17 (1979), pp. 289–320.
36. R.B. Mura and A. Rhemtulla, *Orderable Groups*, Lecture Notes in Pure and Applied Mathematics 27, Marcel Dekker, New York, 1977.
37. B.H. Neumann, On ordered groups, *American Journal of Mathematics* 71 (1949), pp. 1–18.
38. W. Nickel, Matrix representations for torsion-free nilpotent groups by Deep Thought, *Journal of Algebra* 300 (2006), pp. 376–383.
39. H. Rogers, Jr., *Theory of Recursive Functions and Effective Computability*, MIT Press, 1987.
40. D. Segal, *Polycyclic groups*, Cambridge Tracts in Mathematics 83, Cambridge University Press, Cambridge, 1983.

41. A.S. Sikora, Topology on the spaces of orderings of groups, *Bulletin of the London Mathematical Society* 36 (2004), pp. 519–526.
42. E.P. Šimbireva, On the theory of partially ordered groups, *Matematicheskii Sbornik* 20 (1947), pp. 145–178 (in Russian).
43. H.J.S. Smith, On systems of linear indeterminate equations and congruences, *Philosophical Transactions of the Royal Society of London* 151 (1861), pp. 293–326.
44. R.I. Soare, *Recursively Enumerable Sets and Degrees*, Perspectives in Mathematical Logic, Springer-Verlag, Berlin, 1987.
45. D.R. Solomon, Π^0_1 classes and orderable groups, *Annals of Pure and Applied Logic* 115 (2002), pp. 279–302.
46. D.R. Solomon, *Reverse Mathematics and Ordered Groups*, PhD dissertation, Cornell University, 1998.
47. L. VanWyk, Graph groups are biautomatic, *Journal of Pure and Applied Algebra* 94 (1994), pp. 341–352.

MEASURE THEORY AND HILBERT'S TENTH PROBLEM INSIDE $\mathbb{Q}$

Russell Miller

Department of Mathematics
Queens College – C.U.N.Y.
65-30 Kissena Blvd.
Queens, NY 11367, USA
Ph.D. Programs in Mathematics and Computer Science
C.U.N.Y. Graduate Center
365 Fifth Avenue
New York, NY 10016, USA
Russell.Miller@qc.cuny.edu

For a ring R, Hilbert's Tenth Problem HTP(R) is the set of polynomial equations over R, in several variables, with solutions in R. When $R = \mathbb{Z}$, it is known that the jump $\mathbb{Z}'$ is Turing-reducible to HTP($\mathbb{Z}$). We consider computability of HTP(R) for subrings R of the rationals. Applying measure theory to these subrings, which naturally form a measure space, relates their sets HTP(R) to the set HTP($\mathbb{Q}$), whose decidability remains an open question. We raise the question of the measure of the topological boundary of the solution set of a polynomial within this space, and show that if these boundaries all have measure 0, then for each individual oracle Turing machine Φ, the reduction $R' = \Phi^{\text{HTP}(R)}$ fails on a set of subrings of positive measure. That is, no Turing reduction of the jump R' of a subring R to HTP(R) holds uniformly on a set of measure 1.

Contents

1. Introduction

The original version of Hilbert's Tenth Problem demanded an algorithm deciding which polynomial equations from $\mathbb{Z}[X_0, X_1, \ldots]$ have solutions in integers. In 1970, Matiyasevic [5] completed work by Davis, Putnam and Robinson [1], showing that no such algorithm exists. In particular, these authors showed that there exists a 1-reduction from the Halting Problem $\emptyset'$ to the set of such equations with solutions, by proving the existence of a single polynomial $h \in \mathbb{Z}[X, \vec{Y}]$ such that, for each n from the set ω of nonnegative integers, the polynomial $h(n, \vec{Y}) = 0$ has a solution in $\mathbb{Z}$ if and only if n lies in $\emptyset'$. Since the membership in the Halting Problem was known to be undecidable, it followed that Hilbert's Tenth Problem was also undecidable.

One naturally generalizes this problem to all rings R, defining Hilbert's Tenth Problem for R to be the set

$$\mathrm{HTP}(R) = \{f \in R[\vec{X}] : (\exists r_1, \ldots, r_n \in R^{<\omega})\ f(r_1, \ldots, r_n) = 0\}.$$

Here we will examine this problem for one particular class: the subrings R of the field $\mathbb{Q}$ of rational numbers. Notice that in this situation, deciding membership in $\mathrm{HTP}(R)$ reduces to the question of deciding this membership just for polynomials from $\mathbb{Z}[\vec{X}]$, since one readily eliminates denominators from the coefficients of a polynomial in $R[\vec{X}]$. So, for us, $\mathrm{HTP}(R)$ will always be a subset of $\mathbb{Z}[X_1, X_2, \ldots]$.

Subrings R of $\mathbb{Q}$ correspond bijectively to subsets W of the set $\mathbb{P}$ of all primes, via the map $W \mapsto \mathbb{Z}[\frac{1}{p} : p \in W]$. We write R_W for the subring $\mathbb{Z}[\frac{1}{p} : p \in W]$. In this article, we will move interchangeably between subsets of ω and subsets of $\mathbb{P}$, using the bijection mapping $n \in \omega$ to the nth prime p_n, starting with $p_0 = 2$. For the most part, our sets will be subsets of $\mathbb{P}$, but Turing reductions and jump operators and the like will all be applied to them in the standard way. Likewise, sets of polynomials, such as $\mathrm{HTP}(R)$, will be viewed as subsets of ω, using a fixed computable bijection from ω onto $\mathbb{Z}[\vec{X}] = \mathbb{Z}[X_0, X_1, \ldots]$.

We usually view subsets of $\mathbb{P}$ as paths through the tree $2^{<\mathbb{P}}$, a complete binary tree whose nodes are the functions from initial segments of the set $\mathbb{P}$ into the set $\{0, 1\}$. This allows us to introduce a topology on the space $2^{\mathbb{P}}$ of paths through $2^{<\mathbb{P}}$, and thus on the class $\mathbf{Sub}(\mathbb{Q})$ of all subrings of $\mathbb{Q}$. Each basic open set $\mathcal{U}_\sigma$ in this topology is given by a node σ on the tree:

$\mathcal{U}_\sigma = \{W \subseteq \mathbb{P} : \sigma \subset W\}$, where $\sigma \subset W$ denotes that when W is viewed as a function from $\mathbb{P}$ into the set $2 = \{0, 1\}$ (i.e., as an infinite binary sequence), σ is an initial segment of that sequence. Also, we put a natural measure μ on **Sub**($\mathbb{Q}$): just transfer to **Sub**($\mathbb{Q}$) the obvious Lebesgue measure on the power set $2^{\mathbb{P}}$ of $\mathbb{P}$. Thus, if we imagine choosing a subring R by flipping a fair coin (independently for each prime p) to decide whether $\frac{1}{p} \in R$, the *measure* of a subclass $\mathcal{S}$ of **Sub**($\mathbb{Q}$) is the probability that the resulting subring will lie in $\mathcal{S}$.

It is also natural, and in certain respects more productive, to consider Baire category theory on the space **Sub**($\mathbb{Q}$), as an alternative to measure theory. Here we will focus on measure theory. For questions and results regarding Baire category theory on subrings of $\mathbb{Q}$, we refer the reader to the forthcoming [6]. Due to the common subject matter of that article and this one, there is a substantial overlap between the introductions and background sections of the two papers, which we trust the reader to forgive. Naturally, we have also made every effort to maintain the same notation across both papers.

2. Background

2.1. *Measure theory and Cantor space*

The topological space $2^{\mathbb{P}}$ of all paths through $2^{<\mathbb{P}}$, which we treat as the space of all subrings of $\mathbb{Q}$, is obviously homeomorphic to *Cantor space*, the space 2^ω of all paths through the complete binary tree $2^{<\omega}$. We assign to the basic open set

$$\mathcal{U}_\sigma = \{W \subseteq \mathbb{P} : (\forall n < |\sigma|)\ [n \in W \iff \sigma(n) = 1\}$$

the measure $2^{-|\sigma|}$, as in the standard Lebesgue measure on 2^ω, and extend this measure to all Lebesgue-measurable subsets of $2^{\mathbb{P}}$. Thus we have a natural measure on the space of all subrings R_W of $\mathbb{Q}$.

All sets $W \subseteq \omega$ satisfy $W \oplus \emptyset' \leq_T W'$, and for certain W, Turing-equivalence holds here. Those W for which $W \equiv_T W \oplus \emptyset'$ are said to be *generalized-low*, and it turns out that this is the standard situation, according to both measure and Baire category.

Lemma 2.1: *(Folklore) The class*

$$\boldsymbol{GL}_1 = \{W \in 2^\omega : W' \equiv_T W \oplus \emptyset'\}$$

of generalized-low sets is comeager and has measure 1 *in* 2^ω. *However, there is no single Turing functional* Φ_e *for which the subclass*

$$\boldsymbol{GL}_{1,e} = \{W \in 2^\omega : W' = \Phi_e^{W \oplus \emptyset'}\}$$

has measure 1.

We express the content of the second statement by saying that although $W' \leq_T W \oplus \emptyset'$ on a set of measure 1, the reduction is *nonuniform*. This contrasts with the situation in Baire category, where a single Turing reduction does succeed on a comeager set.

We give the full proof of the measure-theoretic statements in Lemma 2.1 even though they are well known. They will be illustrative when we come to consider measures of boundary sets of polynomials.

Proof: First, for an arbitrary rational $\varepsilon > 0$, we describe a Turing reduction Φ_e for which $\mu(\mathbf{GL}_{1,e}) \geq 1 - \varepsilon$. This will prove that $\mathbf{GL}_1$ has measure 1.

With an oracle $W \oplus \emptyset'$, on input x, the functional Φ_e simultaneously performs two searches. At each stage s, it checks the first s strings $\sigma \in 2^{<\omega}$. Whenever it finds a σ for which $\Phi_{x,s}^{\sigma}(x)\downarrow$, it enumerates this σ into its set $S_{x,s}$. Simultaneously, for the sth rational r, it asks its $\emptyset'$ oracle whether

$$(\exists t\ \exists \langle \sigma_0, \ldots, \sigma_n \rangle \in (2^{<\omega})^{<\omega})\ [\mu(\cup_i \mathcal{U}_{\sigma_i}) > r\ \&\ (\forall i \leq n) \Phi_{x,t}^{\sigma_i}(x)\downarrow].$$

If this statement is false, it enumerates that r into its set $R_{x,s}$. (We start with $r = 1$ in $R_{x,0}$, since the statement must be false for this r.)

These searches continue until we reach a stage s at which some r in $R_{x,s}$ has the property that $\mu(\cup_{\sigma \in S_{x,s}} \mathcal{U}_\sigma) + \frac{\varepsilon}{2^{x+1}} \geq r$. Since $\mu(\cup_{\sigma \in S_x} \mathcal{U}_\sigma)$ (where $S_x = \cup_s S_{x,s}$) must equal the infimum of the subset $\cup_s R_{x,s}$ of $\mathbb{Q}$, it is clear that this process eventually halts. When it does, we use the W-oracle to check whether any of the finitely many σ already in $S_{x,s}$ is an initial segment of W. If so, we conclude that $\Phi_x^W(x)\downarrow$; if not, we conclude that $\Phi_x^W(x)\uparrow$.

Of course, this conclusion will not always be correct. However, it fails only on the class of those W for which $\Phi_x^W(x)\downarrow$ but no initial segment of W lies in the finite set $S_{x,s}$ we had enumerated by the stage s at which the process halted. The class of all these W must have measure $\leq \frac{\varepsilon}{2^{x+1}}$, since $S_x = \{W : \Phi_x^W(x)\downarrow\}$, which has measure $\leq r$, and at least $r - \frac{\varepsilon}{2^{x+1}}$ of this set has initial segments in $S_{x,s}$. Since the process gives the wrong answer (on input x) only for a class of measure $\leq \frac{\varepsilon}{2^{x+1}}$, the class of all W such that, for that W and some x, it gives the wrong answer is a class of measure $\leq \varepsilon$, as required.

Now we prove the nonuniformity, by constructing (the index x of) a specific program. Fix an effective numbering $\sigma_0 = \langle\rangle$, $\sigma_1 = \langle 0\rangle$, $\sigma_2 = \langle 1\rangle, \ldots$ of all of $2^{<\omega}$. The function Φ_x, on arbitrary input, halts if and only if there exists an n such that its oracle has initial segment $\sigma_n \widehat{\ } 1^{n+2}$ (that is, σ_n followed by $(n+2)$ consecutive 1's). Notice that the class $\mathcal{A}$ of all W such that $\Phi_x^W(x)\downarrow$ has measure $\leq \sum_n 2^{-|\sigma_n|-n-2} \leq \sum_n 2^{-(n+2)} = \frac{1}{2}$. It follows that, if a functional Φ_e computes W' from $W \oplus \emptyset'$ uniformly on a class of measure 1, then there must be sets W for which $\Phi_e^W(x)\downarrow = 0$. However, the initial segment $\sigma \subseteq W$ used in this computation is equal to σ_n for some n, and then $\Phi_e^V(x)$ must give the same output 0, incorrectly, on each V in the basic open set $\mathcal{U}_{\sigma_n \widehat{\ } 1^{n+2}}$. Since each basic open set has positive measure, we see that Φ_e fails, on a class of positive measure, to compute W' from $W \oplus \emptyset'$ correctly. □

The proof of Lemma 2.1 showed that $\mathbf{GL}_1$ had measure 1 by showing that, for each $\varepsilon > 0$, a single Turing functional could compute W' from $W \oplus \emptyset'$ correctly on a set of measure $> 1 - \varepsilon$. This near-uniformity may sound like a useful fact, but in fact the next lemma can be adapted to show that it has to be the case in order for that lemma to be true at all.

Lemma 2.2: *(Folklore) Suppose $\mu\{W : A \oplus W \geq_T B\} > r$. Then there exists a single Turing functional Ψ such that $\mu\{W : \Psi^{A\oplus W} = B\} > r$.*

Proof: The idea is that we glue finitely many functionals together, using initial segments of the different W to choose which one to run. Formally, set $r + \varepsilon = \mu\{W : A \oplus W \geq_T B\}$. Since there are only countably many Turing functionals in all, there must exist finitely many functionals $\Psi_0, \ldots, \Psi_m$ such that

$$\mu\{W : (\exists e \leq m)\ \Psi_e^{A\oplus W} = B\} > r + \frac{\varepsilon}{2}.$$

But then, for each $e \leq m$, there must also exist finitely many initial segments $\sigma_{e0}, \ldots, \sigma_{ek_e}$ such that

$$\sum_{e\leq m} \mu\{W : (\exists j \leq k_e)\ [\sigma_{ej} \subseteq W \ \&\ \Psi_e^{A\oplus W} = B]\} > r + \frac{\varepsilon}{4},$$

and such that all these σ_{ej} are pairwise incomparable. So the desired functional Ψ, given any oracle $V \oplus W$, checks to see whether any of the finitely many σ_{ej} is an initial segment of W: if so, it runs Ψ_e on its oracle, while if not, it simply diverges. □

Lemma 2.3: *(Folklore) Let $A \not\geq_T B$. Then the class $\mathcal{C} = \{W : A \oplus W \geq_T B\}$ has measure* 0.

Proof: If $\mu(\mathcal{C}) > 0$, then there is some σ for which $\frac{\mu(\mathcal{C}\cap\mathcal{U}_\sigma)}{\mu(\mathcal{U}_\sigma)} > \frac{1}{2}$. (See e.g. Lemma 3.1a of [4] for a proof of this result, which is standard. The broader principle is the *Zero-One Law,* which states that all measurable subsets of 2^ω invariant under Turing equivalence have measure either 0 or 1.) Let $\delta = \mu(\mathcal{U}_\sigma) = \frac{1}{2^{|\sigma|}}$, and pick $\varepsilon > 0$ so that $\mu(\mathcal{C} \cap \mathcal{U}_\sigma) = \frac{\delta}{2} + \varepsilon$. Now by Lemma 2.2, there is a single Turing functional Φ such that $\Phi^{A \oplus W} = B$ for all W in a subclass of $\mathcal{U}_\sigma$ of measure $\frac{\delta}{2} + \frac{\varepsilon}{2}$. But then we could compute B directly from A: $B(x) = n$ iff there exist finitely many pairwise-incomparable strings $\tau \supseteq \sigma$ of total measure $> \frac{\delta}{2}$ for which $\Phi^{A\oplus\tau}(x)\downarrow = n$. □

In [11], Stillwell went much further, proving the decidability of the entire almost-everywhere theory of the Turing degrees under meet, join, and jump. (The class of pairs of sets A and B for which the meet $A \wedge B$ is defined has measure 1, so it is reasonable to discuss the almost-everywhere theory with $\wedge$ as a binary function.)

2.2. *Subrings of the rationals*

Now we turn to background results specifically about subrings of $\mathbb{Q}$. For all $W \subseteq \mathbb{P}$, we have the Turing reductions

$$W \oplus \text{HTP}(\mathbb{Q}) \leq_T \text{HTP}(R_W) \leq_T W'.$$

Indeed, each of these two Turing reductions is a 1-*reduction.* For instance, the Turing reduction from $HTP(R_W)$ to W' can be described by a computable injection which maps each $f \in \mathbb{Z}[\vec{X}]$ to the code number $h(f)$ of an oracle Turing program which, on every input, searches for a solution $\vec{x}$ in $\mathbb{Q}$ to the equation $f = 0$ for which the primes dividing the denominators of the coordinates in $\vec{x}$ all lie in the oracle set W. The reduction $W \leq_T \text{HTP}(R_W)$ is simple: $p \in W$ if and only if $(pX - 1) \in \text{HTP}(R_W)$. The reduction from $\text{HTP}(\mathbb{Q})$ to $\text{HTP}(R_W)$ uses the fact that every element of $\mathbb{Q}$ is a quotient of elements of R_W, so that $f(\vec{X})$ has a solution in $\mathbb{Q}$ if and only if $Y^d \cdot f(\frac{X_1}{Y}, \ldots, \frac{X_n}{Y})$ has a solution in R_W with $Y > 0$. (Here d is the total degree of f, so that Y^d suffices to cancel all denominators.) Since the Four Squares Theorem ensures that every nonnegative integer is a sum of four squares of integers, we may express the condition $Y > 0$ by a polynomial: if a rational y is positive, then there is a solution in $\mathbb{Z}$ to:

$$h(y, U_1, \ldots, U_4, V_1, \ldots, V_4) = y(1 + V_1^2 + \cdots + V_4^2) - (1 + U_1^2 + \cdots + U_4^2).$$

Conversely, any solution in $\mathbb{Q}$ to $h(y, \vec{U}, \vec{V}) = 0$ forces $y > 0$. It follows that when $y > 0$, this polynomial has a solution in every subring of $\mathbb{Q}$, while when $y \leq 0$, it has no solution in any subring. Therefore we may use it within any subring we like, to define the positive elements there. From all this we see (for arbitrary W) that $f \in \text{HTP}(\mathbb{Q})$ if and only if the following polynomial lies in $\text{HTP}(R_W)$:

$$\left(Y^d \cdot f\left(\frac{X_1}{Y}, \ldots, \frac{X_n}{Y}\right)\right)^2 + (h(Y, \vec{U}, \vec{V}))^2.$$

Recall that the *semilocal* subrings of $\mathbb{Q}$ are precisely those of the form R_W where the set W is cofinite in $\mathbb{P}$, containing all but finitely many primes. It will be important for us to know that whenever R is a semilocal subring of $\mathbb{Q}$, we have $\text{HTP}(R) \leq_1 \text{HTP}(\mathbb{Q})$. Indeed, both the Turing reduction and the 1-reduction are uniform in the complement. This result, stated formally below, essentially follows from work of Julia Robinson in [9]. For a proof by Eisenträger, Park, Shlapentokh, and the author, see [2].

Proposition 2.4: *(See Proposition 5.4 in* [2]*) There exists a computable function G such that for every n, every finite set $A_0 = \{p_1, \ldots, p_n\} \subset \mathbb{P}$ and every $f \in \mathbb{Z}[\vec{X}]$,*

$$f \in HTP(R_{\mathbb{P}-A_0}) \iff G(f, \langle p_1, \ldots, p_n \rangle) \in HTP(\mathbb{Q}).$$

That is, $HTP(R_{\mathbb{P}-A_0})$ is 1-reducible to $HTP(\mathbb{Q})$ for all semilocal $R_{\mathbb{P}-A_0}$, uniformly in A_0.

The proof in [2], using work from [3], actually shows how to compute, for every prime p, a polynomial $f_p(Z, X_1, X_2, X_3)$ such that for all rationals q, we have

$$q \in R_{\mathbb{P}-\{p\}} \iff f_p(q, \vec{X}) \in \text{HTP}(\mathbb{Q}).$$

Therefore, an arbitrary $g(Z_0, \ldots, Z_n)$ has a solution in $R_{\mathbb{P}-A_0}$ if and only if

$$(g(\vec{Z}))^2 + \sum_{p \in A_0, j \leq n} (f_p(Z_j, X_{1j}, X_{2j}, X_{3j}))^2$$

has a solution in $\mathbb{Q}$.

3. The Boundary Set of a Polynomial

For a polynomial $f \in \mathbb{Z}[\vec{X}]$ and a subring $R_W \subseteq \mathbb{Q}$, there are three possibilities. First, f may lie in $\text{HTP}(R_W)$. If this holds for R_W, the reason is finitary: W contains a certain finite (possibly empty) subset of primes generating the denominators of a solution. For this reason, the set $\mathcal{A}(f) = \{W : f \in \text{HTP}(R_W)\}$ is open: for any solution of f in R_W and any $\sigma \subseteq W$ long enough to include all primes dividing the denominators in that solution, every other $V \supseteq \sigma$ will also contain that solution.

The second possibility is that there may be a finitary reason why $f \notin \text{HTP}(R_W)$: there may exist a finite subset A_0 of the complement $\overline{W}$ such that f has no solution in $R_{\mathbb{P}-A_0}$. For each finite $A_0 \subset \mathbb{P}$, the set $\text{HTP}(R_{\mathbb{P}-A_0})$ is 1-reducible to $\text{HTP}(\mathbb{Q})$, by Proposition 2.4; indeed the two sets are computably isomorphic, with a computable permutation of $\mathbb{Z}[\vec{X}]$ mapping one onto the other. We write

$$\mathcal{C}(f) = \{W \subseteq \mathbb{P} : (\exists \text{ finite } A_0 \subseteq \overline{W})\ f \notin \text{HTP}(R_{\mathbb{P}-A_0})\}$$

for the set of W where this second possibility holds. $\mathcal{C}(f)$ is another open set, for the same reasons that $\mathcal{A}(f)$ is open.

The third possibility is that neither of the first two holds: W may not lie in $\mathcal{A}(f) \cup \mathcal{C}(f)$. Now one can computably enumerate the collection of those σ such that the basic open set $\mathcal{U}_\sigma = \{W : \sigma \subseteq W\}$ is contained within $\mathcal{A}(f)$. The set $\text{Int}(\overline{\mathcal{A}(f)})$ is similarly a union of basic open sets, and these can be enumerated by an $\text{HTP}(\mathbb{Q})$-oracle, since $\text{HTP}(\mathbb{Q})$ decides $\text{HTP}(R)$ uniformly for every semilocal ring R. The *boundary* $\mathcal{B}(f)$ of f remains: it contains those W which lie neither in $\mathcal{A}(f)$ nor in $\text{Int}(\overline{\mathcal{A}(f)})$. This set $\mathcal{B}(f)$ will be the focus of much of the rest of this article. Topologically, it is indeed the boundary of $\mathcal{A}(f)$, since it contains exactly those points which lie neither in the interior of $\mathcal{A}(f)$ (namely $\mathcal{A}(f)$ itself) nor in the interior of its complement. Therefore $\mathcal{B}(f)$ is always closed. In computability theory, $\mathcal{B}(f)$ is a Π^0_2 subset of $2^{\mathbb{P}}$, and indeed is $\Pi_1^{\text{HTP}(\mathbb{Q})}$, since with an $\text{HTP}(\mathbb{Q})$-oracle one can enumerate its complement $(\mathcal{A}(f) \cup \mathcal{C}(f))$.

To reduce the computability discussion to first-order, one can say of nodes σ that it is Σ^0_1 for $\mathcal{U}_\sigma$ to be contained within $\mathcal{A}(f)$, while it is $\text{HTP}(\mathbb{Q})$-decidable whether $\mathcal{U}_\sigma \subseteq \mathcal{C}(f)$. However, no $\mathcal{U}_\sigma$ can be contained within any $\mathcal{B}(f)$. Indeed, if $\mathcal{U}_\sigma \not\subseteq \mathcal{C}(f)$, then some $\tau \supseteq \sigma$ must have $\mathcal{U}_\tau \subseteq \mathcal{A}(f)$. It follows that, in Baire category theory, $\mathcal{B}(f)$ is nowhere dense, as shown in [6], and therefore the union

$$\mathcal{B} = \bigcup_{f \in \mathbb{Z}[\vec{X}]} \mathcal{B}(f)$$

is meager. Since meager sets often (but not always) are of measure 0, and vice versa, we will ask below whether $\mathcal{B}$ has measure 0 (or equivalently, whether every $\mathcal{B}(f)$ has measure 0). Theorem 3.4 will suggest the importance of this question.

3.1. *Examples of* $\mathcal{B}(f)$

A boundary set $\mathcal{B}(f)$ can be empty, but need not be, and we now give a specific example where it is nonempty. The basic idea is to use the polynomial X^2+Y^2-1. Of course, this polynomial has two trivial solutions $(0,1)$ and $(1,0)$ in $\mathbb{Z}$, so we modify it: our actual f has as its solutions those rationals (x,y) with $x^2+y^2=1$ and $x>0$ and $y>0$. This is readily accomplished using the Four Squares Theorem. Technically, the polynomial f uses twelve other variables as well, but it has a solution in R_W iff R_W contains positive rationals (x,y) with $x^2+y^2=1$.

Now if this f lies in HTP(R_W), we may write each solution in R_W as $(\frac{a}{c},\frac{b}{c})$, where a,b,c are all nonzero integers with no common factors and $c>1$. Every prime p dividing c must lie in W. For each such p, we have $a^2+b^2\equiv 0 \bmod p$. But p cannot divide both a and b (lest it be a common factor), and so easy arithmetic yields

$$\left(\frac{a}{b}\right)^2 \equiv -1 \bmod p.$$

This forces $p \not\equiv 3 \bmod 4$. It follows that f has no solutions in any subring R_V for which V contains only primes congruent to 3 modulo 4.

On the other hand, it is known that every prime $p\equiv 1 \bmod 4$ is a sum of two squares of integers. Poonen pointed out that, writing $p=m^2+n^2$, this yields

$$\left(\frac{m^2-n^2}{p}\right)^2+\left(\frac{2mn}{p}\right)^2=\frac{(m^2+n^2)^2}{p^2}=1.$$

With p prime, we know $mn\neq 0$ and $m\neq \pm n$, so this is a solution to f in the subring $R_{\{p\}}$. It follows that f has solutions in every subring R_W for which W contains any prime $\equiv 1 \bmod 4$, and only in such subrings. (The only remaining prime that could divide c is 2, in which case 4 divides c^2. But if $a^2+b^2=c^2\equiv 0 \bmod 4$, then a and b must both be even, giving them a common factor with c.)

It now follows that $\mathcal{A}(f)$ has measure 1, since the probability is 1 that an arbitrary W contains at least one prime $\equiv 1 \bmod 4$. Hence $\mathcal{C}(f)$, being open of measure 0, must be empty. But we saw above that $\mathcal{A}(f)\neq 2^{\mathbb{P}}$, so $\mathcal{B}(f)\neq\emptyset$, although $\mu(\mathcal{B}(f))=0$. In particular, every subring W which contains no prime $\equiv 1 \bmod 4$ has the defining property for being in the boundary set of f: no initial segment $\sigma \subset W$ determines whether or not R_W contains a solution to f.

Next, imitating the foregoing proof, we show that for each odd prime q, there is an infinite decidable set V of primes such that R_V contains no nontrivial solutions to $X^2 + qY^2 = 1$. (Here the trivial solution is $(1, 0)$, which can be ruled out by a messier polynomial, just as above.)

Lemma 3.1: *Fix a prime $q \equiv 3 \bmod 4$, and let x and y be positive rational numbers with $x^2 + qy^2 = 1$. Then every prime factor of the least common denominator of x and y is a square modulo q.*

For a prime $q \equiv 1 \bmod 4$, the situation is a little more complicated. If $x^2+qy^2 = 1$ and $y \neq 0$ and a prime p divides the least common denominator of x and y, then one of the following holds:

- *$p \equiv 1 \bmod 4$ and p is a square modulo q.*
- *$p \equiv 3 \bmod 4$ and p is not a square modulo q.*

Proof: We proceed similarly to the $q = 1$ case done above. Suppose that $q \equiv 3 \bmod 4$ and a, b, c are positive integers, with no common factor, satisfying $a^2 + qb^2 = c^2$. Thus $\left(\frac{a}{b}\right)^2 \equiv -q \bmod p$ for every prime p dividing c. If $p \equiv 1 \bmod 4$, then -1 is also a square mod p, so q is a square mod p, and by quadratic reciprocity p must be a square mod q. Likewise, if $p \equiv 3 \bmod 4$, then -1 is not a square mod p, so q is not either; but with both p and q congruent to 3 mod 4, quadratic reciprocity now shows that p is again a square mod q. (The number-theoretic results here may be found in any standard text on the subject, e.g., [10].)

When $q \equiv 1 \bmod 4$, a similar analysis, with careful use of quadratic reciprocity, gives the result stated in the lemma. □

One could use Lemma 3.1 to build infinitely many distinct polynomials (with different prime values of q) such that there is an infinite set V for which R_V contains no solution of any of those polynomials, yet for which no finite subset of the complement of V suffices to establish that any individual one of those polynomials fails to have a solution in R_V. (Saying the same thing: every one of these polynomials has solutions in every HTP-generic subring of $\mathbb{Q}$.) In particular, for every prime q and every nonzero $j \in \mathbb{Z}$, we have

$$\left(\frac{j^2 - q}{j^2 + q}\right)^2 + q \cdot \left(\frac{2j}{j^2 + q}\right)^2 = 1,$$

so that $X^2 + qY^2 = 1$ has a nontrivial solution in each subring with the prime factors of $(j^2 + q)$ inverted. Given a finite set $\{r_1, \ldots, r_k\}$ of primes which we may not invert, just take $j = \Pi_{r_i \neq q} r_i$; then no r_i divides $(j^2 + q)$, and so the semilocal ring $R_{\mathbb{P}-\{r_1,\ldots,r_k\}}$ contains the above solution to the polynomial.

3.2. *HTP-genericity*

Recall that $\mathcal{B}$ denotes the union of all boundary sets $\mathcal{B}(f)$, over all polynomials $f \in \mathbb{Z}[\vec{X}]$. For a set W to fail to lie in $\mathcal{B}$, it must be the case that for every polynomial f, either $f \in \text{HTP}(R_W)$ or else some finite initial segment of W rules out all solutions to f. This is an example of the concept of *genericity*, common in both computability and set theory, so we adopt the term here. With this notion, we can show not only that $\text{HTP}(R_W) \leq W \oplus \text{HTP}(\mathbb{Q})$ for all $W \in \overline{\mathcal{B}}$, but indeed that the reduction is uniform on $\overline{\mathcal{B}}$.

Definition 3.2: A set $W \subseteq \mathbb{P}$ is *HTP-generic* if $W \notin \mathcal{B}$. In this case we will also call the corresponding subring R_W HTP-generic.

Proposition 3.3: *HTP(R_W) is Turing-reducible to $W \oplus$ HTP($\mathbb{Q}$) uniformly on the set $\overline{\mathcal{B}}$. That is, there exists a single Turing reduction Φ such that, for every HTP-generic set W, $\Phi^{W \oplus HTP(\mathbb{Q})} = HTP(R_W)$.*

Proof: Given $f \in \mathbb{Z}[\vec{X}]$ as input, the program for Φ simply searches for either a solution $\vec{x}$ to $f = 0$ in $\mathbb{Q}$ for which all primes dividing the denominators lie in the oracle set W, or else a finite set $A_0 \subseteq \overline{W}$ such that the HTP($\mathbb{Q}$) oracle, using Proposition 2.4, confirms that $f \notin \text{HTP}(R_{\mathbb{P}-A_0})$. When it finds either of these, it outputs the corresponding answer about membership of f in HTP(R_W). If it never finds either, then $W \in \mathcal{B}(f)$, and so this process succeeds for every W except those in $\mathcal{B}$. □

Theorem 3.4: $\mu(\mathcal{B}) = 0$ *if and only if there exists a Turing reduction of HTP(R_W) to $W \oplus$ HTP($\mathbb{Q}$) which succeeds uniformly on a class of measure 1. Moreover, if these equivalent conditions hold, then:*

(1) The measures $\mu(\mathcal{A}(f))$ and $\mu(\mathcal{C}(f))$ are HTP($\mathbb{Q}$)-computable uniformly in f. (This claim is precisely defined later in this subsection.)
(2) There is no *Turing reduction of W' to HTP(R_W) which succeeds uniformly on a class of measure* 1.

Proof: Proposition 3.3 proves the forwards direction of the equivalence immediately. For the backwards direction, suppose that $\text{HTP}(R_W) = \Phi^{W \oplus \text{HTP}(\mathbb{Q})}$ for every W in a class $\mathcal{D}$ of measure 1. Fix any polynomial

f, and any set $W \in \mathcal{B}(f)$. Then $f \notin \mathrm{HTP}(R_W)$, but we claim that $\Phi^{W \oplus \mathrm{HTP}(\mathbb{Q})}(f)$ cannot halt and output 0. If it did, then fix n large enough that $\Phi^{(W \restriction n) \oplus \mathrm{HTP}(\mathbb{Q})}(f) \downarrow = 0$. Then also $\Phi^{V \oplus \mathrm{HTP}(\mathbb{Q})}(f) \downarrow = 0$ for every $V \supset W \restriction n$. However, since $W \in \mathcal{B}(f)$, there exists some $V \supset W \restriction n$ for which $f \in \mathrm{HTP}(R_V)$. Fix $m \geq n$ large enough that $f \in \mathrm{HTP}(R_{V \restriction m})$. Then, for every $U \supset V \restriction m$, we have both $f \in \mathrm{HTP}(R_U)$ and $\Phi^{U \oplus \mathrm{HTP}(\mathbb{Q})}(f) \downarrow = 0$, contradicting the assumption that Φ succeeds uniformly on a class of measure 1. Thus we must have either $\Phi^{W \oplus \mathrm{HTP}(\mathbb{Q})}(f) \uparrow$ or $\Phi^{W \oplus \mathrm{HTP}(\mathbb{Q})}(f) \downarrow \neq 0$ whenever $W \in \mathcal{B}(f)$. But since Φ succeeds uniformly on a class of measure 1, this means that every class $\mathcal{B}(f)$ has measure 0, and hence so does the countable union $\mathcal{B}$ of these classes.

Now suppose that the two equivalent conditions hold. (1) will follow from the more general Theorem 3.6 below. For (2), we simply note that by Lemma 2.1, no single Turing functional can compute W' from $W \oplus \emptyset'$ uniformly on a class of measure 1. Since we are assuming that $\mathrm{HTP}(R_W)$ can be computed from $W \oplus \mathrm{HTP}(\mathbb{Q})$ (hence from $W \oplus \emptyset'$) uniformly on a class of measure 1, there cannot possibly exist a further reduction of W' to $\mathrm{HTP}(R_W)$ which also succeeds uniformly on a class of measure 1. (The intersection of two classes of measure 1 also has measure 1, of course.) □

For the next theorems, we simplify our notation by writing

$$\alpha(f) = \mu(\mathcal{A}(f)) \qquad \beta(f) = \mu(\mathcal{B}(f)) \qquad \gamma(f) = \mu(\mathcal{C}(f))$$

for each $f \in \mathbb{Z}[\vec{X}]$. Of course $\alpha(f) + \beta(f) + \gamma(f) = 1$. Notice, however, that we have very little *a priori* information about the values in the images of these functions. Complexity bounds exist, since $\alpha(f)$ is always a left-c.e. real number, and $\gamma(f)$ is always left-c.e. relative to $\mathrm{HTP}(\mathbb{Q})$, but it is not clear whether these real numbers need to be algebraic over $\mathbb{Q}$. The author is not aware of any polynomials f for which $\alpha(f)$ fails to be a dyadic rational, nor of any for which $\beta(f) > 0$. Likewise, it is open whether the set $\mathcal{C}(f)$ must always be a finite union of basic open sets $\mathcal{U}_\sigma$. (Our examples in Subsection 3.1 do show that $\mathcal{A}(f)$ can fail to be a finite union of basic open sets.)

In order to deal with these functions from ω (or from $\mathbb{Z}[\vec{X}]$) into $\mathbb{R}$, we will say that we can *compute* a real number r (such as $\alpha(f)$) if we can enumerate both the strict lower cut and the strict upper cut in $\mathbb{Q}$ defined by r. (Notice that if r itself is rational, this means that r will never appear in either cut.) Of course, an enumeration $E = \cup_s E_s$ of the non-strict lower cut of r quickly yields an enumeration of its strict lower cut: we may assume $|E_s| \leq s$ and take $E'_s = E_s - \{\max(E_s)\}$.

A *uniform enumeration* of the strict lower cuts of a sequence of real numbers $\langle r_i \rangle_{i\in\omega}$ (such as $\langle \alpha(f)\rangle_{f\in\mathbb{Z}[\vec{X}]}$) is a single procedure that, on input i, enumerates the strict lower cut of r_i; similarly for strict upper cuts. To *compute the sequence* $\langle r_i\rangle_{i\in\omega}$ *uniformly* means to enumerate both strict upper and lower cuts for the sequence uniformly. Finally, a function θ from ω into $\mathbb{R}$ (such as α, β, or γ) is *computable* if the sequence $\langle\theta(i)\rangle_{i\in\omega}$ is uniformly computable.

The next theorems show the potential power of being able to compute α, β, and γ. Of course, if it turns out that $\mu(\mathcal{B}) = 0$, then β would be very readily computable, and α and γ would both be HTP($\mathbb{Q}$)-computable.

Theorem 3.5: *The following are equivalent.*

(1) The strict upper cuts of the measures $\alpha(f)$ of polynomials f are computable uniformly in f.
(2) $\alpha(f)$ is computable uniformly in f.
(3) There is a single Turing functional Ψ such that, for all rational $\varepsilon > 0$,

$$\mu(\{W \subseteq \mathbb{P} : (\forall f)\ HTP(R_W)(f) = \Psi^W(\varepsilon, f)\}) \geq 1 - \varepsilon.$$

Proof: The equivalence of (1) and (2) is immediate, since we can enumerate the lower cuts of $\alpha(f)$ uniformly in f simply by searching for solutions to f in $\mathbb{Q}$. The argument for (2) $\Longrightarrow$ (3) is similar to the proof of Lemma 2.1. Given ε and the nth polynomial f in $\mathbb{Z}[\vec{X}]$, we first approximate $\alpha(f)$ by finding an $r \in \mathbb{Q}$ with $r < \alpha(f) \leq r + \frac{\varepsilon}{2^{n+1}}$. (Since $\alpha(f)$ could be 0, we allow $r < 0$ here, in which case the next step is trivial.) Next, find finitely many solutions to f in $\mathbb{Q}$ such that the class of subrings containing these solutions has measure $\geq r$. If W contains any of these solutions, then output "yes" (that is, $f \in$ HTP(R_W)); otherwise output "no." This output will be correct except on a class of measure $\leq \frac{\varepsilon}{2^{n+1}}$, and so our functional computes HTP(R_W) correctly except on a class of measure $\leq \sum_n \frac{\varepsilon}{2^{n+1}} = \varepsilon$.

It remains to show that (3) $\Longrightarrow$ (1). Here we use the uniformity of the procedure Ψ with respect to ε. Given an f, we wish to enumerate the upper cut of $\alpha(f)$. For each $s > 0$ in turn, set $\varepsilon = \frac{1}{s}$, and search for strings σ and naturals t such that $\Psi_t^\sigma(\frac{\varepsilon}{3}, f)\downarrow$. By assumption, we will eventually find a t and a finite set of strings $\sigma_0, \ldots, \sigma_k$ such that all $\Psi_t^{\sigma_i}(\frac{\varepsilon}{3}, f)\downarrow$ and $\mu(\cup_{i\leq k}\mathcal{U}_{\sigma_i}) \geq 1 - \frac{2\varepsilon}{3}$. Let $r = \mu(\cup_{\sigma(i)=1}\mathcal{U}_{\sigma_i})$, and enumerate all rationals $> r + \varepsilon$ into the upper cut of $\alpha(f)$. Now according to the computation by Ψ, at least $(1 - r - \frac{2\varepsilon}{3})$-much of $2^{\mathbb{P}}$ lies outside of $\mathcal{A}(f)$. Ψ may have been wrong about as much as $\frac{\varepsilon}{3}$ of these, but must have been correct on the remaining ones, whose measure is at least $(1 - r - \varepsilon)$. So our enumeration did nothing incorrect. Moreover, if $q \in \mathbb{Q}$ satisfies $q > \alpha(f)$, then once we reach an s with $\varepsilon = \frac{1}{s} < \frac{q-\alpha(f)}{2}$, we will have $r \leq \alpha(f) + \frac{\varepsilon}{3}$, and hence $r + \varepsilon \leq \alpha(f) + \frac{4\varepsilon}{3} < q$. Thus every $q > \alpha(f)$ will eventually be enumerated into the upper cut given by our procedure. This completes the proof. □

For the next theorem, we consider more than just the characteristic function of $\mathrm{HTP}(R_W)$. For each polynomial f and each $W \subseteq \mathbb{P}$, define

$$\chi(f, W) = \begin{cases} 2, & \text{if } W \in \mathcal{A}(f); \\ 1, & \text{if } W \in \mathcal{B}(f); \\ 0, & \text{if } W \in \mathcal{C}(f). \end{cases}$$

Theorem 3.6: *The following are equivalent.*

(1) The strict lower cuts of the measures $\beta(f)$ of the boundary sets $\mathcal{B}(f)$ of polynomials f are enumerable uniformly in f using an HTP($\mathbb{Q}$)-oracle.
(2) $\alpha(f)$, $\beta(f)$, and $\gamma(f)$ are all HTP($\mathbb{Q}$)-computable uniformly in f.
(3) There is a single Turing functional Θ such that, for all rational $\varepsilon > 0$,

$$\mu(\{W \subseteq \mathbb{P} : (\forall f)\ \chi(f, W) = \Theta^{W \oplus HTP(\mathbb{Q})}(\varepsilon, f)\}) \geq 1 - \varepsilon.$$

Proof: We can enumerate the strict lower cut of $\alpha(f)$, of course, just by searching for solutions of f in $\mathbb{Q}$, and with an HTP($\mathbb{Q}$)-oracle we can similarly enumerate the strict lower cut of $\gamma(f)$. Assuming (1), this allows us to use an HTP($\mathbb{Q}$)-oracle to enumerate the strict lower cuts of $\alpha(f) + \beta(f)$ and of $\gamma(f) + \beta(f)$. But $\gamma(f) = 1 - \alpha(f) - \beta(f)$, so we can then enumerate the strict upper cut of $\gamma(f)$, and similarly for the strict upper cut of $\alpha(f)$. Finally, the strict upper cut of $\beta(f)$ is just that of $(1 - \alpha(f) - \gamma(f))$. Thus (1) $\Longrightarrow$ (2).

Assuming (2), the program for Θ is along similar lines to that of Ψ in Theorem 3.5. Given ε, the nth polynomial f, and an oracle for $W \oplus \mathrm{HTP}(\mathbb{Q})$, it uses the oracle to compute an r with $r < \alpha(f) < r + \frac{\varepsilon}{2^{n+2}}$, then finds a finite union of basic open sets, of measure $\geq r$, all contained within $\mathcal{A}(f)$. If W lies within one of these basic open sets, then it outputs 2 (that is, it concludes, correctly, that $f \in \mathrm{HTP}(R_W)$), and otherwise it continues with the procedure below. Thus, for this f, the class of W with $f \in \mathrm{HTP}(R_W)$ and $\Theta^{W \oplus \mathrm{HTP}(\mathbb{Q})}(f) \neq 2$ has measure at most $\frac{\varepsilon}{2^{n+2}}$.

If the program concludes (possibly incorrectly) that $f \notin \mathrm{HTP}(R_W)$, then it goes back to its $HTPQ$-oracle to approximate $\gamma(f)$, and finds an r' with $r' < \gamma(f) < r' + \frac{\varepsilon}{2^{n+2}}$. Using the entire oracle $W \oplus \mathrm{HTP}(\mathbb{Q})$, it then enumerates strings σ such that $f \notin \mathrm{HTP}(R_{\mathbb{P} - \sigma^{-1}(0)})$ until it has found a finite union of basic open sets, of measure $\geq r'$, contained within $\mathcal{C}(f)$. If

W lies within any of these basic open sets, then the program outputs 0 (correctly, since $W \in \mathcal{C}(f)$); while otherwise it outputs 1. Thus, for this f, the class of W with $f \in \mathcal{C}(f)$ and $\Theta^{W\oplus\mathrm{HTP}(\mathbb{Q})}(f) \neq 0$ has measure at most $\frac{\varepsilon}{2^{n+2}}$, and so the class of those W such that the program output is incorrect has measure $\leq \frac{\varepsilon}{2^{n+1}}$. (The outputs 2 and 0 are always justified, so the only possible errors are output 1 with either $W \in \mathcal{A}(f)$ or $W \in \mathcal{C}(f)$. These are the two classes described above, each of measure $\leq \frac{\varepsilon}{2^{n+2}}$.) It follows that, apart from a class of measure $\leq \sum \frac{\varepsilon}{2^{n+1}} = \varepsilon$, the $\Theta^{W\oplus\mathrm{HTP}(\mathbb{Q})}$ outputs the correct answer for every f. This proves (3).

The proof that (3) $\implies$ (1) is entirely analogous to that in Theorem 3.5. To enumerate the strict lower cut of $\beta(f)$, using an HTP($\mathbb{Q}$)-oracle, we search for strings σ and naturals s, t for which $\Theta_s^{\sigma\oplus\mathrm{HTP}(\mathbb{Q})}(\frac{1}{t}, f)\downarrow= 1$. When (and if) we find such strings $\sigma_0, \ldots, \sigma_n$ (for any single t), we conclude that $\mu(\beta) \geq \mu(\cup_{i\leq n} \mathcal{U}_{\sigma_i}) - \frac{1}{t}$, and we enumerate all rationals less than this value. As in Theorem 3.5, this process enumerates precisely the strict lower cut of $\beta(f)$. □

4. Questions

The obvious question arising from this article is number-theoretic: does there exist a polynomial $f \in \mathbb{Z}[\vec{X}]$ with $\beta(f) > 0$? Equivalently, is $\mu(\mathcal{B}) > 0$? (Recall that $\beta(f)$ denotes the measure of the boundary set $\mathcal{B}(f)$ of the polynomial f, with $\mathcal{B} = \cup_f \mathcal{B}(f)$.) The useful analogy here is to the class $\mathcal{A} = \{W : x \in W'\}$ defined (for a specific index x) in Lemma 2.1. That class $\mathcal{A}$ had measure $\leq \frac{1}{2}$, and could readily have been built to have arbitrarily small positive measure. However, the interior of its complement was empty, and so the boundary of $\mathcal{A}$ had measure $\geq \frac{1}{2}$, and could have been made to have boundary of measure arbitrarily close to 1. The question is whether one can build a polynomial f for which the set $\mathcal{A}(f)$ acts the same way as the $\mathcal{A}$ from the lemma, with small measure itself but with $\mathcal{C}(f)$ empty, so that $\mathcal{B}(f)$ must have positive measure. (Of course, the question of possible values of $\beta(f)$ does not require $\mathcal{C}(f) = \emptyset$; this simply seems like the easiest way to address it.)

A stronger version of this question appears in [2]. There Eisenträger, Park, Shlapentokh, and the author ask (in essence) whether a polynomial f could have the properties that $\mathcal{C}(f) = \emptyset$, yet that there also exists $\varepsilon > 0$ such that for every $W \in \mathcal{A}(f)$, there is an n for which $\frac{|W\cap\{0,\ldots,n\}|}{n+1} > \varepsilon$. (One might as well assume here that W contains only the elements necessary to cause a single solution of f to appear in R_W.) The reasons for posing this question are explained there. For any f with these properties, the set $\mathcal{A}(f)$ would imitate the set $\mathcal{A}$ in our Lemma 2.1: not necessarily ending with a long string of 1's, but at least avoiding any string with too many 0's.

Several other questions are listed in Subsection 3.2. These include whether $\alpha(f)$ is always dyadic, or always rational, or always algebraic, or always computable; and also whether the set $\mathcal{C}(f)$ is always a finite union of basic open sets. The computability of the function α remains open: this is one of the equivalent conditions in Theorem 3.5. Notice that the ability to compute $\alpha(f)$ uniformly in f does not automatically confer the ability to decide HTP($\mathbb{Z}$): $\alpha(f)$ can equal 1 even when $f \notin$ HTP($\mathbb{Z}$), as in the first example in Subsection 3.1, for instance. Likewise, Theorem 3.6 proves three conditions to be equivalent, but leaves open the question of whether or not those conditions actually hold.

Acknowledgments

The author was partially supported by Grant # DMS – 1362206 from the National Science Foundation, and by several grants from the PSC-CUNY Research Award Program. This work grew out of research initiated at a workshop held at the American Institute of Mathematics and continued at a workshop held at the Institute for Mathematical Sciences of the National University of Singapore. The author wishes to acknowledge useful conversations with Bjorn Poonen and Alexandra Shlapentokh.

References

1. Martin Davis, Hilary Putnam, and Julia Robinson. The decision problem for exponential diophantine equations. *Annals of Mathematics*, 74(3): 425–436, 1961.
2. Kirsten Eisenträger, Russell Miller, Jennifer Park, and Alexandra Shlapentokh. As easy as $\mathbb{Q}$: Hilbert's Tenth Problem for subrings of the rationals, to appear in *Transactions of the American Mathematical Society*.
3. Jochen Koenigsmann. Defining $\mathbb{Z}$ in $\mathbb{Q}$. *Annals of Mathematics*, 183(1): 73–93, 2016.
4. Stuart Kurtz. Randomness and Genericity in the Degrees of Unsolvability. Ph.D. thesis, University of Illinois at Urbana-Champaign, 1981.
5. Yu. V. Matijasevič. The Diophantineness of enumerable sets. *Dokl. Akad. Nauk SSSR*, 191: 279–282, 1970.
6. Russell Miller. Baire category theory and Hilbert's Tenth Problem inside $\mathbb{Q}$, in *Pursuit of the Universal: 12th Conference on Computability in Europe, CiE 2016*, eds. A. Beckmann, L. Bienvenu & N. Jonoska *Lecture Notes in Computer Science* **9709** (Berlin: Springer-Verlag, 2016), 343–352.

7. Bjorn Poonen. Hilbert's Tenth Problem and Mazur's conjecture for large subrings of $\mathbb{Q}$. *Journal of the AMS*, 16(4): 981–990, 2003.
8. Bjorn Poonen. Characterizing integers among rational numbers with a universal-existential formula. *American Journal of Mathematics*, 131(3): 675–682, 2009.
9. Julia Robinson. Definability and decision problems in arithmetic. *Journal of Symbolic Logic*, 14: 98–114, 1949.
10. Jean-Pierre Serre. *A Course in Arithmetic.* Graduate Texts in Mathematics, Springer, Berlin, 1996.
11. John Stillwell. Decidability of the "almost all" theory of degrees. *Journal of Symbolic Logic*, 37(3): 501–506, 1972.

Printed in the United States
By Bookmasters